STILL
TURNING
A HISTORY OF
AERMOTOR WINDMILLS

Number Twenty-seven: Tarleton State University

Southwestern Studies in the Humanities

T. Lindsay Baker, General Editor

A list of other books in this series may be found at the back of this book.

STILL TURNING

A HISTORY OF AERMOTOR WINDMILLS

CHRISTOPHER C. GILLIS

Foreword by T. Lindsay Baker

TEXAS A&M UNIVERSITY PRESS

College Station

This paper meets the requirements of ANSI/NISO Z39.48-1992 (Permanence of Paper).
Binding materials have been chosen for durability.
Manufactured in the United States of America

LIBRARY OF CONGRESS CATALOGING-IN-PUBLICATION DATA

Gillis, Christopher C., author.
  Still turning: a history of Aermotor windmills / Christopher C. Gillis; foreword by
T. Lindsey Baker. — First edition.
      pages cm — (Tarleton State University southwestern studies in the humanities; no. 27)
  Includes bibliographical references and index.
  ISBN 978-1-62349-335-6 (cloth: alk. paper) —
  ISBN 978-1-62349-336-3 (ebook)
  1. Aermotor Windmill Company—History. 2. Aermotor Windmill Corporation—
History—21st century. 3. Aermotor Windmill Corporation—History—20th century.
4. Aermotor Company—History—20th century. 5. Aermotor Company—History—
19th century. 6. Pumping machinery industry—United States. 7. Windmills.
8. Water-power. I. Title. II. Series: Tarleton State University southwestern
studies in the humanities; no. 27.
  HD9705.5.P854 A474 2015
  338.7'6214530973—dc23

                              2015012119

In memory of Stan "The Windmill Man" Anderson (1925–2013),
who unselfishly shared so much of his knowledge about Aermotor windmills
with historians and collectors alike.

CONTENTS

ILLUSTRATIONS

FOREWORD

The most commonly seen name emblazoned on windmill vanes in towns and fields worldwide is Aermotor. Created by the combined geniuses of a professional engineer and a marketing master, the Aermotor appeared on the American market in the 1880s and within a decade dominated sales in the United States. By 1900 the Chicago-based manufacturer was exporting them to six continents, where they remain in use today. Becoming the most widely sold wind pumps in North America, their silhouette symbolizes successful rural life in demanding environments.

Author Christopher Gillis here tells the story of Aermotor windmills and the people who built and used them. His is the first book-length academic account of any wind pump that brought life to areas of the earth's surface that otherwise might never have been settled.

Gillis came to wind power history as an American high school pupil traveling to Belgium as part of a foreign-exchange program. There he discovered European windmills and began a serious study. His university graduation thesis examined the use of factory-made American windmills as devices for pumping groundwater in developing countries. Eventually, as a professional editor, he turned his attention to wind-electric generation, authoring two books on this topic.

I learned of Christopher from his 1994 article in *Old Mill News* on an Ohio windmill manufacturer. We became personally acquainted after he was approached by the maker of Aermotor windmills to prepare a history of the company, and I offered him use of my substantial files on wind power history. Though the economic recession of the early twenty-first century caused his project for the company to fail, I suggested to Gillis that he consider instead writing an academic book on Aermotor. Through these circumstances the book came to be included in the Tarleton State University Southwestern Studies in the Humanities series.

In the course of his research, Christopher Gillis uncovered far more about the history of Aermotor windmills, their makers, and their users than I or anyone else ever imagined. He located and spoke with aged employees of the manufacturers, he interviewed people from companies worldwide that now make the old Chicago product, and he found heretofore unknown written sources. He skillfully combined these materials to prepare an engaging story that places a genuinely human face on one of the icons of rural life, the Aermotor windmill.

—*T. Lindsay Baker*, Tarleton State University, Stephenville, Texas

PREFACE

When I was seventeen years old, I had the fortunate opportunity to study abroad in the Belgian province of East Flanders. It was on a crisp autumn day in 1988 when I set out on my bicycle to further explore the countryside around the town of Ronse. Next to a rustic farmhouse I encountered the remains of a windmill. I stopped and peered inside the squat tower made of stone and covered with a form of plaster, and I found it was empty. The wooden sails of the windmill had disappeared decades earlier. Always curious about how things worked, I set out to learn more about Belgium's centuries-old windmill heritage. This included an introduction to the Dierckx family, who kindly drove me around from village to village in the eastern Belgium province of Limburg to visit operating windmills. I also reviewed copies of a weekly Flemish-language milling publication, *De Belgische Molenaar*, for additional historic insights into the country's industry. (*De Belgische Molenaar* is no longer published and has since been superseded by the longtime monthly sister publication *Levende Molens*.) Needless to say, I was hooked and I took every occasion thereafter to visit, photograph, and obtain information about Belgium's windmills. Even after I returned home to Maryland in the summer of 1989, my interest in this technology continued.

I then attended Mount St. Mary's College in Emmitsburg, Maryland, and during the summer after my sophomore year was awarded a grant from the National Endowment for the Humanities to research the socioeconomic impacts of windmill and water mill technologies on medieval France. It was then that I learned mills were more than just facilitators of flour production for daily bread but had come to replace many former man- and animal-powered tasks across early Europe, such as crushing oil seeds and mineral ores, sawing timber, and draining marshy land.

Having grown up in Maryland, I had limited firsthand exposure to American water-pumping windmills. The majority of them had rapidly faded from the landscape by the late 1900s. Availability to electricity coming through power lines rendered water-pumping windmills unnecessary for most mid-Atlantic farmers by the middle of the twentieth century. Their water needs were easily met by high-output electric pumps. Those windmills that remained mostly became inoperable, frozen in place by their rusting gears and having their wind wheels blown to pieces by decades of storms. In most cases, only the tower remained, since it made an excellent place upon which the farmer could erect his television antenna.

During my last year of undergraduate studies at Mount St. Mary's in 1992–93, I was awarded a fellowship to research the socioeconomic benefits of deploying low-tech solar, wind, and other renewable energy technologies in least-developed countries. That's when I truly began to understand the mechanics and history of the American-made water-pumping windmill. As part of my research, I contacted the remaining US windmill manufacturers for product information—Aermotor, Dempster, and Heller-Aller. I credit Max Kelley, president of the Heller-Aller Company in Napoleon, Ohio, for sparking my passion in the history of the American windmill. Mr. Kelley always made time to answer my telephone calls and written correspondence in detail. He also introduced me to Dr. T. Lindsay Baker, American windmill historian, author, and publisher of the *Windmillers' Gazette*. Even after completing my fellowship thesis, I realized I still had much to learn about the windmill industry. Routine communication with Dr. Baker and reading of the *Windmillers' Gazette* in the years that followed my graduation from college and the launch of my career as a business journalist and editor continued to fuel my interest in the history of windmills, particularly those used to generate electric power. I published two books on the history of wind–electric power generation, in 2008 and 2011.

At the start of 2012, Dr. Baker, who by now served as the W. K. Gordon Endowed Chair in Texas Industrial History at Tarleton State University, approached me about writing a scholarly history on the most recognized mechanical windmill in North America—and the world for that matter: the Aermotor. In 2008 the windmill company had reached its 120th year in continuous business and had already outlasted all former US competitors with the exception of Dempster, which finally exited the windmill manufacturing business in 2013. I embraced the opportunity to write this book about Aermotor and how it emerged as an iconic symbol of American agriculture. However, I was well aware that the task before me would be arduous, as Aermotor's historic record after repeated moves over the past sixty years was scattered to the wind. I was already aware that many documents had been destroyed or thrown out by the company during subsequent moves of production and sales. Yet some of Aermotor's former managers had the foresight to save certain hallmark documentation, such as the company's meeting minutes between 1903 and 1958 and the stock ledger from its days in Chicago, from the burn barrels. Today these documents are housed in private collections and public archives throughout the Southwest and Midwest of the United States. Through the kindness and trust of many individuals, I gained access to these prime research materials. While it is impossible to mention every individual whom I encountered while crafting this book, I will do my best to thank as many of them as possible.

First, Dr. Baker served as my mentor, editor, and guide at every step of the book's research and development. He unselfishly opened his voluminous research files in Rio Vista, Texas, to me to collect significant materials related to Aermotor's history, such as scholarly and trade journal articles, letters, brochures, sales catalogues, and various forms of advertising. His wife, Julie Baker, made me feel at home while I spent days at a time during my visits hunkered down among stacks of research materials.

Stanley A. Anderson of Skiatook, Oklahoma, to whom this book is dedicated, offered me significant insight into Aermotor's mid-twentieth-century Chicago plant operations, as well as its Midwest branch offices, relationships with its water well drilling customers, and changes in ownership. On page after page of yellow-lined legal notepad paper, he answered in writing my numerous questions with great detail. I regret that he passed away before the book's publication. However, he kindly read several of my chapter drafts and offered his input prior to his death in October 2013.

I would also like to express my gratitude to other past and current Aermotor executives and employees who shared with me their experiences at the company, namely H. Beck Atkinson, Angel Fire, New Mexico; David P. Suey Sr., Beatrice, Nebraska; Kees Verheul, Port O'Connor, Texas; and Bob Bracher, Calvin Schovajsa, Michael Guy Morrow, and Jesse Zwiebel, of San Angelo, Texas.

Other individuals who shared with me their private collections of Aermotor materials, company memories, and windmill knowledge included Bob and Francine Popeck, Batavia, Illinois; Mark S. Welsh, the Second Wind Windmill Service, Fort Worth, Texas; Alan and Judi Spear, Batavia, Illinois; Jack and Ray Roberts, Roberts Pump & Supply Co., Grand Island, Nebraska; R. Nolan Clark, Amarillo, Texas; Neal Yerian, Windmill-Parts, Westfield, Indiana; Austin Dempsey, Batavia Enterprises, Batavia, Illinois; Chuck Jones, Benton, Kansas; Mike Brigolin, Columbus, Michigan; Marcia Scholes Baldwin, Twin Lakes, Wisconsin; Paul A. Behrends, Paul's Windmill Service, Foosland, Illinois; Kevin Moore, Rock Ridge Windmills, Cloverdale, California; Ken and Sharen O'Brock, O'Brock Windmill Distributors, North Benton, Ohio; Norman H. Marks, Aero Manufacturing Co., Geneva, Nebraska; Larry A. Chilton, Vintage American Windmills, Georgetown, Texas; Mark Durham, Gicon Pumps & Equipment, Abernathy, Texas; Craig Donges, Berlin Center, Ohio; Susan Wegehoft Scott, Conway, Arkansas; Vivian R. Whalen, Hartford, Wisconsin; Carlos Fernandez-Bueno, Potomac Wind Energy, Dickerson, Maryland; Barbara Raney, Tulsa, Oklahoma; Lee Raney, Frisco, Texas; Patricia Tucker, Broken Arrow, Oklahoma; Robert R. Nigh Jr., Tulsa, Oklahoma; Craig Runyan, Williamsburg, New Mexico; David Vana, Davana LLC, Bloomingdale, New York; and Mary Ellen Rickert, Rochester, New York.

I received research assistance from a number of overseas windmill experts and historians as well, including Frans Brouwers, *Levende Molens*, Ekeren, Belgium; Etienne Rogier, Toulouse, France; Tom Conlon, Iron Man Windmill Co., Wuhan, China; Helen Walter, *The Windmill Journal*, Morawa, Australia; Geoffrey J. Moore, W. D. Moore & Company, O'Connor, Australia; and Theo Andrag, Agrico (formerly P. Andrag & Sons), Belleville, South Africa. Their efforts to answer my many questions helped me present the influence and prowess that Aermotor had in the global spread of mechanical windmill technology.

In addition, I would like to thank the research staff and librarians at the University of Chicago Library's Special Collections, Chicago Historical Society, Iowa

State University's Special Collections, and the Bentley Historical Library at the University of Michigan. On a more personal note regarding assistance from historical societies and academic institutions, I remain grateful to Carla Hill, Batavia Depot Museum, Batavia, Illinois; Reverend Elias Yelovich, Phillips Library, Mount St. Mary's University, Emmitsburg, Maryland; Steve Rosengard, Museum of Science and Industry, Chicago; and Coy Harris and his staff at the American Wind Power Center, Lubbock, Texas.

The preparation of the photos and line drawings used to illustrate this book would not have been possible without the kind technical assistance of my friend and photographer Walter J. Leskuski Jr. Many hours during the evenings were spent in front of two laptop computers at his kitchen table analyzing and discussing which images provide the reader the best visual presentation for the story of Aermotor.

Finally, I owe a tremendous debt of gratitude to my wife, Theresa, and children, Christopher Jr. and Elizabeth Ann. Without their support, this book would not have been possible. They endured my solo trips in search of historic information about the Aermotor Company, windmill-related discussions around the kitchen table, and numerous hours in the evenings after work holed up in my home office quietly reviewing files upon files of documentation, conducting phone interviews, corresponding with sources, and writing the chapters of this book. The tediousness of this project required my utmost attention.

Overall, it is my hope that I have captured the majority of the rich history of this windmill company and its place in a once-vibrant and dynamic American industry within the following pages. I have endeavored during the past three years to leave no stone unturned in my pursuit of the historic record, but at the same time realizing there may be some additional facts and details about Aermotor and its charismatic founders, La Verne W. Noyes and Thomas O. Perry, that may be uncovered by other historians and windmill enthusiasts in the future. I hope this book encourages that endeavor.

STILL
TURNING
A HISTORY OF
AERMOTOR WINDMILLS

CHAPTER ONE

Raising Water

To live and flourish, humans, animals, and plants all require water. Getting enough of this necessary liquid is often challenging. This situation has led people over thousands of years to develop an entire science around the management of this resource, known as hydrology. Some of the first efforts involved civilizations living along rivers, such as the Euphrates and Tigris in Mesopotamia (modern-day Iraq), the Hwang Ho, or Yellow River, in China, and the Nile in ancient Egypt, all areas receiving annual floods. Besides serving as a source of water, these powerful rivers brought with them nutrient-rich silt that sustained repeated years of crop production. To manage these high-water periods best, early engineers developed and constructed dams, levees, and canals, which relied on natural forces such as gravity, to channel water to and from rivers and tributaries safely and efficiently. As these civilizations evolved and learned from each other, the science of water management subsequently spread across the region encompassing Europe, the Middle East, and Asia. In the New World, ancient South and Central American civilizations understood the value of water control as well. Fixed stone channels and pipelines routed water through cities for both drinking and sanitation purposes.

In many smaller communities of the ancient world, obtaining fresh water for drinking, cleaning, and irrigation was arduous. Streams and rivers may not have been readily accessible due to difficult terrain. Daily treks were often made on foot by women and children toting heavy vessels to water sources sometimes miles away. This laborious work was further compounded by the sheer weight and sloshing of the water in the transport containers. Five gallons of water, for example, weighs about forty pounds. Day in and day out, this physical drudgery led to injuries and sometimes even death. However, without this water, communities would be forced to move or face certain death. To break the cycle of searching for water, some communities enlarged the heads of springs that reached the surface to gain greater access to water. Others constructed stone- or brick-lined cisterns, both above and below ground, to collect and store rainwater. However, some people learned they could access unseen water sources just below ground by digging holes, or wells.

Drawing of a Persian-style windmill at Nashtifan in the southern part of the Khorasan Razavi province in what is today Iran. Nashtifan ("storm's sting") is an area known for its strong wind. Courtesy Stephen James Govier, Suffolk, England.

Some early Middle Eastern and Asian civilizations developed ways to lift larger quantities of water from streams or cisterns with less human effort. About 3000 B.C., people living along the Mesopotamian river valley developed a leverage-based device called a shaduf, which used a pole balanced on a post stuck in the ground next to a water source such as a stream or spring. One end of the pole had a bucket, while the other, a counterweight. When the bucket was pushed down into the water, the counterweight helped lift the bucket of water and empty it into a trough. More than 2,000 years later, more sophisticated lift pumps emerged, including waterwheels fitted with clay pots or wooden compartments for scooping water, and tethers with a series of buckets that ran across a pulley. Unlike the simple lever, these pumps allowed for more continuous and larger lifts of water. Power for these water lifts often came from beasts or humans, whichever was more readily available.

It's uncertain which civilization can lay claim to being the first to use the wind as a mechanical power source on land, when this event actually occurred, and what its impetus may have been. According to fragmentary historic accounts, the earliest use of wind power in the form of a windmill may be linked to ancient Persia, near the modern-day border of Pakistan and Afghanistan. These early mills, used to break down grains into meal, operated in a horizontal position. They had sets of lightweight sails made from woven reeds that were spaced in a cylindrical fashion around a vertical wooden shaft that turned a set of millstones. The early Persian horizontal windmills are mentioned in a text from A.D. 950 referring to the tales of the three Banu Musa brothers of Baghdad between A.D. 850 and 870. It's possible that these windmills operated in the region even earlier than that. Around 1300, Mohammed Al Dimashqi provided a detailed description of a horizontal windmill in his book *Nukhbat-al-Dahr* (*Stories of the Centuries*). It stood in a building with vertical slots in the walls that channeled the wind toward the wheel.[1]

How the windmill first appeared in Europe is also sketchy. Some historians believe windmills first came to the continent with the crusaders when they returned from the Middle East. This story likely originated in 1690 when Antoine Furestiere published his universal dictionary. Furestiere mentioned that a crusader brought the windmill back with him to Europe. This account, while often embraced by many early wind power historians, may be farfetched since it is known that European Christians made pilgrimages to Palestine as early as A.D. 950. The first crusade did not begin until 1096. In 1190 Ambrosius described how the Arabs were surprised to witness crusaders, under the command of Filips Augustus and Richard "the Lionheart," erecting a windmill in Acre, now Syria.[2]

Since the mid-twentieth century, more historians have embraced the notion that early European windmill developers applied the wind to the preexisting principles of water mills.[3] The oldest documented mention of a European windmill was in 1183. In a document written by the Count of Flanders, Filips van de Elzas (1168–1191), a decree was made that no one could build a windmill. The document pertains to mills in Wormhout in present-day Belgium.[4] This demonstrates that windmills were known in Flanders at the time. Researchers have also found documents with references to windmills in Normandy thought to date to 1090, but they have since been determined to come from a later date. Changing dates on early documents was often done for tax purposes. In some cases, windmill dates were misinterpreted in translation of early documents from Latin. Some researchers translated the Latin word *molendinum* too often to mean windmill. The Count van de Elzas added the words *quod vento movetur* ("driven by the wind") in his landholding records.[5] Other old documents have also been found referencing windmills. Early examples include 1185 at Weedly, Sussex, and Dinton, Buckinghamshire, both in the United Kingdom; 1192 at Cabtatre on the Somme in France; and 1197 at Zonnebeke, near Ypres, Belgium.[6] By the thirteenth century, the area covering the present-day Belgian provinces of East and West Flanders contained more than one hundred documented windmills.

The earliest European windmills were "post" mills, meaning that the mill structure was mounted on top of a sunken timber post. A heavy crossbeam that balanced on top of the post served as a bearing on which the housing was turned into and out of the wind. Millwrights in time developed trestles above the ground to support the posts, as opposed to burying them, allowing them to last longer. The mill housing, which contained the millstones and related machinery, was accessed by a ladder. From the ground, the miller could adjust the cloth stretched over the wooden framework of sails according to the force of the wind. A pole directly behind the mill housing allowed him to direct the sails into or out of the wind.

It was not long before other types of windmill structures were developed in Europe, most prominently stone and wooden tower mills. Although these towers

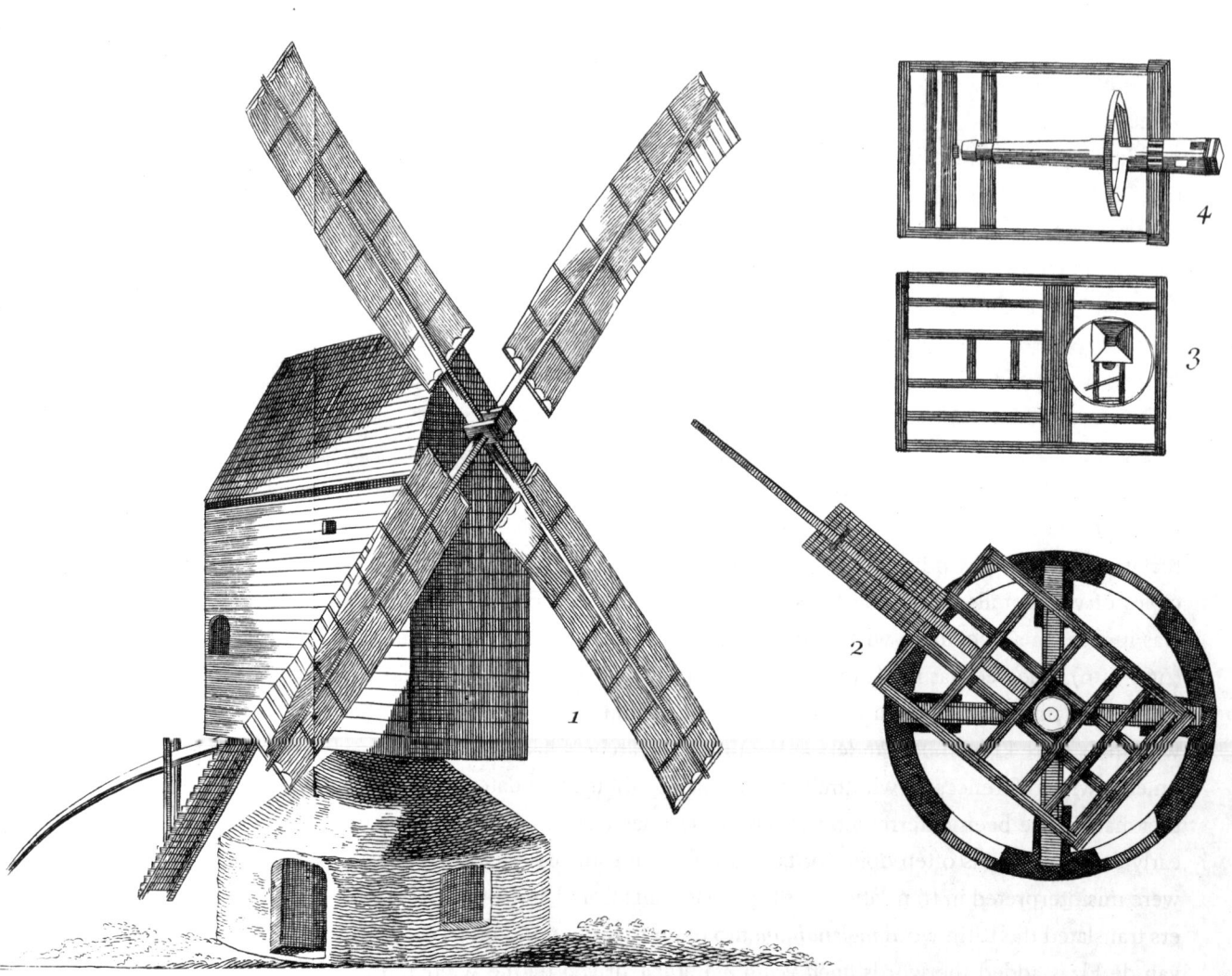

A traditional post mill was manually turned into wind on a wooden post. From *Le Spectacle de la Nature* (Paris, 1751), plate 7.

were stationary, their caps through which the sails protruded could be rotated by the miller using a pole, either from the ground or along a catwalk around the middle of the tower, depending on the height of the structure. Some stone tower mills in northern Europe reached up to one hundred feet tall with wall thicknesses at the base of more than two feet. Popular wooden tower mills in northern Europe, often called "smock" mills because from a distance they resembled a man's work smock, had octagonal sides that sloped outward at the bottom. The structure was faced with boards or, in some cases, thatch. Along the Mediterranean, tower mills were generally much shorter and simpler in design and often had stone towers. Some of these southern European windmills were built with nonadjustable sails that pointed all the time toward prevailing winds.

The oldest known water-pumping windmill dates from the early fourteenth century. This mill, called "Hoosmolen," was mentioned in a 1316 document and stood in an area known as the Bourgoyen-Ossemeersen in Ghent, Flanders.[7] Its purpose was to drain the area so that it could be used for agriculture. A century later a "poldermolen," or drainage mill, was noted in the Netherlands. In 1407 the count of Holland ordered an investigation of a "water-pumping mill" built by Floris van Alkemade and Jan Grietensoen in the neighborhood of Alkmaar.[8] From the sixteenth century onward, the need for more land caused a boom in the construction of drainage mills in this part of Europe.

At first these drainage mills could not empty the lowest waterlogged areas. Then by using a Dutch system, called "molengangen," water could be pumped out of extremely low areas in stages. A molengang is a series of drainage mills, whereby the first and lowest-situated windmill pumps water to a higher elevation where it is then pumped again to the next level and so on until the water can flow into a river or canal.[9]

Large waterlogged areas in Flanders and the Netherlands were made habitable by the work of hundreds of drainage mills over the last few centuries. These regions included Schiphol (meaning, in Dutch, "hole in which ships disappeared"), a former lake that connected to the sea, and in Flanders the area known as De Moeren near the Belgian/French border. De Moeren was a marshland pumped dry with the help of twenty-six windmills in the beginning of the seventeenth century.[10] Another impressive group of drainage mills at Kinderdijk-Elshout in the Dutch area of Alblasserwaard, which had been ravaged by deadly floods over the centuries, has become a symbol of flood control and land reclamation from the sea and was placed on the United Nations Educational, Scientific and Cultural Organization's World Heritage List in 1997.[11]

Starting in the early 1600s, water-pumping windmills were pressed into service to drain marshlands in Great Britain, especially in the eastern region of England known as the Fens. A number of patents were issued during the reign of Elizabeth I (1533–1603) for windmills to be used in pumping.[12] Walter Blith, in his 1652 book *The English Improver Improved*, depicted a water-pumping windmill with a waterwheel that lifted and relocated water. Despite use of these early windmills, draining the Fens proved far more challenging from both technical and social perspectives. Some drained areas would easily re-flood during the next rainy period. The windmills simply could not keep up with the volume of the returning water.[13] Working against these windmills were the changes to the marshes themselves. Once the wetlands dried out, the natural peat that had helped keep the landscape intact decayed with air contact or was physically removed and sold for fuel, leaving the land susceptible to worse flooding than before. The indigenous people, who once made their living from the Fen marshes by trapping and fishing, despised the intrusion of the windmills and made every effort to disable or destroy them.[14]

Drainage mills in Britain and Europe were constructed similarly to non-water-pumping tower mills. They typically consisted of the moveable tower cap from which the sails protruded, mounted atop a fixed tower. The wind wheel shaft turned a vertical wooden shaft positioned in the center of the structure that in turn meshed with another cogged gear to drive the mechanism for lifting the water. The oldest of these drainage mills used a wheel to scoop water to a higher

A hollow post mill (left) near Kinderdjik, southern Holland, with a side wheel for scooping water from a marsh into a drainage ditch. A smock mill (right) at Schermer, northern Holland, which used an Archimedes-style screw to lift water. Courtesy Jean Rogier and Etienne Rogier, Toulouse, France.

level, usually limited to about two meters. Later, many of these scoop wheels were replaced by Archimedes screws, which allowed for a greater pumping height, about four meters. It is estimated that a historic drainage mill with sufficient wind could transfer up to sixty cubic meters of water per minute.[15]

The earliest drainage mills did not have any self-regulation for sail speed. Their sails, usually four in number and sometimes more, were made from long wooden beams and overlaid with wooden lattice work and cloth for catching the wind. The sails on an axle were attached to a central hub at the top of the mill, which connected to a system of internal cogged wooden gears and shafts to drive the waterwheel or screw below. The operator was responsible for arranging the cloth sails in such a way that the mill was working at the highest rate of efficiency without endangering the physical structure from centrifugal force or from fire caused by excessive friction on the wooden bearings. One of the more interesting drainage mills developed in the fifteenth century was the so-called *wipmolen* (derived from the Dutch word *wippen,* or to "throw"), which had a distinctively large windsail sweep—in the range of ninety feet—but had a much smaller, squared building containing only the brake wheel and gearing for the drainage equipment at the bottom of the structure. These windmills were used mostly in the lowlands of West and North Holland.[16] Taking into account the changeable weather conditions that in the Low Countries (Flanders and the Netherlands) could give "four seasons" in a single day, the miller was always busy changing the spread of cloth on the sails and turning the mill structure to face the variable winds.[17]

The fantail of the Woolpit post mill in Suffolk, England. In the late 1700s, English blacksmith Edmund Lee invented a wooden multi-sailed device mounted to the back of a post mill or behind the cap of a tower or smock mill to mechanically assist in turning a windmill's sails into the wind. Courtesy Stephen James Govier, Suffolk, England.

The Chinese developed their own distinctive water-lifting windmills that often used a vertical-axis and a lightweight, wood-framed wheel with evenly spaced vertical cloth sails. When the sails turned, the power transferred to a gear mechanism that in turn ran a rope with small buckets for lifting and dumping water into fields surrounded by earthen berms. These windmills were common in coastal areas where rice farming required movement of water during growing stages. As in Europe, historians found the earliest recorded references to these mills dating to the twelfth century, but their origin may date even earlier.[18]

Besides either draining or irrigating lands for farming, windmills also found a use in ancient salt production. The extraction of salt from ground deposits or evaporation of seawater dates back more than 4,500 years. Historic accounts of who first learned to remove salt from nature and its earliest uses are up for debate. However, many scholars cite the Chinese for using salt for both medicinal and preservation purposes as far back as 2500 B.C. For hundreds of years, windmills have been used in coastal regions of Europe, North Africa, and the Middle East to pump briny water into drying pools. After the water evaporates, the leftover crystallized salt residue can then be raked, gathered, and crushed.[19]

For centuries millwrights have explored various sail designs and positions to secure the most power from the wind. But it was not until the eighteenth century that scientists and engineers gained a better understanding of how the wind interacts with windmill sails. In 1745 English blacksmith Edmund Lee (?–1763) invented a fantail, which was a wooden multisailed device mounted on the back of a post mill at ground level or directly behind the cap of either a tower or smock mill at a right angle to main sails in the front. The fantail remained at the edge of the wind while the main sails were turning. If the wind changed direction, the fantail was set in motion and through gearing guided the main sails back into the wind, saving the miller from having to redirect the mill's sails manually with every change in wind direction. Years would pass before Continental millers would begin adopting this innovation.[20]

In 1759 English physicist and engineer John Smeaton (1724–1792) presented groundbreaking research to the Royal Society of London in which he explained that the power available from the wind is proportional to the cube of the wind speed, meaning that as the wind speed doubles the power from the wind increases eight times. He conducted nineteen sets of experiments on windmill sails, including various structures, positions, and quantities of surface. Smeaton was well aware of the various sail designs used by both English and Dutch windmill builders at the time. Blades were generally angled consistently along the length of the sail. However, Smeaton's research concluded that the most efficient windmill sails should be angled at about 18 degrees where they attach to the rotor shaft and contour to 7 degrees at the tip, giving them varied pitch and a more swept appearance.[21]

Smeaton in his experiments used traditional cloth-covered wooden lattice sails. It was not until the early 1770s that this general appearance of windmill sails began to change. Scottish millwright Andrew Meikle (1719–1811) in 1772 introduced the variable-spring sail. In his design the standard sailcloth was replaced by a series of small wooden shutters, all connected to a single rod. The miller could move this rod to adjust the position of the shutters depending on the wind velocity. In time a system of springs and weights was introduced essentially to allow for the self-regulation of the shutter position and thus windmill operating speed. When the wind increased, the shutters would open, and conversely close when the velocity decreased. This improvement meant the miller was able to manage the sails with less work, but proper adjustment could be carried out only when the mill was stopped.[22] In 1789 English millwright Stephen Hooper of Yorkshire devised a sail that replaced the wood shutters with a cloth that could be unrolled more or less as required by the force of the wind. However, Hooper's "roller reefing" sail system was not automatic, was deemed too complicated, and never became widespread.[23]

By the end of the eighteenth and start of the nineteenth centuries, numerous millwrights continued their search for the most efficient windmill sails. In 1807 English millwright William Cubitt (1785–1861) devised the so-called "patent" windmill sail. Although based on Meikle's variable-spring sail, Cubitt introduced a combination of connections and weights going to the center of the main shaft of the wind wheel. This improvement allowed the miller to adjust the position of the shutters without stopping the mill.[24]

About 1840 French wheelwright Pierre Théophile Berton (1803–1861) was asked to make some repairs on a windmill. His son (with the same name, 1827–1894), who assisted him on the job, noticed the mill had very little power in low-wind speeds. He also witnessed how the sails on the mill needed to be adjusted continuously to match the variable winds. This inspired the young Berton to design a windmill sail with wood boards set parallel to the sail stock, which increasingly

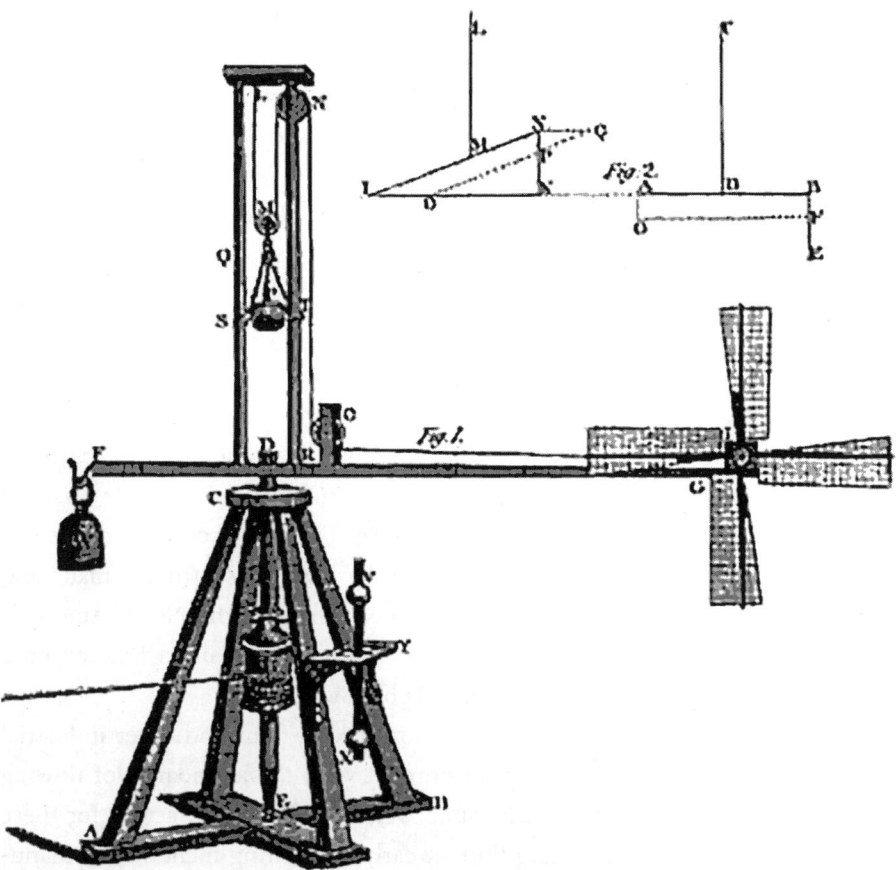

English physicist John Smeaton used this experimental device to determine the relationships between wind speed, peripheral rotational speed of sails, and achievable maximum power. He presented his groundbreaking research to the Royal Society in 1759. From J. Smeaton, "An Experimental Enquiry Concerning the Natural Powers of Water and Wind to Turn Mills and Other Machines Depending on Circular Motion" (London, 1759), plate 6.

fanned outward as the wind lessened without stopping the mill. In 1841 he patented this improvement and installed it on a windmill in Anjou, France, in 1852. These sails, however, weren't used much in northern France or Flanders due to a general lack of interest among millers for them.[25]

During the 1700s and early 1800s, a time of industrialization in Europe, the windmill remained a viable power source, and this continuing role helped drive some subsequent innovations. Windmills were built larger to generate more horsepower for jobs other than grinding flour or pumping water. They found use in a variety of industrial jobs, including crushing ores, oil seeds, salt and oak bark (for leather tanning); sawing timber; making paper, snuff, pottery, cement, fertilizers, and paints; and assisting in mining and quarry operations.[26] Although coal-fired steam engines had come into operation by the mid-1700s, it took decades for these engines to evolve further and proliferate within Europe's industrial landscape. By the mid-1800s steam power had effectively marginalized the windmill. Even so, the wind continued to prove its worth as a mechanical power source over steam engines in rural areas until the early twentieth century and the spread of electricity. As meteorologist Samuel B. Goslin wrote in 1879, the wind "needs no gauges and constant watching to feed with water, or draw the fire in the event of a short supply of water: it can now, as it was in days gone by, when men were far from the present skill and enlightenment, as found in Europe, be utilised for draining, irrigating, grinding, and many other purposes *where the introduction of steam would be a decided waste of money for fuel*, and the working at an immense cost by European and other skilled drivers."[27]

Europeans exported their windmill technology to colonized lands throughout the world. In the early seventeenth century, the Dutch settled in what is known today as Lower Manhattan, New York, and erected a windmill. In 1621 a post windmill was built on the Flowerdew Hundred estate twenty miles outside the Jamestown settlement in Virginia. The windmill survived the Indian uprising in March 1622. The last historic reference to it was in 1624 when the estate was sold by Sir George Yardley to Abraham Piersey.[28] Over the next 250 years, numerous European-style windmills were erected along the East Coast, from Canada and Maine to Florida, and eventually deep into the interior of the North American continent with the migration west. There were even colonial-era Russian-built windmills at Fort Ross in northern California by 1814.[29]

Windmills were less prominently used for making flour and other industrial applications in the New World than in Europe. With the abundance of flowing water along many coasts, water mills often were simply more practical for these activities. Windmills, however, did find an early and lasting niche in the manufacture of salt in coastal America. With the outbreak of the Revolutionary War in 1776, some remote coastal communities, such as those on Martha's Vineyard and Cape Cod, Massachusetts, found themselves cut off from the mainland by the presence of British warships. One of the commodities that quickly ran short was salt for preserving fish, drying leather, and making medicines. The old local process of evaporating salt water by boiling it, leaving a residue of salt crystals, proved to be too inefficient to support the demand. In 1776 Captain John Sears of Dennis, Massachusetts, developed a new system using large wooden vats to hold seawater, which utilized the sun to evaporate the liquid. The salt left inside the vats then could be scraped off. This method was soon combined with a water-pumping windmill developed by Major Nathaniel Freeman of Harwich, Massachusetts, and with the increase in consumer demand following the Revolutionary War, the concept rapidly spread throughout Cape Cod and neighboring areas. Freeman's windmills, which sat on top of wooden stands, consisted of four wood-lattice sails covered in canvas to catch the wind and drive a small pump.[30] The system generally consisted of one or two wind-powered pumps, a series of wooden pipes, and evaporation vats.[31] These salt works required little maintenance, and the salt exposure actually helped preserve the wooden pipes, vats, and workers' hand tools. It's estimated that by 1802 there were 136 separate wind-powered salt works on the Massachusetts coast, producing more than 40,000 bushels (at about 80 pounds a bushel) of salt annually. The number of operators had increased to five hundred by the start of the Civil War, and they were able to produce more than a million

A typical wood-constructed windmill once used for pumping brine for salt-making in New England. From Albert Cook Church, "The Padanaram Salt Works," *New England Magazine* 41, no. 2 (October 1909): 488.

bushels of salt per year. This industry on Cape Cod withered away in the early 1900s, as it was unable to compete with other types of large-scale salt operations elsewhere in the United States and Canada.[32]

Salt manufacturing employing simple wind pumps was similarly set up along the coast of California shortly after the signing of the Treaty of Guadalupe Hidalgo in 1848, which placed the territory within the United States. The population exploded soon thereafter as a result of the 1849 Gold Rush.[33] Sea-salt production sites emerged along the San Francisco Bay area. Erected low to the ground and dotting the flat landscape were numerous large, multibladed wooden windmills using the prevailing winds off the ocean to elevate seawater to drying pools. The towns of Hayward and Mt. Eden, in particular, had at the height of their sea-salt industry about seventeen companies producing upward of 17,000 tons of salt annually.[34]

A downside to living around America's coastal marshes was often insufficient access to clean drinking water. Most surface water became mixed with salt water due to regular tidal flows. Early inhabitants either had to collect rain or buy fresh water from vendors who transported it in barrels on wagons from further inland. Some early villages were able to tap fresh water from underground using rudimentary pumps. In the Maryland eastern shore town of Crisfield before the Civil War, for example, a simple windmill was installed to pump water from a 200-foot-deep well into a large cypress tank. A local historian wrote, "The water was sold to the general public at two cents a bucket. Drinking water consumed on the premises was free . . . The windmill was owned and operated by Captain Augustus Maddox who made a comfortable living from his water enterprise. It was a novel sight to see the throngs of people lined up at the old windmill for water during a summer drought."[35]

In the early 1800s, increasing numbers of Americans began to venture deeper into the interior of the continent in search of economic and agricultural opportunities. Despite occasional clashes with Indian tribes and the harshness of Mother Nature, these early westbound settlers found an abundance of timber for shelter, food sources (both natural and cultivated), and rivers, streams, and springs with fresh water near which they built their dwellings. However, the farther one moved past the Mississippi River and onto the Great Plains and into the arid prairie and desert West, readily accessible surface water became increasingly difficult to find. It often became necessary for the homesteader or stock raiser to hand-dig a well. The process of locating a suitable well site included both observations of the geography and a bit of luck. Some farmers employed a water-finding technique known as "dowsing," in which an individual used a V- or Y-shaped twig. The process went like this:

> The forks of the twig are about 18 to 24 inches in length; the user holds one in each hand; the twig is held with the point downwards, and when the operator passes over the spot where water exists, the twig simultaneously and entirely independent of the operator, moves upward, in the form of a semi-circle.[36]

Dowsing is found among many cultures throughout the world, even though many have questioned its value in locating water.

It was not uncommon for well holes to be about four feet wide and to exceed fifty or more feet in depth. Deadly cave-ins were always a concern, and sometimes toxic buildups of carbon dioxide, or "damp," in the bottom of the well could overcome the digger.[37] Dug wells were generally lined with rock or wooden boards and covered at the top to prevent contamination. An early twentieth-century authority wrote, "Even a small fishworm may cause a considerable disturbance in the water, and toad will render it unfit for drinking for many days."[38] Consumption of water from contaminated wells was considered a facilitator of diseases such as typhoid fever, which was common in rural areas.[39]

Other tools were developed to tap water deeper underground. One of those tools, the hand auger, employed a piece of one-inch-diameter pipe to which was attached a cutting edge at one end and a perpendicular handle at the other. One

Men drinking water from a hand-dug well in the early 1890s. Author's collection.

An Elizabethtown, Pennsylvania, well driller's postcard advertising the health benefits of drilled wells versus traditional hand-dug wells, ca. 1875. Author's collection.

to two men turned the handle, allowing the edged end to dig into the earth. When the dirt filled the top of the cutting edge, the auger was pulled up and emptied, and the process was repeated. Pipe would be added to keep the wall of the bored hole from caving. Another common tool used to reach water required one or two men to hammer a pipe fitted on one end with a drive point and strainer. As each section was sunk, another piece of pipe was added until the drive point and strainer reached the water-bearing layer. Both manual processes proved sufficient in areas with high water tables or well depths no deeper than twenty to twenty-five feet. Once the pipes in the well were secure, a hand-driven suction pump was installed on top to bring the water to the surface.[40]

In areas where water tables were deeper, such as on the Great Plains, large percussion drills and hammers were required to "drive" the wells. Supported by a wooden derrick or tower, the percussion method used a cable to which was suspended a heavy steel drill bit. The bit would be dropped into the earth, so that with each strike it bored a hole, simultaneously smashing through any rock. The bit would then be lifted out of the hole by a winch turned by either a few men or a couple of horses. In the case of the hammer method, a heavy weight was suspended by a cable and repeatedly dropped on the top end of a pipe in order to drive it into the ground. For either boring method, water was poured down the holes to help float rock fragments and dirt, forming a muddy mixture or slurry. A long tubular bucket with a foot-valve was lowered into the hole, and when the valve-end struck the bottom of the well, it opened and allowed the bucket to fill with the slurry. When it was time to pull the bucket to the surface, the foot-valve closed, keeping the slurry in the bucket. This process of either drilling or hammering and then clearing out the hole was repeated over and over, often at one- to two-foot intervals, until the water table was reached. Driven wells were lined with iron casing, while hammered wells used the driven pipe to case the hole.[41]

By the late nineteenth century, more powerful steam-powered equipment inspired by the burgeoning oil and gas industry entered service and allowed water wells to reach depths from several hundred feet to more than a thousand feet. These expensive, one- to two-ton machines generally consisted of an engine, several drums for winching and controlling ropes, and a derrick to support the drills or hammers for driving the well. The apparatuses were constructed on wheeled carriages to transport them from job site to job site. While the work remained difficult and dirty, well drillers were generally well paid for their services, often charging their customers on a per-foot-drilled basis.[42]

An array of manual methods was developed to lift water from wells. For open hand-dug wells, water could be raised from the depths with a bucket attached to a rope. To make it easier to lift the bucket, windlasses and pulley mechanisms were added. Mechanical hand pumps were also developed in large numbers in the 1800s to bring water more easily to the surface. Most of these pumps were made of wood or cast iron and included a handle, which was then connected to a "sucker" rod that ran down the pipe to the well where at the end was a cast iron or brass cylinder containing a plunger pump. The upright cylinder was fitted with a check valve in its lower end, which allowed water to enter on the upstroke of the plunger. On the down stroke, the valve mounted to the plunger opened, allowing water to stay inside the cylinder. With the next upstroke the plunger's check valve, now closed, pushed the water out of the cylinder and into the drop pipe, while at the same time the cylinder's check valve, now opened, refilled the cylinder. With each subsequent plunger stroke, a column of water was forced up the drop pipe until it eventually reached the surface where it could be either used or put in a storage tank. To put water from the well under pressure through pipes

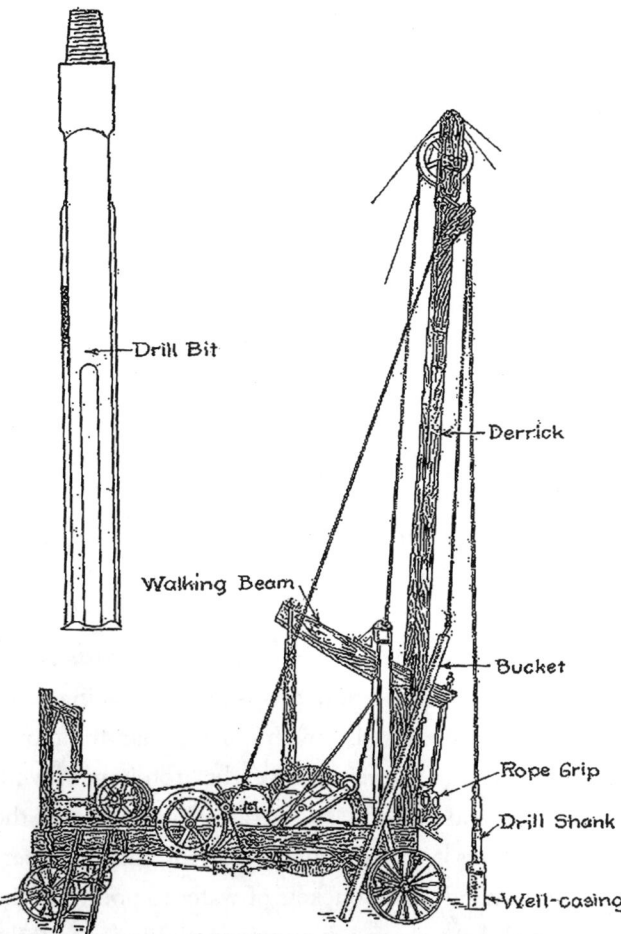

Drill Bit

Derrick

Walking Beam

Bucket

Rope Grip

Drill Shank

Well-casing

A typical professional well driller's rig from the late nineteenth century. From Harry C. Ramsower, *Equipment for the Farm and the Farmstead* (Boston, 1917), 154.

or move it to higher elevations, it could be run through a device called a "stuffing box" or "packer head."[43] Some hand pumps became quite sophisticated in an effort to reduce physical exertion on the pump handle and drive more water more quickly to the surface from the well. One of the more unusual lifting devices, used over topped cisterns or hand-dug wells, employed a rotary handle with a geared wheel on which was hung a loop of chain with tiny buckets spread out in intervals. With a turn of the handle, the chain ran down through a pipe submerged in the well water and came up parallel to the pipe in the open to carry water to the top. As the chain passed over the top of the geared wheel, the water in the buckets emptied into the delivery spout. A practical guide to farming published in 1870 viewed traveling pump salesmen and their wares useful, but recommended caution: "these pumps have been mostly in the hands of pump peddlers, who have charged enormous prices for putting in the pumps; though to their credit, ready to warrant a supply of water, and, so far as we know, have carried out their contracts faithfully. But the high price charged has been ample to protect them against loss."[44]

Increasing numbers of livestock, particularly cattle, placed tremendous pressure on available surface water sources in the arid West. It has been estimated that between 1850 and 1860 the combined herds of dairy, working oxen, and other cattle in the United States and its then territories increased 33.5 percent, or from nearly 18.5 million head to more than 25.5 million head.[45] While people on small farms could use hand pumps to supply their cows, hogs, sheep, and poultry with water, large cattle operations on the open ranges had to move their herds around continuously, often many miles, to find suitable water supplies, such as rivers, streams, and watering holes. Competition between cattlemen for water was fierce, and during the hot, dry summer months it intensified. As western lands eventually became compartmentalized through ownership and with the introduction of barbed wire in the 1870s and 1880s, access to sufficient water became an even bigger problem.[46] "Since a cow would not walk over fifteen miles for water in a day, the man who controlled a spring or never-failing water hole monopolized the grass for seven and a half miles around it. In the 1860s and 1870s, farseeing cattlemen busied themselves securing all the streams and isolated springs for cattle range," explained Nebraska historian Everett Dick.[47] Less powerful ranchers and farmers and late-arriving homesteaders quickly found themselves holding land with no readily available surface water.

Starting in the late 1850s, the railroads from the East and West Coasts began to penetrate deeper into the heartland, linking new towns and generating commerce through their ability to haul tons of freight much quicker than traditional horse or oxen drawn wagon transport. However, these early "iron horses," with their steam engines, required water, and lots of it, to operate. Without water, the engines would seize. In the earliest days it was not uncommon for train engineers to stop along streams and ponds to hand-scoop buckets of water to pour into the boilers.[48] Operators quickly learned that to keep their engines running efficiently, especially across the arid West, a dense network of stored water tanks needed to be erected along the tracks.

Closely following these two trends for large-scale water use was a group of inventors, engineers, and salesmen who seized upon the need for commercially manufactured water-pumping windmills. These wind pumps needed to be relatively easy to erect, durable, and able to run with minimal maintenance. Most importantly, however, they needed to "self-govern" themselves in variable winds. Without this functionality, windmills could be easily torn apart by high winds.

The pursuit of self-governing water-pumping windmills had its origins in both Europe and America. Traditional-style European windmills were labor-intensive in that they required the physical presence of a skilled miller to operate. A self-governing windmill should be able to run anytime, day or night, adjusting itself to changing wind directions and speeds, even turning off, so to speak, when the winds became too strong for safe operation. Numerous claims can be made from patent records and other documentation about when and who developed the first self-governing windmill.

French sculptor and engineer Amedée Durand (1789–1873) began work in the late 1820s on designing a self-governing windmill. In 1836 he published the article "Note sur un Moulin à vent s'orientant et se réglant de lui-même" ("Note on a windmill that directs itself in the wind and regulates its speed"), and he put one of these windmills into service in the early 1840s. Durand's work was further highlighted in an illustrated article, published in 1864 by the *Gazette du Village*. One of these windmills was installed on top of the city hall to supply water to the Gerberoy (Oise) commune, while another was put up at Montbron in Charente around 1850.[49] According to John Walters and Régis Girard, who researched the history of French water-pumping windmills, "Durand wind-engines were distinguished by canvas sails spread over wooden frames, giving an appearance similar to many mills that can be seen working in Southern Europe from Portugal to Crete. However, the blade disk 'trailed' behind the main vertical pivot and a system of levers attached to the sails allowed them to rotate axially to shed the wind. A weight-and-chain mechanism returned the sails to their original position when the gusts abated."[50] Durand had sold at least forty of his windmills by the 1860s. During this period he practically had no competitors in France.[51]

One of the earliest patents in the United States for a water-pumping windmill was issued on September 6, 1833, to James Kerr of Maury County, Tennessee. The mill consisted of a swiveling framework on top of which sat a wind wheel made from six sheet-iron blades and backed by a vane to direct the device into the wind. The windmill design flaw was that it could not regulate the speed of its blade. A wind could have easily destroyed this machine.[52] Another experimenter, Allen S. Ward, a teacher at the Shawnee Indian Manual Labor School in eastern Kansas, claimed in 1850 to have built a windmill that "turns to face the wind, & the sails are so arranged with springs that they yield to the force of the wind when it blows hard, so that it will not run much faster in a hurricane than in a light breeze."[53] In January 1856 inventor Frank G. Johnson of Brooklyn, New York, obtained a patent for his self-regulating windmill. Johnson included in his windmill design iron weights and springs to control the blades' position, "causing them, whenever their velocity is too great, to be more or less turned edgewise to the wind, and *visa versa*."[54] The windmills could also be easily taken down and moved for erecting at another location, and sold at prices ranging from $50 to $800 depending on size.[55] While there was significant desire for water-pumping windmills, many people distrusted the claims made for their performance. An author writing in an 1860 issue of *Scientific American* about water needs in Texas wrote, "There is a million dollars lying waiting here for the first man who will bring us . . . a windmill, strong, durable, and controllable."[56]

As agriculture in northern California expanded beyond San Francisco Bay into the Sacramento and Stockton areas during the 1850s, homemade and locally produced windmills were frequently used to water gardens and livestock. In a typical instance, the *History of Contra Costa County, California*, published in 1882, refers to two brothers, John and Joseph Pulsifer, who settled along the banks of the San Joaquin and, during the dry periods, "arranged a windmill and a pump for raising water from the tules, making one of the finest gardens in 1851–52 within ten miles."[57] The design of this windmill, or whether it was manufactured or homemade, is unclear. However, records indicate that William Isaac Tustin of Benicia, outside Stockton, had developed a water-pumping windmill as early as 1849.[58] His

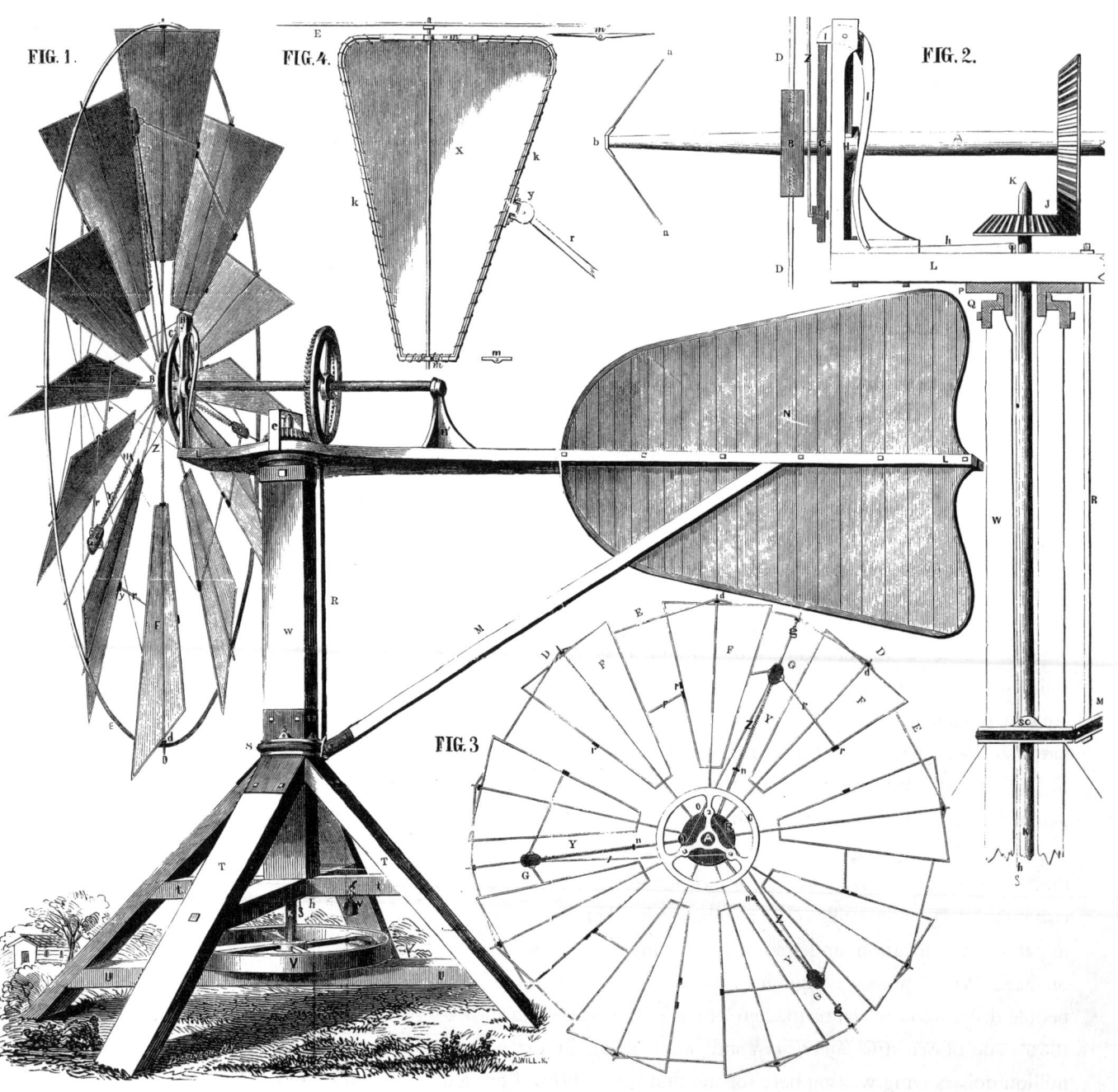

FIG. 1.

FIG. 4.

FIG. 2.

FIG. 3

An engraving of Johnson's windmill, a wind machine for both pumping and power work that was manufactured and sold on both the Atlantic and Pacific coasts in the 1850s. From Frank G. Johnson, *The Wind as a Motive Power* (New York, 1856), n.p.

machine consisted of a wheel of wood paddles erected upon a four-post tower made of heavy timber. Tustin had become California's largest windmill manufacturer by the early 1870s, but his business eventually would be overtaken by larger manufacturers.[59]

One of the most successful windmill entrepreneurs in the history of the business was Connecticut mechanic Daniel Halladay, who was inspired by pump repairman John Burnham to design and build a self-governing mill. Halladay, who accepted the challenge, was reportedly skeptical about whether anyone would want to use it. He quipped, "I can invent a self-regulating windmill that will be safe from all danger of destruction in violent windstorms; but after if I should get

Patented June 11, 1872.

William Isaac Tustin of Benicia, California, developed a commercial water-pumping windmill in 1849 and became the largest windmill manufacturer in the territory by the early 1870s. From *Langley's San Francisco Directory* (San Francisco: Francis, Valentine & Co., 1880), 33.

it made, I don't know of a single man in the world who would want one."[60] He would certainly prove himself wrong, for in time Halladay would become a dominant, early windmill manufacturer that others would attempt to emulate.

On August 29, 1854, the US Patent Office granted Halladay a patent for a four-bladed self-governing windmill. A broad wooden vane kept the blades facing into the wind. Halladay's "improved governor" consisted of an iron lever mechanism attached to a hub, which kept the one-and-half-inch-wide blades at an angle best suited to catch the wind. As the velocity of the wind increased, the centrifugal hub reduced the pitch of the blades automatically to prevent the wind wheel from spinning out of control, destroying itself and the water pump.[61] He further

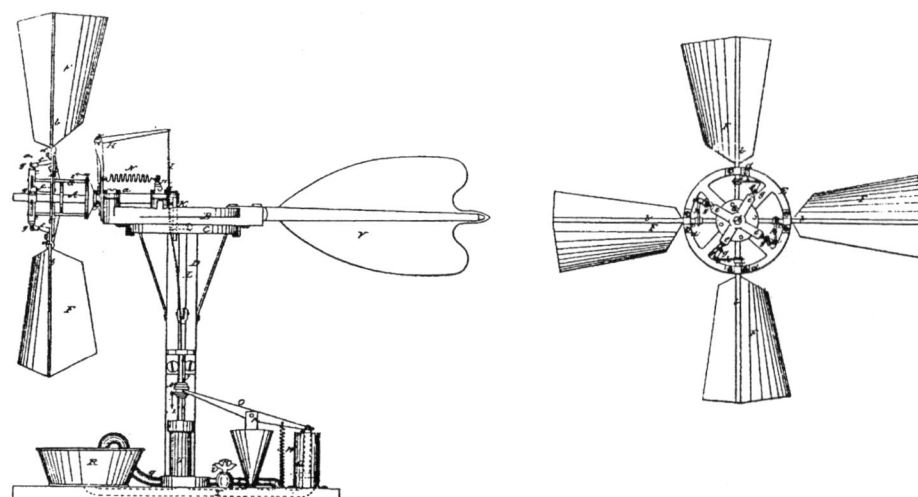
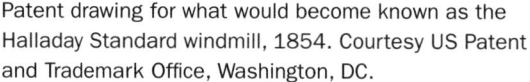

Patent drawing for what would become known as the Halladay Standard windmill, 1854. Courtesy US Patent and Trademark Office, Washington, DC.

Daniel Halladay, who in his older years developed one of the most commercially successful self-regulating windmills in the mid-1850s. Courtesy Batavia Depot Museum, Batavia, Illinois.

emphasized in his patent application that the windmill is "particularly designed for farmers' use and in other cases where a moderate power is required. They are not expensive to manufacture, and may be moved with little difficulty from place to place in order to perform different sorts of work required to be done."[62]

It's inconclusive, but some historians occasionally speculate that both Halladay and Tustin were influenced in their designs by already operating salt-manufacturing water pumps, since both men were in proximity to these thriving industries.[63] Geographer Terry G. Jordan, for example, wrote in a 1973 article on the evolution of US windmills that "Halladay's major contribution, and a very important one, lay in making his mill self-adjusting to changes in wind direction. In most other respects, Halladay and Burnham simply experimented with relatively minor modifications of a machine already well known to them."[64] However, these two men actually took the windmill to another level by adding a governing mechanism to regulate its speed of operation.

With patent in hand, Halladay joined Burnham in forming a manufacturing company based in South Coventry, Connecticut, in 1854. Halladay was designated superintendent of the new operation, and Burnham became the general sales agent. The Halladay Wind Mill Company promised its customers an operating cost of no more than $5 a year.[65] The first commercial Halladay windmill was erected on the Long View farm of H. A. Grant of Enfield, Connecticut, in 1854 and served as his primary source for water on the property until 1872. "[D]uring the eighteen or nineteen years that it was in constant use, day and night, (except when my water tank was full), the entire cost of repairs, exclusive of Sperm oil for lubricating, was not over $15; and with one or two dollars expenditure the old Mill would do satisfactory work for the next five years. I consider it the cheapest and most durable piece of machinery I have ever had on my farm," Grant wrote in a testimonial letter on September 18, 1876.[66]

Factory of the U.S. Wind Engine and Pump Company, Batavia, Illinois, during the late nineteenth century. Courtesy Batavia Depot Museum, Batavia, Illinois.

Burnham quickly realized that the future windmill market was in the West and relocated to Chicago in 1857 to form the U.S. Wind Engine and Pump Company for sales, while manufacturing stayed in South Coventry under the Halladay Wind Mill Company. By 1859, the company sold mills with wheel sizes from eight feet in diameter to twenty feet. The prices ranged from $120 for the smallest mills to $475 for the biggest.[67] However, Burnham became frustrated by the shipping delay that occurred between South Coventry and the western states. In 1862 the U.S. Wind Engine and Pump Company purchased the Halladay Wind Mill Company and relocated manufacturing to Batavia, Illinois, just west of Chicago. The company built a factory along the Fox River and benefited from an abundance of skilled Swedish immigrants in the town.[68] Within fifteen years of its relocation from South Coventry to Batavia, the company had cash capital of more than $100,000 and about 150 men making windmills and pumps.[69]

Batavia, often called the "windmill city," is located about thirty-five miles west of Chicago. The town gradually became a hub for other windmill manufacturers not long following the arrival of the U.S. Wind Engine and Pump Company. This activity was further propelled by the opening of a second channel for the Fox River in the early 1870s, giving additional access to water power. The Chicago, Burlington & Quincy Railroad, which connected to the North-Western Railroad, brought in raw materials and shipped out finished goods.[70] At its height, there were upward of six windmill factories operating in Batavia, in addition to other substantial businesses that manufactured paper, wagons and carriages, and agricultural equipment, flour, and lumber, among other products. Many of the factory buildings were constructed of cream-colored limestone taken from quarries in the vicinity.[71]

Other windmill manufacturers similarly migrated west to be closer to the largest user market, setting up factories throughout the states bordering the Great Lakes. Census records registered sixty-nine windmill manufacturers in 1880, and of those twenty-three were located in Illinois—the biggest concentration of any state.[72] While many numbers have been bantered around, the census records suggest the industry peaked at seventy-seven manufacturers in 1890.[73] The size of these companies ranged from small machine shops to large agricultural implement makers, which included windmills as part of their sales portfolio. Even some of the early mail-order catalogue firms such as Sears, Roebuck and Company and Montgomery Ward offered windmills for sale. Many large windmill makers became farm implement producers and sellers in their own right. For example, U.S. Wind Engine sold a range of hay-moving equipment to farmers.

A few years after the relocation to Batavia, Halladay redesigned his wooden wind wheel away from the individual paddle-shaped blades to the so-called "Rosette" form, which consisted of narrowly spaced sections of thin wooden blades.[74] Windmill historian T. Lindsay Baker explained that this new style of wind wheel used a centrifugal governor to draw the hinged wheel sections away from increasing winds as the rate of rotation increased, and returned them to the flat wind wheel position as the winds decreased. Like Halladay's earlier mills, a rigid vane assembly kept the wheel facing toward the wind.[75] Throughout the late nineteenth and early twentieth centuries, other US windmill manufacturers developed wheels with blades that varied their pitch according to the velocity of the wind. The two most notable examples included the "Hazen" windmill of F. W. Winters and Company of Fairbault, Minnesota, and the Tucker Automatic windmill by Temple Pump Company of Chicago.[76]

Early commercial windmills tended to fall within four categories of design. They included the sectional wheel with centrifugal governor and rigid vane, solid wheel with side-vane governor and hinged vane, a solid wheel with swiveling vane parallel to the wheel, and vaneless (sectional wheel with no vane). Each type, as was proved over time, had its own advantages and disadvantages. For example, it has been observed that the vaneless windmill, which relies on the pressure of the wind on the wheel to keep it in the correct position, is best suited for some applications and in locations with sustained strong winds.[77]

The Eclipse, another commercially successful windmill introduced in the mid-1860s, initially used a wheel with four wooden paddle-shaped blades and a side-vane governor. The windmill was designed by Reverend Leonard H. Wheeler, a missionary who lived among the Ojibway Indians in northern Wisconsin starting during the 1840s. He thought that a lightweight, self-governing windmill would be a useful tool for the Indians to grind their grain and pump water. It is said that Wheeler began working on his windmill concept while convalescing after a fall in which he broke his wrist:

> [T]o keep his mind engaged while the fracture was healing he drafted with his uninjured hand the plan of his invention, using a jack-knife and a board laid out across his chair. The said day that his wrist was broken, his son, who had started on a journey to St. Paul, 200 miles away, was brought back with a broken leg caused by a falling tree. The father had a good pair of legs and the son a pair of sound hands, and as soon as they were able

they made the wood parts of their wind-mill. The government black-smith nearby became interested in the project, and made the necessary mountings. April 26, 1866, the new invention was put in operation and worked successfully at first, but a storm soon after tore it to pieces. This led to a deeper study of the problem, and Wheeler then conceived the idea of a "side vane" set against wind, with the wheel pivoted so a strong wind would blow it around at an angle or entirely out of the wind. In two months a mill of the self-regulating type was in operation. The inventor, however, was failing fast in health, and that fall he moved to Beloit, Wis., surrendering his missionary work. In the spring of 1867 a wealthy banker from the east, a relative, visited them, and he recognized what had never occurred to Wheeler, the value of a patent on the invention. At his solici-tation Wheeler began on a model, but his strength only permitted him to work a few minutes at a time, and it was not finished for two months. It was then sent on to Washington, however, and a patent was granted in due time. Manufacturing was begun soon after at Beloit, laying the foundation for the business of the Eclipse Wind Engine Company.[78]

The railroads were some of the biggest customers for Halladay and Eclipse wind-mills. One of the early large orders to the U.S. Wind Engine and Pump Com-pany was for seventy windmills for the Union Pacific Railroad to supply water to its locomotives. By the mid-1870s, more than one hundred railroads, large and small, had begun using the Halladay windmills. Most of these companies paid for these windmills through monthly installments, thus remaining the property of U.S. Wind Engine until the purchase contracts were met.[79] At the outset the com-pany made little effort to sell windmills to cattlemen west of the Mississippi until the later 1870s.[80] Early sales efforts for Eclipse mills were similarly focused on rail-roads, especially between 1873 and 1880, with orders coming from such firms as the Chicago & Northwestern; Chicago, Milwaukee & St. Paul; Illinois Central; Great Northern; Atchison, Topeka & Santa Fe; Chicago, Burlington & Quincy; Chicago, Great Western; Chicago, Rock Island & Eastern Illinois; Wabash; and Missouri, Kansas & Texas, to name a few.[81]

For many farmers and ranchers, the windmill was a welcome investment and an early example of how to start mechanizing agricultural activities previously performed by muscle power. Before the windmill, the farmer or a member of his family was responsible each day—in many cases for hours at a time—for hand-pumping or lifting water to the surface from a well for household use, gar-den and crop irrigation, and livestock watering. The windmill helped to supplant this burdensome manual task. Excess water from the windmill pump could also be stored in tanks for future use and even channeled to where it was needed on the property. Historian Walter Prescott Webb stated that the attractiveness of windmills was their ability to be broken down for shipping and assembled quickly on site with basic hand tools; interchangeable and durable parts; and ability to self-regulate in the wind.[82] Yet these early machines required routine

Original Eclipse windmill of 1867 as it later appeared in a pageant at Beloit, Wisconsin, in 1912. Courtesy *Windmillers' Gazette*, Rio Vista, Texas.

maintenance, particularly lubrication of their exposed metal parts. On ranches where windmills were frequently abundant and spread across hundreds of acres, a farmhand or cowboy generally was designated to check the mills and make sure the moving parts were properly greased. This was essential maintenance for early windmills with open-gear systems exposed to airborne dirt and moisture. Sprawling ranches, like the XIT in Texas, which had over 350 windmills across its property, would have individual cowboys whose sole job was to ride from windmill to windmill to apply grease and/or make repairs. The work was surely dangerous at times, both on the ground and in the air. If injured by falling off a rickety tower and breaking a leg, for example, a man might not be found for days, risking death in what could be harsh conditions. Some of these windmill men became quite skilled in their work and in time became independent businessmen as freelance "windmillers." [83]

In an 1884 presentation to the Engineer's Club of St. Louis, member James W. Hill observed that solid-wheel windmills in the range of ten to thirty feet in diameter, like the Eclipse, had also carved a prominent place in the Midwest landscape,

A Halladay Standard pumping windmill shown in the style after the paddle-shaped blades were replaced with thin wooden blade sections nailed to slotted wooden rims. From U.S. Wind Engine and Pump Company, *Special Geared Wind Mill Circular* (Chicago: Culver, Page, Hoyne & Co., ca. 1878), n.p.

not just at railroad stops and on ranches but in towns for municipal water purposes. He further described how a fourteen-foot-diameter Eclipse windmill and a related tank and underground piping installed at Arkansas City, Kansas, in 1881 quickly made a return on its investment: "We estimate that this mill is pumping from 18,000 to 20,000 gallons of water every 24 hours. We learned that these works have saved two buildings from burning, and that the water is being used for sprinkling the streets, and being furnished to consumers at the following rates per annum: Private homes, $5; stores, $5; hotels, $10; livery stables, $15. At these very low rates, the city has an income of $300 per annum. The approximate cost of the works was $2,000. This gives 15 per cent interest on the investment, not deducting anything for repairs or maintenance which has not cost $5 per annum so far."[84]

In addition to pumping water, the large windmill manufacturers of the 1870s supplied farmers and businesses with so-called "power" mills for grinding grain, cutting fodder, shelling corn, sawing wood, and even washing clothes. These windmills were more costly than their pumping counterparts, and some of the machines could be set up to perform both pumping and power work. U.S. Wind Engine in 1871 sold power mills ranging in price from $350 to $3,500.[85] Some of these power mills, based on U.S. Wind Engine designs, incorporated much larger wind wheels. J. A. Wheeler erected one of these mills at Freeborn, Michigan, just after the Civil War, which had a fifty-eight-foot-diameter wind wheel, based on his own patent, that provided power to turn one set of four-foot-wide millstones for flour and another set of three-foot-wide stones for corn and feed. Wheeler wrote to U.S. Wind Engine in March 1870, claiming that two months earlier the mill had ground 8,400 bushels of feed and that February, 2,000 bushels of wheat. Wheeler's brother operated a similar U.S. Wind Engine–based windmill for manufacturing flour and feed at Owatonna, Minnesota, but this unit had a sixty-two-foot-diameter wind wheel.[86]

These early US windmill makers were fierce competitors, displaying their products at county fairs and big city expositions; advertising in numerous farm journals, both in the United States and abroad; and deploying small armies of salesmen to travel the countryside with working scale models and descriptive literature to entice dealers to sell their windmills. It was not uncommon for disgruntled managers and workers to break away from one firm and either join another or set up a new business altogether. This led to windmill companies taking legal action against each other. One colorful example, which played out in the courtroom and the press in 1869, involved a patent infringement case between Continental Wind Mill Company in New York City and Empire Wind Mill Manufacturing Company of Syracuse, New York. A federal circuit court ruled in 1871 that Empire owned the patent rights of Continental's former employee Addison P. Brown, at that time with Empire. During the case, both Continental and Empire threatened to sue customers to dissuade them from doing business with either company.[87] With so many experimenters and entrepreneurs seeking patents for "improved" windmill designs during the 1850s through to the 1880s, there were constant threats to established manufacturers of product infringements and outright thefts of designs. William Isaac Tustin in an advertisement of his Economy Wind-Mill, published in 1880, warned, "Persons infringing on this patent will be prosecuted to the full extent of the law."[88]

Even during tests and placement of windmills at public events, stirred-up hostilities could quickly result in public rants. Such was the case for U.S. Wind Engine in the aftermath of a test of windmill efficiency by the Pennsylvania State Agricultural Society on its exhibition grounds on September 18–27, 1884. The society sent invitations to fourteen major windmill manufacturers of the day. Five agreed to participate, and one of those, U.S. Wind Engine, only grudgingly accepted after some back-and-forth discussions with the organizers over the size of the windmill it could enter. The other entries in the test were the Leffel's Improved Iron Wind Mill, made by the Springfield Machine Company of Springfield, Ohio; the Perkins Solid Wheel windmill, made by the Perkins Wind Mill Company of Mishwauke, Indiana; the Woodmanse Solid Wheel windmill, made by the Woodmanse Wind Mill Company of Freeport, Illinois, and exhibited by S. W. Kennedy of Philadelphia; and the Manvel Solid Wheel windmill, made by B. S. Williams and Company of Kalamazoo, Michigan. The test required the use of windmills with ten-foot wind wheels to be "erected in the same manner as if they were being put up in the ordinary course of business."[89] During the test, the No. 2 Halladay Standard

Large roof-mounted Halladay Standard windmill used to power equipment. Author's collection.

Wind Mill from the U.S. Wind Engine Company pumped 44,538 gallons, followed only by the Woodmanse windmill at 29,742 gallons. (The Springfield company dropped out of the test after two days.) The society noted in its test results that the U.S. Wind Engine mill had a wheel diameter of ten feet, four inches (or an additional four square feet of wind surface) and that "[w]hile this mill pumped a large quantity of water, in the strong winds experienced during the test, had the winds been light, it is questionable as to its being able to perform the proportional large duty required of it."[90] U.S. Wind Engine retorted that this is "not correct and proves the committee ignorant of scientific principles," and further stated the position of its windmill compared to the others was not as "favorable" due to the location of a building "several hundred" feet away, a wind obstacle. "It would have seemed fairer to all concerned if the committee had reported the speed indicators poor, frail machines and let it go at that," U.S. Wind Engine said.[91] Interestingly, U.S. Wind Engine would unashamedly tout the results of this test in its sales literature.[92]

By 1870, U.S. Wind Engine advertised that it made fourteen sizes of windmills. "Our ordinary size mills, in ordinary depth wells, with large reservoirs, will fully supply 500 head of cattle during the entire year," it asserted in an 1871 catalogue.[93] In general, windmills weren't insignificant investments, especially for small farmers, with costs ranging from $110 to several hundred dollars apiece depending on size, tower height, and any additional equipment. This was in addition to the cost of sinking a well. George A. Carter, a windmill installer for U.S. Wind Engine, who started with his father in 1867, described the difficulty of selling farmers windmills in a letter to the *Farm Implement News* twenty years later:

> When we began the sale of mills it seemed too much to tell a farmer that we could set him up a machine which would pump all the water to supply his stock, take care of itself in a storm, and automatically stop itself when the tank was full, and resume work when any water was removed from the tank. Consequently, it was up-hill business introducing it at first, but the plan soon became recognized as a complete success for the purpose, and its adoption by the best farmers has constantly increased.[94]

Yet some farmers resisted buying factory-made windmills and opted to construct their own. These windmills, although heavier and clumsier in appearance than their commercial counterparts, could be built from a variety of scrap materials lying around the farm, such as lumber, roofing tin, iron strapping from barrels and wagons, worn-out farm implement parts, and other basic hardware such as nails, bolts, and wire. Professor and geologist Erwin Hinckley Barbour, with the assistance of three University of Nebraska students, traveled on horseback and in spring wagons along the Platte River to study and photograph these home-made windmills in 1897 and 1898.[95] Even among these windmills, which were more commonly found on farms in western Kansas and Nebraska, the basic designs could be placed in several categories—jumbo (wind rustler), battle-axe, and merry-go-rounds—and cost from $1.50 to $150 according to estimates at the time. The jumbo had about an eight-foot-long horizontal shaft with four paddle blades attached to it. The wind wheel was nestled into a stationary, ground-mounted wooden box with an open top, allowing the tops of the blades to catch the wind. The shaft would turn a small iron crank at one end that drove the pump rod up and down. The battle-axe used a set of four, six, eight, or more slightly turned wooden paddles to create the wind wheel, which was attached to a horizontal shaft set above the ground on wooden posts. Unlike the jumbo, the battle-axe blades were exposed completely to the wind. Its horizontal shaft was similarly attached to a simple metal crank to drive the pump located underground.

Two men inspect a jumbo-style homemade windmill, ca. 1898. The windmill was built by J. S. Brown, proprietor of the Midway Nurseries, Kearney, Nebraska, at a cost of only $1.50. Photograph by Erwin Hinckley Barbour, courtesy US Geological Survey Photo Archives, Denver, Colorado.

Both the jumbo and battle axe were constructed to face prevailing winds.[96] The merry-go-round, on the other hand, was more sophisticated, and a vertical axis with shutter-like sails was reminiscent of certain early Chinese windmills. Lucius Wilcox described in his book *Irrigation Farming* a merry-go-round windmill with a twenty-four-foot-diameter wheel with evenly spaced vertical wooden fans, measuring six feet high by four feet wide: "The fans are free at one edge, and, like a flag floating from the mast, they swing edgewise against the wind, this being the line of least resistance. The moment the center is past each fan in turn swings back against the immovable arms and exposes its 24 square feet of surface to the impact of the wind. Half the fans are thus continually in the wind and half out of it."[97] Throughout the Midwest, there were also unique "shop-made" windmills made with vertical wheels fashioned with turbine-style wooden blades as well as mills "reconstructed" from a menagerie of commercially made wood and metal parts.[98]

By the 1880s, with several hundred thousand commercially manufactured windmills installed throughout the United States, the mechanical device had become a symbol of progressive agriculture, individual prosperity, and community sustainability, which may not have otherwise existed in the more rugged, arid locations of the country. Not everyone appreciated the appearance of these windmills. One noted US architect in the late 1870s complained that windmills "should be condemned, for they are sure to obtrude themselves most offensively upon the sight; and to see these awkward, spider-like structures dancing fandangos before our eyes disturbs the repose, and mars the landscape of our otherwise beautiful homes," although he admitted that the water-pumping windmill is "very effective in its results."[99] Despite the naysayers, Alfred R. Wolff wrote in his comprehensive study of windmills published in 1885, "For certain specific purposes, and, primarily among them, for pumping water in moderate quantities, the windmill is not only a thoroughly reliable, but at the same time the most economical prime mover, and, as far as judgment can now be passed, will hold this place for many years to come."[100]

CHAPTER TWO

Perry's "Mathematical Wheel"

From the 1850s to the early 1870s, the majority of commercial windmills in the United States were predominantly made of wood held together with metal rods and other hardware. Many farmers liked the ability to repair their own windmills when parts wore out or became damaged, and wood made this task easier, since it could be either found around the farm or secured from a local lumberyard or cabinetmaker. On hand were also nails, screws, bolts, wire, and even leather to assist with making repairs. A blacksmith could be called on to manufacture or repair iron parts, but the thrifty farmer would surely want to avoid this expensive and time-consuming undertaking to fix his windmill. On top of that, farming operations, especially in arid country, could scarcely go long without a steady flow of water for their households, livestock, and gardens. A broken windmill needed to be quickly fixed and easily brought back into service.

By the early 1870s, a handful of manufacturers began to challenge the predominance of wooden windmills on the market by designing and selling units made with all-metal assemblies. This was no easy feat, since it was challenging to the engineers of these companies to build windmills that were lightweight, rugged, and balanced to use the wind effectively. Furthermore, to make these windmills commercially attractive to farmers and price-competitive with their wooden competitors, the material costs to make them had to be kept in check. Many early manufacturers of all-metal windmills knew this well, since they already had shops with skilled labor and proper tools to fabricate reasonably priced metal agricultural implements and other iron- and steel-based machinery. The rural population of the country by 1870 was just over 18 million, operating about 2.6 million farms.

The first commercially available all-metal windmill in the United States was manufactured by Mast, Foos and Company in Springfield, Ohio. The enterprise, which was incorporated in 1875, started out producing the Anderson boiler, but quickly exited this product line to start manufacturing its so-called Iron Turbine.[1] This windmill traced its origin to the 1872 patent of James A. Risdon of Genoa, Illinois.[2] Inventor Samuel Webb Martin further refined Risdon's design in a series of patents issued in 1878 and 1879.[3] True to its name, the windmill was completely

The Halladay windmill's sectional wind wheel, late 1870s. The etching shows a combination of iron and wood construction for the hub and gearing. From Michigan State Board of Agriculture, *Nineteenth Annual Report of the Secretary of the State Board of Agriculture of the State of Michigan, for the Year Ending August 31st, 1880* (Lansing, 1880), 307.

fashioned out of wrought iron parts from its wind wheel to its vane. The blades, made from No. 24 sheet metal, were distinguished by their bucket-like appearance, which the company attempted to describe elegantly in its sales literature as "solid and smooth, loosing [*sic*] nothing by friction as is the case with rough wheels ... hence the wind does not loose [*sic*] its force upon the bucket in its first contact, but like a stream of water following the curve still using its momentum till its escape from the wheel."[4] To protect the mill against moisture, the metal parts received three coats of paint, and unlike the wooden windmills, the Iron Turbine's parts would not "swell, shrink, rattle and be torn to pieces by the wind."[5] A front-page advertisement in the March 8, 1879, *Dodge City Times* hailed the Iron Turbine as "A machine long needed in this country, and one which every farmer and stock man should have."[6] The company sold three sizes of this windmill: an eight-and-a-half-foot, six-bucket engine; a ten-foot, seven-bucket engine; and a twelve-foot, ten-bucket engine. The 450-pound, ten-foot Iron Turbine, for example, was rated for wells up to 150 feet deep and cost about $85 in the 1880s.[7] Within fifteen years, the company claimed to have sold its windmills across North America as

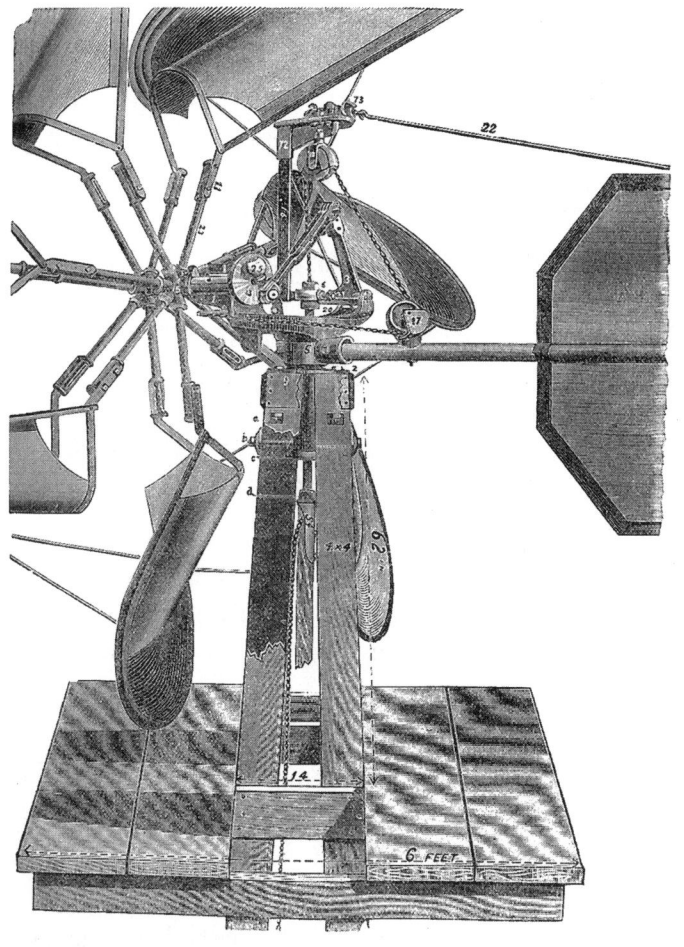

Mast, Foos and Company of Springfield, Ohio, produced one of the first commercially successful all-metal windmills, named the Iron Turbine, in the early 1870s. From Michigan State Board of Agriculture, *Nineteenth Annual Report of the Secretary of the State Board of Agriculture of the State of Michigan, for the Year Ending August 31st, 1880* (Lansing, 1880), 311.

well as in numerous countries, including the United Kingdom, France, Germany, Russia, Australia, New Zealand, and the Sandwich Islands.[8]

At around the same time, another Springfield-based manufacturer, the Springfield Machine Company, developed and sold its own metal windmill wheel. The windmill was first designed by father-and-son inventors Henry Croft and Henry Croft Jr. and then modified by Edward Croft Leffel, whose father, James Leffel, designed and built some of the most sophisticated water-power turbines in the mid-1800s. The Leffel Iron Wind Engine had a wheel composed of blades with a helical shape and measuring about three feet long and two feet wide, which also were made of No. 24 sheet metal. The blades were held in place by curved iron ribs bolted to an iron cone and spider hub. The sixteen-inch wheel bearing was attached to an iron turntable, which contained the mechanism that drove the pump rod. The *Farm Implement News* reported at the time that "[t]his is an excellent machine, and the demand for it is widening largely, especially in hot and dry regions."[9] Unlike the Mast, Foos Iron Turbine, the Leffel had a wooden tail vane.

Another all-metal windmill traces its roots to Ellicott City, Maryland, in the early 1880s, where an inventor named Robert G. Kirkwood patented and manufactured the Kirkwood Iron Wind Engine. Kirkwood's enterprise, which started operations under the name Progress Engine and Machine Works, had plants in Summerfield, Maryland, and Fredericksburg, Virginia. The company's windmill possessed an iron hub mounted to a main shaft. Threaded iron rods were screwed into the hub and fanned out to an inner sheet-metal rim. On the outer surface of this rim were attached the angled sheet-metal blades. The tips of the blades were secured to an outside wire wheel. The ten-foot-diameter Kirkwood windmill had thirty-six blades and weighed about four hundred pounds. The vane assembly was also composed of metal. However, Kirkwood didn't remain in Maryland for long; to be closer to his customers, he decided to relocate manufacturing to Arkansas City, Kansas, in 1887, and the business became the Kirkwood Manufacturing Company.[10]

There were other lesser-known US manufacturers in the early 1880s crafting windmills out of metal, such as the Indianapolis Machine and Bolt Works' Iron Duke windmill. The Plymouth Iron Windmill Company of Plymouth, Michigan, was founded by former watch and clock repairman Clarence J. Hamilton, who patented and built a vaneless all-metal windmill that had a wheel made of broad paddle-shaped fans and was governed into the wind by a side vane and weight. The Hamilton windmill was described in the farm press as having a "good reputation for power and durability."[11] Also during this period, the Columbus Windmill Company in Columbus, Ohio, manufactured Page's patented iron windmill.[12] Even some Canadian manufacturers in the 1880s explored all-metal windmill construction. One of the earliest introduced to that market was the Scientific Iron Wind Mill made by R. McDougal in Galt, Ontario.

In March 1892 the *Farm Implement News* published an article reporting on the emerging trend toward all-metal windmill construction during the preceding twenty years and its direct correlation to the US steel industry. The author wrote: "The cheapening of steel through improvements in the processes for producing it and the growing scarcity of good lumber have brought steel extensively into use in the construction of agricultural implements and machines and caused its application to many purposes for which, a few years ago, only wood was considered adaptable . . . Where lightness of design, rigidity or elasticity, strength and durability are requisites[,] steel is superior to any other material, and its present cheapness permits of its use for every purpose which it is suitable; indeed, this is truly the 'age of steel' in its application to agricultural machinery."[13] The nation's emerging steel industry in the years following the Civil War benefited from an abundance of iron ore unearthed in the Great Lakes region and coal from Appalachian mines. The development of the Bessemer converter made its production comparatively inexpensive. The enormous output from the country's steel plants started to eliminate wood in all types of machinery. Steel conveyed a sense of industrial strength and durability. Yet it was not easy at first to convince customers of the value of steel as construction material for windmills. The *Farm Implement News* article said, "Had steel been as cheap when [windmills such as the Iron Turbine and Leffel] were originally constructed as it is now, undoubtedly steel would have been used instead of iron[,] and steel mills might have taken the market long ago."[14] The iron shafting and rods used in the earliest all-metal windmills were heavier than the thinner, yet sturdier steel components that came later.

Even by the early 1890s, some windmill manufacturers still refused to embrace the so-called "steel craze," citing that due to the need to build steel mills that were light enough to compete with wood, "they will not prove sufficiently durable; that they will rust out if they do not otherwise give out, as no paint yet made will long adhere to and protect the surface; that the wheels do not develop as much power as those made of wood, and the towers if sufficiently strong are too expensive, and that if anything happens to either they cannot be easily repaired."[15] Some long-time manufacturers, such as U.S. Wind Engine and Pump Company and Fairbanks, Morse & Company, by this time had acquiesced and started adding all-metal windmills to their product lines along with their traditional wooden mills. Some of these manufacturers still did not believe consumer demand for steel windmills would last long. However, to their chagrin, they would eventually be proved wrong. As one observer noted in 1892, "in a neighborhood on the Northwestern railway, some ten miles north of Chicago, where five years ago there were three wooden mills now only two wooden and thirty steel mills can be counted within a radius of two miles."[16]

While the debate over the superiority between wood or steel windmills raged, a young inventor and engineer by the name of Thomas Osborne Perry would put the design of windmill wheels to thorough scientific testing, something that had not been done in such detail since the 1750s when English physicist and engineer John Smeaton presented his own study of windmill efficiency research to the Royal Society in London. Thomas Perry was born in Tecumseh, Michigan, on February 28, 1847. His father, Gideon D. Perry, was a retired minister of the Methodist Episcopal Church, who later became involved in Michigan politics as an elected member of both the state house and senate. Together, Gideon Perry and his second wife Mary Osborn Perry had five children between 1844 and 1856. Thomas was their only son.[17] Based on his own writings and expressions from peers, Thomas Perry was a deep thinker, but not boastful. He commented that he "wanted to fight in the Civil War, but it ended too soon," thus the eighteen-year-old entered the University of Michigan at Ann Arbor in 1865 to pursue his education.[18] In his last year of undergraduate studies, he was bestowed the title of "Seer of the Class," or class poet, by his classmates and was noted for his "rich vein of dry wit."[19] Thomas Perry received his Bachelor's of Arts in 1869 and, from 1869 to 1871, taught Latin, Greek and mathematics at a high school in Ypsilanti, Michigan. His students and the community referred to him as "professor," which was customary at the time. He then returned to the university for two years, receiving the degrees of civil and mining engineer in 1872, in addition to a Master's of Arts for the three years he taught school. "As to titles," he reflected nearly forty years later, "I have always called myself a Mechanical Engineer, as my occupation has been mostly mechanical and generally experimental, for some of which I have been paid by various manufacturing concerns, and for which I have sometimes paid pretty dearly myself."[20]

Once he finished graduate studies, Thomas Perry returned to Tecumseh, Michigan, with a passion for "original investigations," in particular on windmills and their performance.[21] In June 1873 he filed his first patent application, titled "Improvements in Windmills," to the US Patent Office, in which he stated:

> My invention relates to that class of windmills which revolve horizon-tally on vertical shafts, and have wings which also revolve on vertical axes; it consists, first, in the adaptation of a vane to regulating the wings, by causing them to revolve on their own vertical axes in such a manner as to present their faces to the wind advantageously in every position; sec-ond, in the application of a governor, so regulating the speed of the mill as to give it an uniform velocity in varying winds, thus securing it against injury, and making it better adapted to driving machinery; third, in vari-ous accessory devices.[22]

Thomas Perry's windmill, which consisted of four vertical wings, already chal-lenged him to investigate the importance of gear ratios to ensure that changing wind wheel speeds from variable winds did not harm the machinery that they powered. Gideon Perry must have supported his son's work, since his signature appeared as one of two witnesses on the patent application. The US Patent Office awarded Thomas Perry his patent on January 20, 1874.[23] While a patent model of this windmill was surely constructed, its fate is unknown. He would go on to take out numerous other patents on pneumatic pumps and for other mechanical movements in the years ahead.

In 1876 Thomas Perry moved to Batavia, Illinois, to become a draftsman for the U.S. Wind Engine and Pump Company, by then one of the preeminent US windmill manufacturers. He acted as a solicitor of patents, or as he described it, "designer and investigator" for the company until 1881. He was then briefly employed by the Eclipse Wind Engine Company in Beloit, Wisconsin. During 1881, for reasons unknown, he returned to his former employer in Batavia.[24] From June 1, 1882, to September 15, 1883, Thomas Perry would conduct a detailed series of dynamometric tests of windmill wheels.

Frederick Haynes Newell, head of the US Geological Survey's Division of Hydrography, recollected that at the time Thomas Perry carried out his experi-ments, "[a]lthough millions of dollars have been invested in the manufacture and purchase of mills and much attention has been given to the mechanical details and the saving in weight and cost, yet comparatively little study has been bestowed upon actual efficiency of various forms and upon their development toward theo-retical ideals."[25] R. L. Ardrey put it more bluntly in his 1894 book: "It may be said that [Perry's windmill] is the invention and design of a mechanical engineer, in sharp contrast with many other inventions in the agricultural implement industry that are the result of a moment's inspiration or a crude experiment, often on the part of an unlettered farmer."[26] In fact, most wind wheels were constructed of nar-row wooden slats set at angles with the plane of the wheel ranging from thirty-five to forty-five degrees as a rule of thumb. "The slats were usually placed as close

Thomas O. Perry, age twenty-five, earned a degree in engineering from the University of Michigan in 1872 and immediately began pursuing his interest in efficient windmill design. Courtesy Bentley Historical Library, University of Michigan, Ann Arbor.

No. 146,548.
Fig. 1.
Fig. 2.
Fig. 3.
Fig. 4.
Fig. 5.
Fig. 6.
Fig. 7.
Fig. 8.
Fig. 9.
Fig. 10.
Fig. 11.
Fig. 12.
Fig. 13.
Gideon D. Perry
Joseph R. Kerry

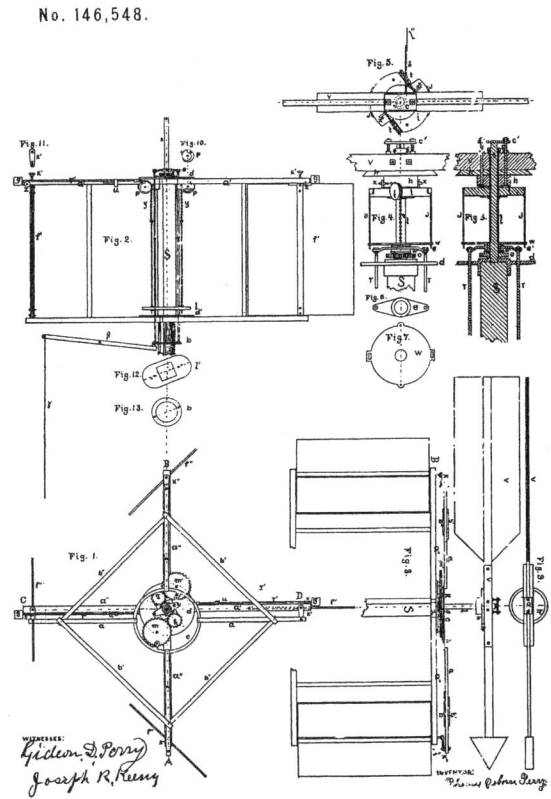

Patent drawing submitted by Thomas O. Perry when he filed for his first windmill patent in June 1873. The windmill consisted of four vertical sails attached to a vertical axis. Perry was challenged to find a gear ratio that allowed the windmill to operate in variable winds without harming the machine. Courtesy US Patent and Trademark Office, Washington, DC.

Owners and employees of the U.S. Wind Engine and Pump Company, Batavia, Illinois, ca. 1880. Courtesy Batavia Depot Museum, Batavia, Illinois.

together as possible without having their projections on the plane of the wheel overlap. The proportions of sail surface and their angles of weather were apparently arrived at without any well-defined purpose. The only experiments made in the United States, so far as could be learned, related to starting forces only. They did not include the measurement of work in foot-pounds," Perry wrote.[27]

Engineer Thomas O. Perry's experiments were conducted in a room measuring thirty-six feet wide by forty-eight feet long and nineteen feet high from the floor to the ceiling. His test apparatus was suspended from roof trusses. It was also necessary for his experiments to have a consistent wind velocity. "We therefore made use of artificial wind, obtained by carrying the wind wheels in a circle against still air on the end of a long sweep suspended beneath the roof trusses of a large room used for setting up tanks. The sweep was made to revolve horizontally around its vertical axis by means of gearing, pulleys, and belts connected with a line shaft driven by an 80-horsepower Reynolds-Corliss steam engine, which furnished the motive power for the works of the United States Wind Engine and Pump Company. The distance from the axis of sweep to center of the wind wheel was 14 feet, so that the velocity of wind against the wind wheel in miles per hour would be indicated very closely by the number of revolutions per minute made by the sweep," Perry wrote. The door to the room was guarded and the windows closed so as to not not disturb the air during testing. He counted the number of turns the sweep made by using an eighty-tooth wheel that revolved at the same rate as the sweep. Thus he was also able to determine the velocity of the wind to within 0.00125 miles per hour.[28] Barometric measurements and air temperature were also recorded during the early stages of testing, but "later we were convinced that the readings of the barometer and thermometer would not always account for variations by the same wheel on different days," Perry said.[29] Perry kept his test wheel diameters to five feet. It's estimated that he made more than 5,000 measurements and tested sixty-one different wheel forms.[30] He concluded from his experiments that it was better to divide the wind wheel between fewer sails, which consequently "reduces the aerial resistance to motion due to the number of edges, and leaves relatively freer interstices for the flow of air between the sails."[31] He also observed that smaller wind wheels—in the range of eight feet in diameter—could be made just as efficient for ordinary water-pumping purposes as the ten-foot and larger ones of the day.[32] One writer in the 1890s said Perry did for the windmill what French engineer and mathematician Jean-Victor Poncelet (1788–1867) accomplished with his improvements to efficiency of the water wheel in the early part of the century.[33]

The conservative management of the U.S. Wind Engine and Pump Company, however, was not sufficiently convinced by Perry's findings to change its windmill products. This decision must have been a blow to the young engineer. His research notes, property of the employer, would not be viewed in total by the public until fifteen years later, when they were published with permission of the company by the US Geological Survey. By this time, many windmill companies, including U.S. Wind Engine, had incorporated principles of Perry's work in their all-metal windmill wheel designs.[34] After presenting his research, Perry quietly left Batavia for Chicago in 1883. One longtime inhabitant of Batavia later reckoned that Perry "did as much as Halladay did to advance the popularity of the windmill, but who knows anything about him? He is just one of our forgotten engineers."[35]

Chicago by the early 1880s was a vibrant city with massive construction projects under way and a substantial manufacturing base. This was a far cry from 1871, when a four-day fire in October destroyed an area of about four miles long and

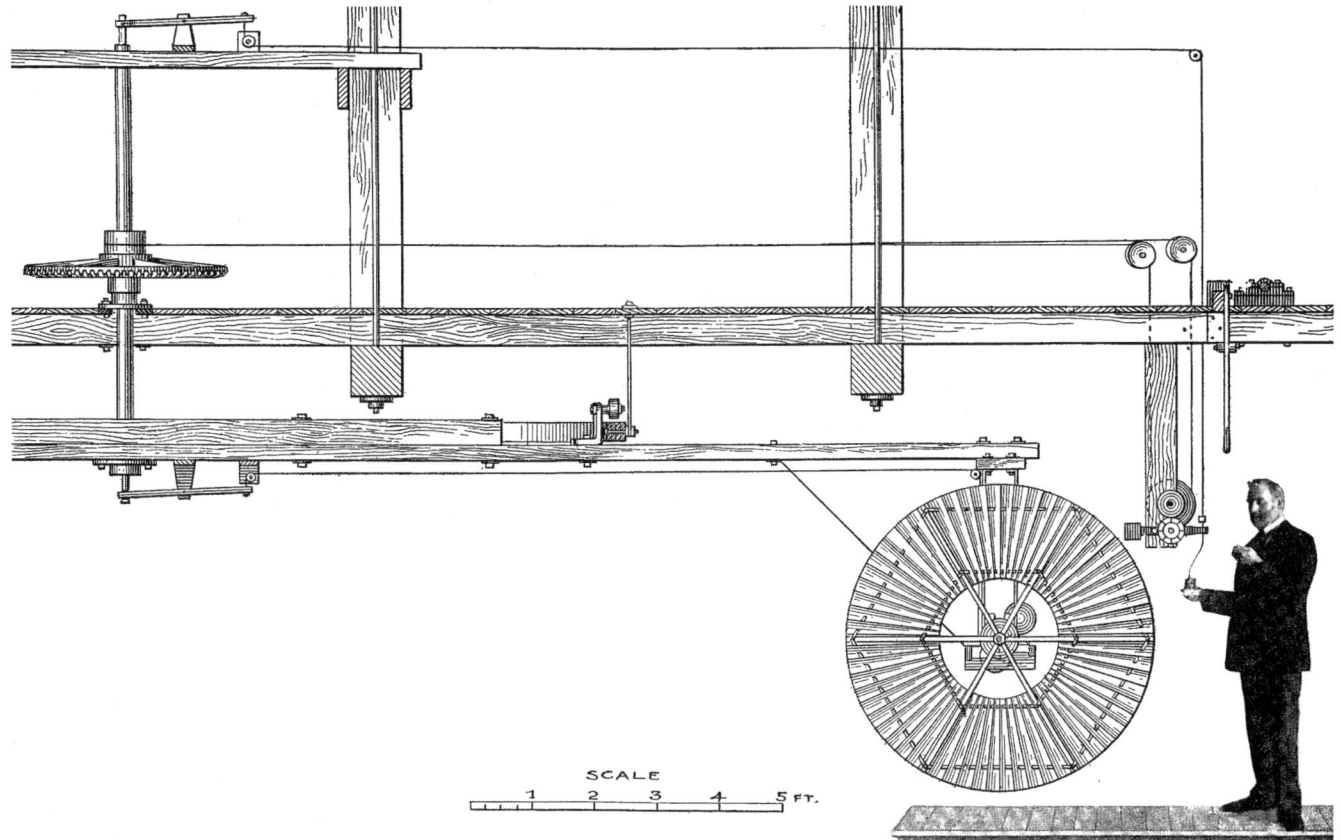

Engineer Thomas O. Perry constructed this apparatus in a large room at the U.S. Wind Engine and Pump Company plant in Batavia, Illinois, and used it to make more than 5,000 measurements and test sixty-one different forms of wind wheels. From the US Geological Survey, *Experiments with Windmills* (Washington, DC, 1899), plate 6.

a mile wide. While much of the business district was incinerated, its large stock-yards and meat-packing plants, lumberyards, and grain facilities, in addition to its railroad tracks, on the South Side remained unharmed. These industries helped Chicago quickly rebuild and increase its population from 300,000 to more than 1 million within a decade of the fire. Large employers like the McCormick Reaper Company and Pullman Car Works set up operations in the city. In addition, there were numerous small manufacturers collectively hiring tens of thousands of people. Chicago's expansive railroad hub also helped the city prosper as its products could be transported in volume throughout the United States. The city suffered, however, a brief setback during the country's mid-1880s economic recession, but recovery was swift and prosperity returned to Chicago at the start of the 1890s.

During his stay in Chicago, Perry partnered with La Verne W. Noyes, another young inventor and ambitious entrepreneur. Perry and Noyes worked together to design a low-down self-binding harvester, which William Deering manufactured for the commercial market. William Deering made another self-binding harvester that contained parts patented by Perry.[36] It's unclear when these two men first met, but it was likely during Perry's employment at U.S. Wind Engine. In addition to windmills, the company sold a variety of other agricultural implements, includ-

At a young age La Verne W. Noyes knew he wanted to be a manufacturer. He attended the Iowa Agricultural College in Ames, graduating with a degree in engineering in 1872. Courtesy Special Collections Department, Iowa State University Library, Ames.

ing hay-carrying barn trolleys and pulleys patented by Noyes in the mid-1870s. After completing the work on the harvester, Perry moved back to his hometown of Tecumseh to continue his windmill experiments, and from 1886 to 1887 he constructed a water-pumping windmill, which he called the "Aërmotor." Some referred to his windmill as the "Mathematical Wheel." Through numerous tests and calculations, Perry designed his windmill to work efficiently in the lightest breezes as well maintain its integrity—both operationally and structurally—in high winds. On Perry's invitation, Noyes traveled from Chicago to Tecumseh to see this windmill in operation and realized there was a significant business opportunity in the making.[37]

La Verne Noyes was born in Genoa, New York, on January 7, 1849, the youngest of four children of Leonard R. and Jane (Jessup) Noyes. The family moved to Springville, Iowa, in 1854, where Leonard Noyes operated a farm.[38] Shortly after the move, in 1856, the family's eldest child, Amanda Malvina, died at the age of seventeen, and the oldest son, Samuel Jessup, who at eighteen years old was part of Company H, Twenty-Fourth Regiment of the Union army's Iowa Volunteers, was killed during the charge at Champion Hill, Mississippi, on May 16, 1863.[39] As a boy, La Verne Noyes showed an interest in mechanics and knew that he wanted to eventually become a manufacturer. He enrolled at the newly started Iowa Agricultural College (Iowa State University) at Ames in 1868 and graduated with a Bachelor's of Science, with a focus on engineering, in 1872. He also had an interest in the literary arts and helped start the school's Crescent Literary Society, which is likely how he came to know his future wife, Ida Elizabeth Smith (class of 1874), of Charles City, Iowa.[40] Immediately after college, Noyes moved to Marion, Iowa, where he entered the agricultural implement business. He received his first patent in April 1875 on a pulley-based hanger for the farm slide-and-swing gate, which he manufactured and sold to wholesale hardware dealers.[41] It was also during this time that he learned the difficulties of business, spending upwards of $5,000 in borrowed money.[42]

For a time during the mid-1870s, Noyes joined forces with a man named Plummer in Springville, Iowa, to begin manufacturing haying tools. In 1876 Noyes patented an improvement to the hay carrier, which allowed for "holding the carriage in position during the loading of the fork and the raising of the same to the top of the mow, and for automatically releasing the said carrier when the loaded fork has reached the desired height, so that it may be run into the barn."[43] By the time Noyes filed for this patent, he had moved to Batavia, Illinois, and had attracted the attention of one of U.S. Wind Engine and Pump Company's founders, Daniel Halladay. In 1876 U.S. Wind Engine had begun selling the Noyes & Plummer Anti-Friction Hay Carrier and would continue to manufacture and sell the device for twenty years. Customers paid $8 apiece for them in 1882.[44] In January 1878 Noyes received a patent for an improvement to horse hay forks. "The object of my invention is to furnish a device for seizing and holding hay, straw, and other loose and fibrous material, and to discharge the same when desired," he wrote in his patent application.[45] Throughout the remainder of the 1870s and early 1880s, Noyes pursued numerous other patents for hay-handling devices, some highly sophisticated. He developed, for example, a field pitching apparatus that was mounted to a long parallel beam supported by a four-timber vertical stand, which allowed farmers to build enormous hay stacks relatively quickly, with one claim stating, "Five tons have been stacked, and the frame moved in forty-five consecutive minutes. Fifty tons have been stacked in a single day."[46] Noyes also patented simpler devices, such as a hay pulley, marketed by U.S. Wind Engine as the "Noyes Pulley," which consisted of a six-inch maple sheave in an iron mount with a swiveling eye. It was noted in U.S. Wind Engine's sales literature in the late 1870s that the pulley's sheave was twice the size of most pulleys and "revolves only about half the number of times in passing the same amount of rope, and hence must last nearly twice as long as a small pulley."[47] Besides the widespread name recognition afforded to Noyes through his relationship with U.S. Wind Engine, the business relationship allowed him to generate earnings to pay off his $5,000 debt at 10 percent interest on all except $800, on which he paid 8 percent for one year.[48]

While his interest in inventing new agricultural implements continued, La Verne Noyes started a new manufacturing venture in the early 1880s based on his invention of a distinctive dictionary stand. It has been said that he received the inspiration for the stand from watching his wife, Ida, a writer and artist of rather petite stature, struggle with heavy dictionaries, atlases, and other unwieldy reference books of the day, some weighing upward of ten pounds apiece. In July 1881 Noyes received a patent on his first portable book holder, which he said "conformed in general shape to the book." The stand opened to a slightly tilted flat surface, cradling the book's spine in the middle; when not in use, the book could be closed and firmly held in place by the stand.[49] The earlier models were not without their flaws. Lightweight cast-metal legs of the stand occasionally broke. Noyes continued to refine the stand's design and usefulness. In 1885 he filed for various patents that added folding cast-iron base legs with caster wheels and a shelf to hold additional books in the middle of the stand in addition to a more robust spring mechanism to hold the book either opened or closed and adjust to any angle for reading.[50] The ability to fold and take apart the stand allowed for easier and cheaper shipping of these stands to customers. His so-called No. 19 stand,

Advertisement for Noyes's dictionary stands. In the 1870s, La Verne W. Noyes patented a number of agricultural implements, which were sold by the U.S. Wind Engine and Pump Company of Batavia, Illinois. However, it was his patented dictionary stands in the early 1880s that generated the first significant income for the young inventor. Author's collection.

which could balance a book up to thirty inches off the floor, sold in 1888 for $3 apiece. He also manufactured a brass stand, which sold for $4.[51] Noyes's dictionary stand business was located on the fourth floor of a building at 42 and 44 West Monroe Street in downtown Chicago. The stands were sold to numerous libraries and schools, in addition to individuals, and promoted through various mass merchandise catalogue retailers and product wholesalers, such as the Chicago-based Sears, Roebuck and Company; and the Redhead, Norton, Lathrop & Co. in Des Moines, Iowa.[52] Noyes estimated that his sales increased by 20 percent between 1885 and 1886 and more than 33 percent by 1887.[53] From sales of the holders he earned net profits between $15,000 and $18,000 a year.[54]

It was the profits from the dictionary stand business and other secured funds that would help Noyes start a windmill business. However, before he could prove the business was worthwhile, the entrepreneur was careful to carry out some windmill tests of his own, which Perry conducted under his direction. According to Perry, Noyes bought two ten-foot windmills "of the prevailing standard, as made by the two leading manufacturers and set them to work for several weeks or months alongside of the first Aermotor which he constructed, pumping water thru water meters borrowed from the Chicago City water department. This he did to satisfy himself that there was no mistake about my claims for the Aermotor."[55]

Besides Perry's innovative metal wind wheel design, he also developed a geared pumping mechanism behind the wheel that would inspire traditional windmill manufacturers to make similar changes to their machines in the 1890s. During his experiments, the engineer realized that his more nimble wind wheel turned faster in the wind than the average wooden-wheel windmill and could start in a breeze of as little as four miles per hour. Windmills until the late 1880s had single-stroke pumps, meaning that for every 360-degree turn of the wind wheel, the pump rod

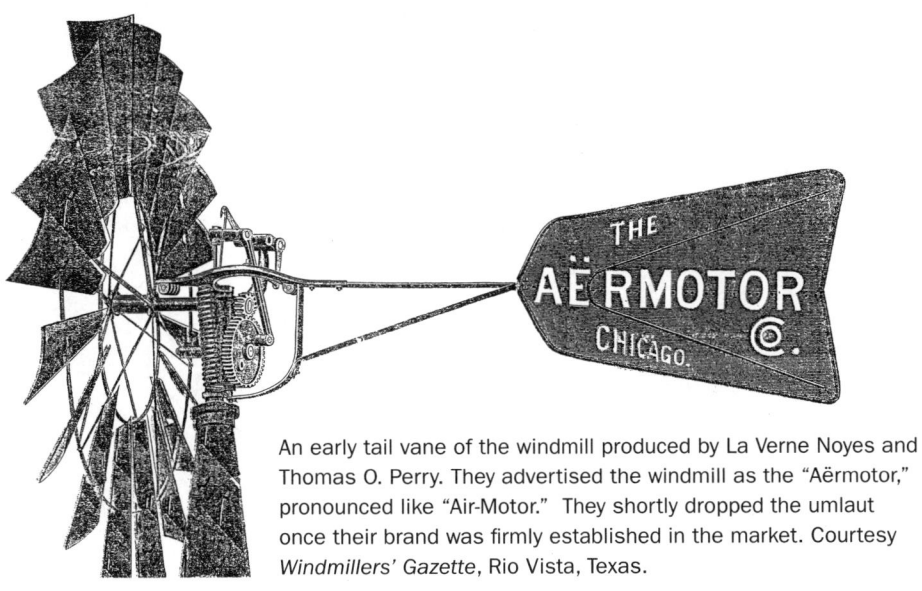

An early tail vane of the windmill produced by La Verne Noyes and Thomas O. Perry. They advertised the windmill as the "Aërmotor," pronounced like "Air-Motor." They shortly dropped the umlaut once their brand was firmly established in the market. Courtesy *Windmillers' Gazette*, Rio Vista, Texas.

would complete a single stroke. The faster the wind wheel turned, the quicker the strokes, causing a jerky and potentially damaging action on the pump rod. Perry developed a system of "back-gearing" to give the windmill a more consistent, smoother operation. Thus he developed a mechanism that essentially divided the windmill's work between two shafts and four bearings instead of one shaft and two bearings, which was found in direct-stroke windmills of the day. In Perry's design, the light metal wind wheel and pinion gear is carried by one shaft, which shifted the heavier load action to the slower-turning shaft with the larger crank gear. This mechanical action allowed for one long pump stroke for approximately every three turns of the wind wheel.[56]

Noyes, after much consideration, decided to name the firm the Aërmotor Company, the name initially prescribed for the windmill by Perry. The company cleverly explained the context behind the windmill's name in its earliest advertising literature: "Since the wind-motor embodies the suggestions and teachings of the researches and experiments of the only competent engineer who has in modern times undertaken a thoroughly scientific and exhaustive investigation of air as a motive power, we feel entitled to coin and appropriate to its exclusive use the name Aermotor, literally meaning Air-Motor. The term windmill, is a misnomer, at best, since in former times it designated the grist-mill which was run by wind as distinguished from one run by water, hand, or other power. Pronounce the name nearly as if Spelled Air-Motor, giving the long sound of A."[57]

CHAPTER THREE

Open-Gear Windmill

In 1888 the newly formed Aërmotor Company may not have looked very much like a threat to the numerous well-established windmill manufacturers already on the US market. New windmill companies sprang up continuously throughout the 1870s and 1880s. Some upstart firms, however, remained niche players, catering to local or regional markets, if they managed to survive that long. The big players' wheels and tail vanes, such as Althouse-Wheeler & Company, Challenge Wind Mill and Feed Mill Company, the Eclipse Wind Engine Company, Perkins Wind Mill & Axe Company, and the U.S. Wind Engine and Pump Company, to name a few, dominated the skylines of the country's rural and suburban landscapes. La Verne W. Noyes, president of Aërmotor, must have known from the outset that he needed more than just his ambition to survive and carve a name for himself in the industry. Fortunately, he already had experience working for one of the giants of the business—U.S. Wind Engine—which gave him the opportunity to understand both its strengths and weaknesses in the market prior to forming his own company. He also retained as his consulting engineer Thomas O. Perry, who is credited with developing the Aërmotor windmill's original all-steel design.[1]

The Aërmotor Company officially started operations on the fourth floor at 42 and 44 West Monroe Street in Chicago, the location of Noyes's successful dictionary stand business. That first year in operation Aërmotor claimed to have manufactured and sold forty-five windmills.[2] In 1889 he relocated operations to a six-story building at 110 and 112 South Jefferson Street, where he took out a five-year lease, believing that the space would suffice for a number of years. That year the firm made 2,288 windmills.[3] Still, this space was not enough, and the company relocated again in November 1890 to a property at Rockwell and Fillmore Streets that had two large buildings surrounded by four acres and was about four miles south from Chicago's city hall. From this location the company was also afforded more direct access to some of the nation's significant rail operators calling at Chicago at the time, such as the Wisconsin Central Railway and the Pittsburgh, Cincinnati, Chicago and St. Louis Railroad.[4] Windmill output from this new factory site took off, with 6,268 units produced in 1890, followed by 20,049 in 1891,

In 1889 Aermotor moved to this building at 110 and 112 South Jefferson Street in Chicago and that year manufactured 2,288 windmills of the style shown on the right. From *Farm Implement News* 11, no. 8 (August 1890): 43.

and a whopping 60,000 by 1892, with claims of producing a complete steel windmill and tower every three minutes during the working day. It's unknown how the 1893 economic recession, the worst in the United States up until that time, dented the company's orders for windmills, but its fierce competitiveness with other makers was not dampened, as expressed frequently in its advertising.[5]

Central to this larger property were two five-story brick buildings, emblazoned with "The Aermotor Co." name along their rooftops. Each building was laid out to ensure smooth and efficient movement of materials as they passed through the windmill manufacturing process. The first building to the left, as it would have appeared from present-day Campbell Street, running in a south-to-north direction, was topped with an Aermotor windmill. It included a first-floor room where workers prepared wood for assembling water tanks. The second floor of this building housed the administrative offices, including Noyes's desk; a company restaurant for office employees, factory foremen, and visitors; and a product display room. The third floor included a drafting room, pattern shop, and tool room, most likely where Perry spent his time. This floor was where the latest product experiments and the building of manufacturing machinery took place. On the fourth floor were the offices for advertising materials, packing, and special machinery used to make wire braces for Aermotor's steel fixed towers. The fifth floor was used for storage and additional experimental work. Starting on the fifth floor of the adjacent building, one entered a stock room of extra parts to meet peak product demands. The fourth floor of this building housed the sheet-metal department for the manufacturing of windmill sails and vanes. The third floor, referred to as the "erecting room," received castings after the machine work was done to them and they were assembled into various machines. The second floor housed Aermotor's machine shop, which contained the equipment to "quickly and cheaply" turn out the company's various windmill parts. The ground floor of the building contained the forge and tower room. "Here is found some very heavy machinery including two gang punches for punching the holes in Steel Towers. One of them, punches at one movement all of the holes wanted in a piece of steel 28 feet long. With these machines hundreds of Steel Towers can be turned out in a single day. Here you will also find the bolt machines, drop hammers, drop presses, some of which were designed especially for the Aermotor work," wrote the company in an 1893 bulletin to its agents.[6] Stepping out of the second building, one entered the company's foundry, which contained the molds for churning out numerous poured castings. Immediately adjacent to this one-story structure was the galvanizing room, where the company estimated that it annually melted about forty tons of zinc mixed with other metals to galvanize its windmill parts and towers. Next to the galvanizing room was found the shipping and packing department, which loaded products onto wagons for delivery to area freight depots. In addition, a track fed railcars, up to twenty at a time, into the storeroom for the larger cross-country shipments. The premise of the plant layout was to operate in a production line fashion. As a worker finished one part, it was moved down the line to another, and so forth, until ultimately all the parts of the mill were crated for shipping. Factory efficiency and minimization of material waste and manpower were essential to this manufacturing model. Many of Aermotor's competitors at the time had trouble keeping up since they operated more like job shops in which material handoffs were more disjointed.[7]

R. L. Ardrey in his 1894 review of the US agricultural implement industry described the significant operational enhancements put in place at Aermotor's Chicago factory:

Machinery must replace slow and tedious hand labor at every point, and machines that have not been used enough to wear them appreciably, must be discarded for something new of greater capacity and efficiency. In this way only can the cost of manufacture be reduced to a minimum, so the factory in question may lead in competition in its industry. The Aermotor factory is remarkable for the improvements that have been made in every department. In the foundry, for example, devices for expediting the moulder's work have been adopted, increasing a workman's product from six to sixteen pieces per day in one case, and from twelve to forty-five in another instance; these devices also improving the quality of the product. In every detail of the work carried on in this factory the same system and organization appears, looking to decreased cost of production and higher quality of the product. As a result of these improvements, and of the advantages inherent in the new principles of the Aermotor, the claim

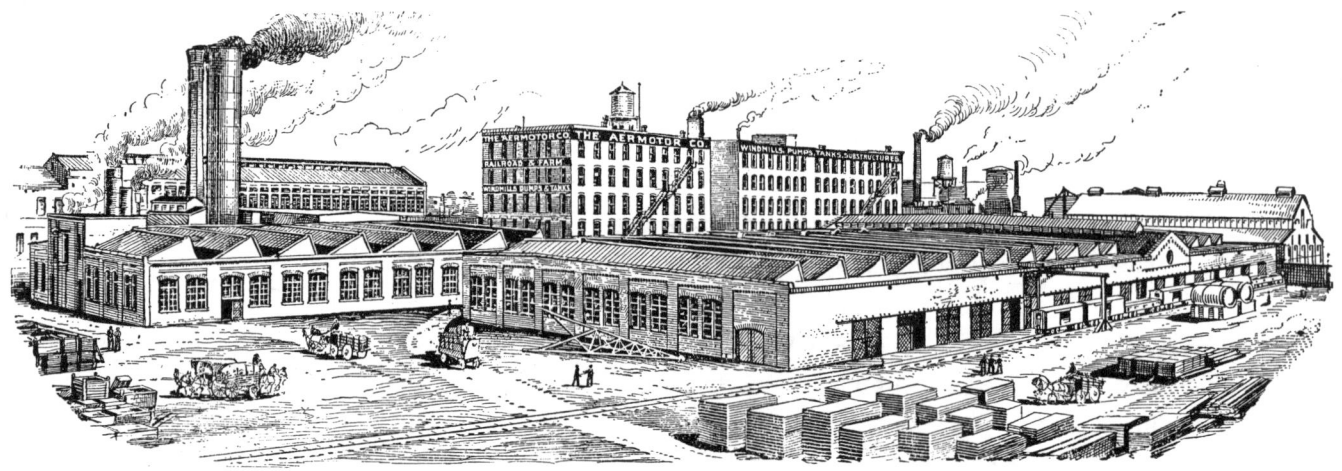

Aermotor relocated to Twelfth and Rockwell Streets in Chicago in 1890 and rapidly built one of the most sophisticated windmill manufacturing and distribution operations in the country. From Aermotor Company, *Twentieth Annual Aermotor Repair List* (Chicago, 1908), n.p.

is made, with reason, that the cost of wind power—the cheapest power known—has been reduced to one-sixth of what it was a few years ago. It has been made possible to greatly increase the use of windmills by bringing the price within the means of a greater number of purchasers.[8]

A major production move for the Aermotor factory was the introduction of galvanizing baths to coat its steel windmill components and preserve them against rust. In its first three years, Aermotor—like many agricultural implement manufacturers—painted its metal products. But it does not take long for changes in temperature and moisture to wear away the paint and for corrosion to start. In 1892 Aermotor invested about $50,000 on a galvanizing system and spent about a year experimenting with the dipping process before making its first galvanized windmills commercially available. The galvanizing process traces its origin to the work of two French chemists, Paul Jacques Malouin and Stanilas Sorel, who conducted their experiments with zinc coatings on steel in the late 1830s and early 1840s. Aermotor decided it was best to galvanize only finished parts so that rivets, punch holes, and bends were thoroughly coated with the material. "When the section of an Aermotor Wheel is all riveted up, completed and cleansed of rust and impurities, it is then immersed in melted zinc and aluminum and left there until it becomes as hot as the metal is and until every crack, cranny, crevice, pore and opening of every sort is filled, closed up and saturated with the molten metal and the whole 23 pieces composing the section become soldered and welded together as one piece," the company explained.[9] Its windmill vanes and tower pieces were given similar treatment. Aermotor used at least forty tons of melted zinc and aluminum annually in the early 1890s to galvanize its windmill parts.[10] The company bragged at the time that maintaining the process was expensive and kept many competitors from offering similar quality galvanized products. However, it was not long before other windmill manufacturers were doing the same.

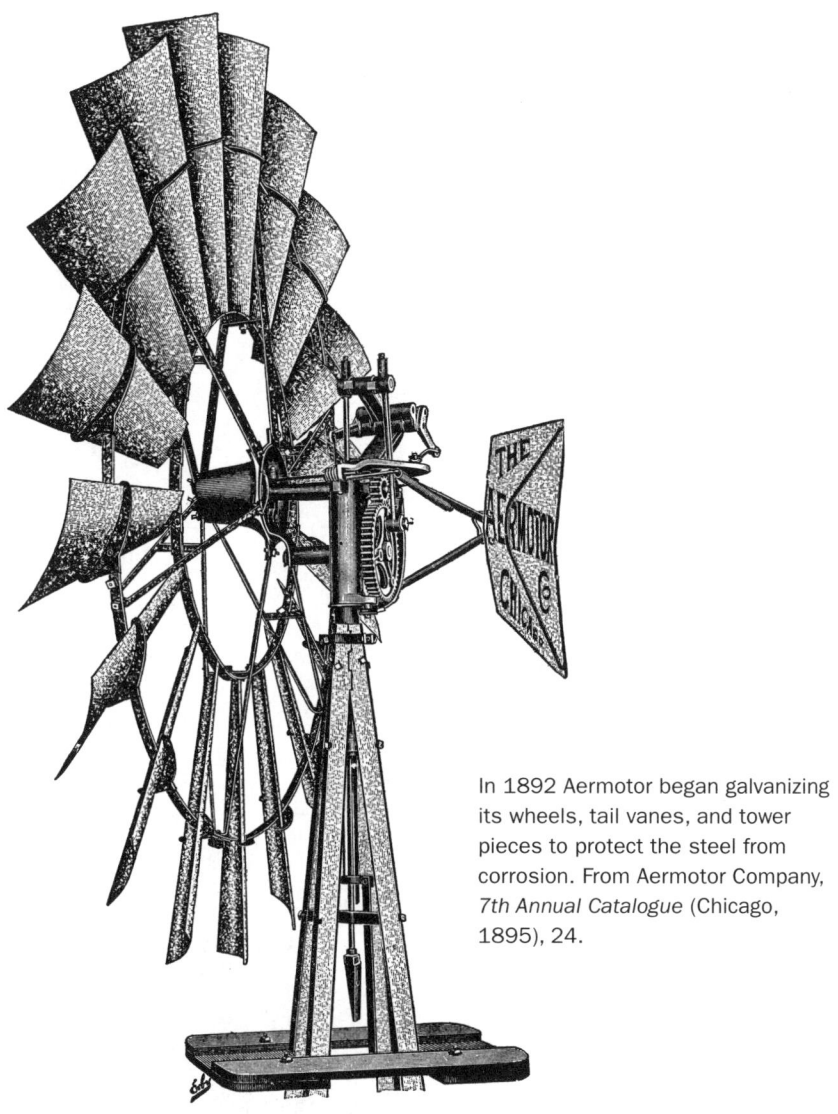

In 1892 Aermotor began galvanizing its wheels, tail vanes, and tower pieces to protect the steel from corrosion. From Aermotor Company, *7th Annual Catalogue* (Chicago, 1895), 24.

At its inception Aermotor set out to make its windmills not only inexpensive but also sensible in design, deployment, and use. However, nothing was perhaps more daunting about owning a windmill than climbing the tower every week or two to grease the exposed gears behind a wind wheel. The poor placement of hands or feet could send a person to the ground, resulting in severe injury or death, not to mention that windmill tower ladders could become slippery under wet and icy conditions and were prone to deterioration if made from wood or exposed iron. Besides, some windmill owners were simply afraid of heights. Perry must have been keenly aware of this problem because a year before Aermotor started operations, the US government approved a patent application for his tilting "windmill-derrick." In his patent application, Perry said the purpose of this invention was "first, to provide a support for sustaining a windmill high in the air while in use, and which shall admit of readily lowering the windmill to the ground or where it can be easily reached, so as to do away with the necessity of climbing whenever it may be desirable to oil the bearings or make repairs, or to

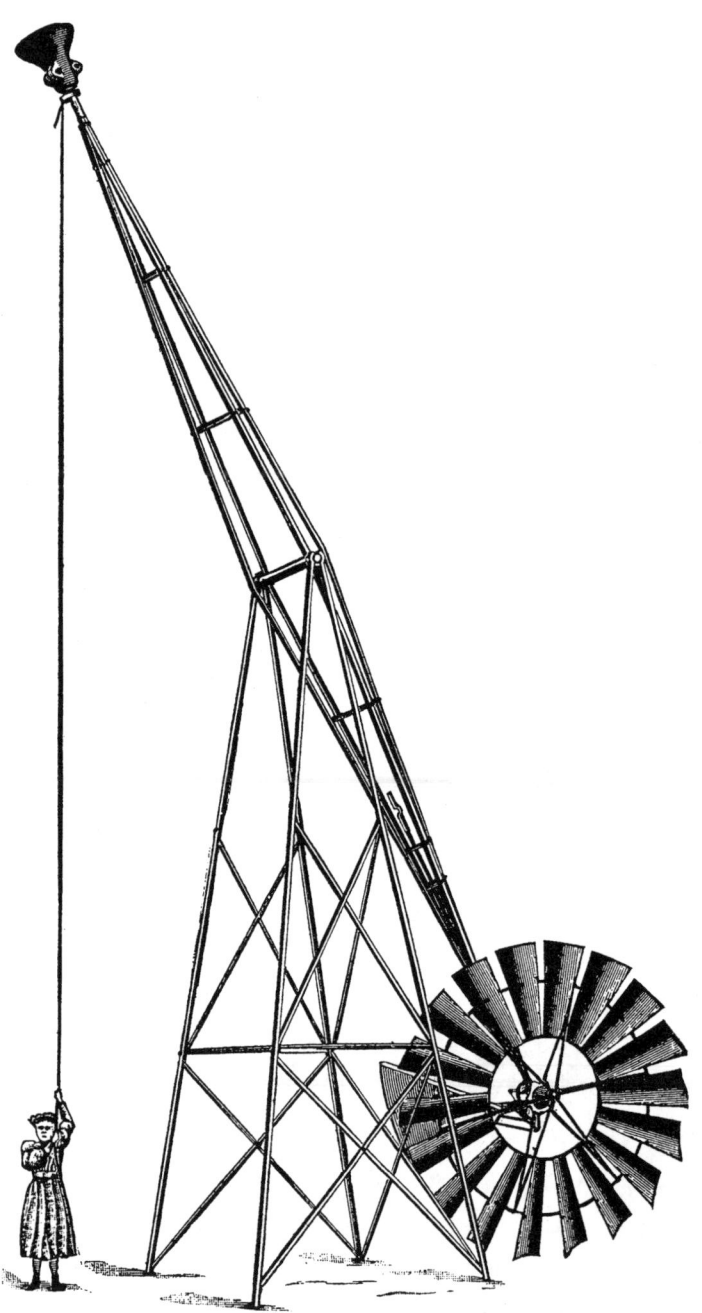

A young girl is shown lowering a forty-foot galvanized steel tilting Aermotor tower for oiling the windmill's gears. From Aermotor Company, *7th Annual Catalogue* (Chicago, 1895), 39.

avoid damage in case of storm, and second, to facilitate the erection of the windmill and reduce cost."[11] Perry's early tilting tower was essentially a single post with a hinge nearly halfway up the structure. The top portion of the tower was lowered to the ground by releasing a catch and by pulling on a cable. By the time Aermotor started manufacturing windmills in 1888, Perry had developed a more sophisticated wooden tilt tower with four legs that anchored to the ground and offered a clamping mechanism to lower the wind wheel to the ground. On the lower end of the movable portion was an iron counterweight.[12]

The tilting tower was an immediate hit with Aermotor's customers, and many thousands were sold within the first three years of starting operations. "The idea

Aermotor pumping windmill on a wooden tilting tower, ca. 1905. Courtesy *Windmillers' Gazette*, Rio Vista, Texas.

was at first laughed at as impracticable, but thousands of Aermotors are now sold to parties who would not use a windmill if it were necessary to climb a tower to oil or look after it," one publication reported in 1891.[13] Aermotor's tilting towers were sold to the customer complete with precut and predrilled marked timber and the necessary hardware for assembly. All one needed was a hammer and wrench to assemble the tower. The towers were designed for both eight- and twelve-foot-diameter windmills and came in heights of thirty-eight, forty, fifty, and sixty feet. In 1891 the company's wooden tilting towers could be purchased starting at $20 for the thirty-eight-foot model with the eight-foot wind wheel and up to as much as $72 for the sixty-foot variant that supported the heavier twelve-foot wheel.[14] To highlight its ease of use, Aermotor often showed in its advertising and brochures an adolescent girl operating it by tugging on a line attached to the weight.[15]

By 1892 Aermotor had supplanted wood with all-steel components for both its tilting and newly introduced fixed steel windmill towers. The company claimed that its steel towers were not more expensive than traditional wooden towers and more durable, weighed one-third that of their wood counterparts, presented one-fifth the surface "to the grasp of the storm," and was aesthetically appealing.[16] The problem the company immediately encountered, however, with its steel towers was insufficient supply of angle iron. Aermotor, thus, decided "at an expense of several hundred dollars" to build a machine to bend flat bars of steel into the proper angles. A test fixed tower was attached to the outside wall of one of its brick buildings alongside two wooden towers of similar design and height, all parallel to the ground. "A horizontal strain was put upon the wood towers through spring balances made for the purpose, that indicated the exact strain which it took to break the towers—a similar strain was put upon the Steel Tower and was steadily increased until its anchor posts pulled, at a very much higher strain than

Open back-geared Aermotor pumping windmill on a steel tower being raised in one piece at unidentified location, ca. 1900. Courtesy *Windmillers' Gazette*, Rio Vista, Texas.

sufficed to wreck both wooden towers. This was a two inch round-cornered angle," the company said.[17] Aermotor commented at the time that most competitors' steel towers used two-inch angled steel for all their windmill sizes, threatening the integrity of the tower as the windmills became bigger and heavier, whereas Aermotor only used the two-and-half-inch angle for towers supporting its eight-foot windmills, scaling up to heavier three-and-a-half-inch and four-inch steel corner posts for those towers that supported its sixteen-foot mills.[18] However, cost was a significant differentiator between Aermotor's tilting and fixed steel towers, with the smallest, thirty-foot tilting tower for an eight-foot windmill priced at $48 and the eighty-foot-tall tilting tower for the twelve-foot windmill costing as much as $288 in 1895, or twice as much as the fixed variants.[19] Sales for the steel tilting towers gradually dropped off, with the last reference in Aermotor's sales literature to them in 1899.[20] Yet a number of these tilting windmills remained in use well into the early 1900s.

Aermotor's fixed steel tower design for pumping windmills remained relatively unchanged by the company for many decades, proving their durability. The 1918 model remains in production to this day. One of those design hallmarks was the way the company dovetailed the angle-iron steel at the top of the tower.[21] Much

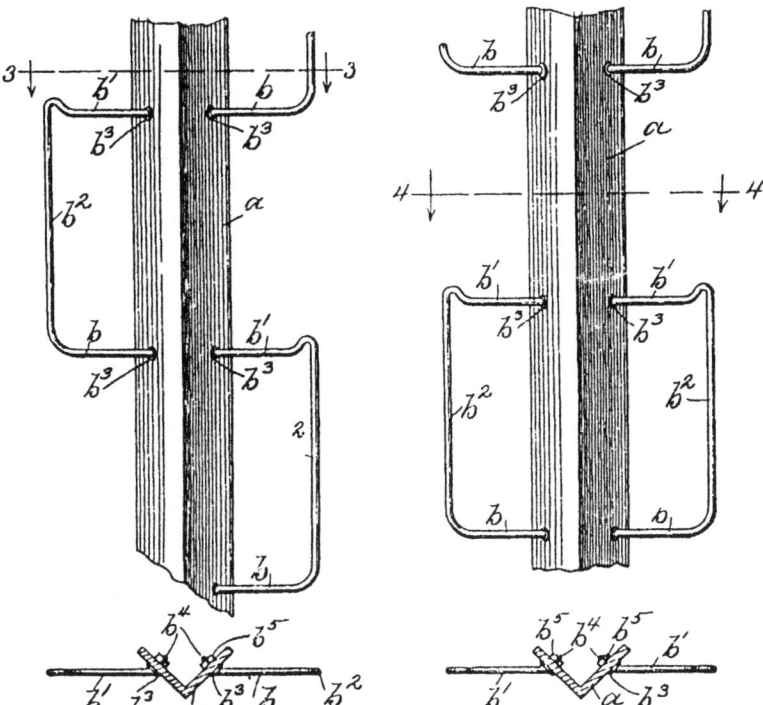

Patent drawing of La Verne W. Noyes's looped step ladder for steel windmill towers, 1897. Courtesy US Patent and Trademark Office, Washington, DC.

thought was also put into the tower turntable, on which the windmill rotated, to avoid wind obstruction. Other significant and lasting attributes of Aermotor's early tower design included the steel crossties that kept the tower sections and the overall structure firmly in place as well as the stay in the center of the tower to prevent lateral bending or buckling of the pump rod during its downward motion into the well.[22] However, the most recognizable Aermotor tower design to this day is the ladder for scaling the tower. The company's earliest tower steps appeared much like some of its competitors' towers. They used a single corner post of the tower to bolt evenly spaced steel steps upturned slightly on the ends at the width of an average man's foot and hand.[23] Another early design prescribed by the company included a series of one-inch-by-one-inch pieces of angled steel turned up at the ends so that the flat portions were the width of an average man's foot and hand.[24] At the top of the tower, just below the leading edge of the wind wheel, was placed a square, wooden platform, large enough for a person to stand for oiling or checking the windmill's gears. Then in 1897 Aermotor patented a looped step, which Noyes described in his patent application as being "manufactured at small cost and readily placed in position [with threads and nuts into the tower leg], and, furthermore, to provide a construction which will afford a vertical hand-hold for the person climbing the ladder."[25] These ladders were sold with all Aermotor fixed steel towers by 1899 and praised by users for offering "greater security" during climbs.[26] These loop steps remain in production for Aermotor towers to the present day.

Another late 1890s tower innovation for Aermotor was the introduction of a three-post or "tripod" tower. A three-legged tower by its very nature can offer a more stable footing than four legs, and Aermotor, as it attempted to reason the structural benefits with its customers, said that it is "almost impossible to anchor four posts exactly on a level; and when you do, they will not remain level long if the ground is soft. The trouble comes from the leg which is anchored high. It buckles and it tends to push the tower over, putting the other legs on a strain. It becomes worse than no leg at all."[27] Another benefit to these towers was manufacturing costs, which allowed Aermotor to price them 20 percent less than their four-legged counterparts.[28] Internally, the company's management believed customers would gravitate to the three-post tower, eventually phasing out sales for four-legged models. However, sales of these towers never eclipsed the traditional four-post tower for a number of reasons, including the general perception that three legs are weaker than four and the inconvenience of climbing a tower with legs erected at a steep, sixty-degree angle from ground level.[29] In the early 1900s the company continued to refine and strengthen the tripod tower, lengthening the ground-connected tower cross-members and allowing for less obstruction around the pump area for people and livestock. These towers could effectively be erected closer to houses and buildings.[30]

Farmers, who were eager to tap shallow wells or nearby streams to deliver greater volumes of water to their crops, would couple their windmills to various suction pumps, and Aermotor's windmills were no exception. In 1893 a farmer in California connected his sixteen-foot Aermotor to a triplex mechanism, which with three side-by-side pumps was capable of drawing 8,000 gallons of water an hour. Aermotor claimed its twelve-foot windmill, coupled with a smaller triple-acting pump, could just as easily deliver 4,000 gallons of water an hour.[31]

The Aermotor Company also matched some of its competitors in the early 1890s by offering the so-called "suburban outfits," which were essentially windmill towers supporting elevated wooden water tanks. To accomplish this, a piece of gas pipe was securely flanged from the ground to the bottom of the tank, allowing the pump rod to lift the water up and into the tank. These tanks also included pipes to prevent overflows.[32] To keep the water clean, tanks were usually covered. The water was fed through a pipe by gravity to wherever it needed to go. Since suburban tanks carried enormous weight—a gallon of water weighs about eight pounds—it required extremely sturdy and balanced windmill towers. Aermotor took advantage of the strength of its fixed steel towers to offer these outfits. Instead of the tank in the tower, some farmers installed a windmill next to a house or barn and fed the pumped water into an attic tank where it could be better protected from winter freezing.[33]

The company also sold numerous tools, pump parts and accessories, sheet-metal livestock tanks, and tank heaters to its windmill customers.[34] Even the tools used in its factory could be purchased, including its unique four-wheel steel trucks. These trucks had sixteen-inch wheels—two in the middle with one in front and back. The truck turned on its two main center wheels. To accomplish this, one end wheel was on the floor, the other was off the ground about an inch. Aermotor said the trucks were handy for moving lengthy materials such as steel

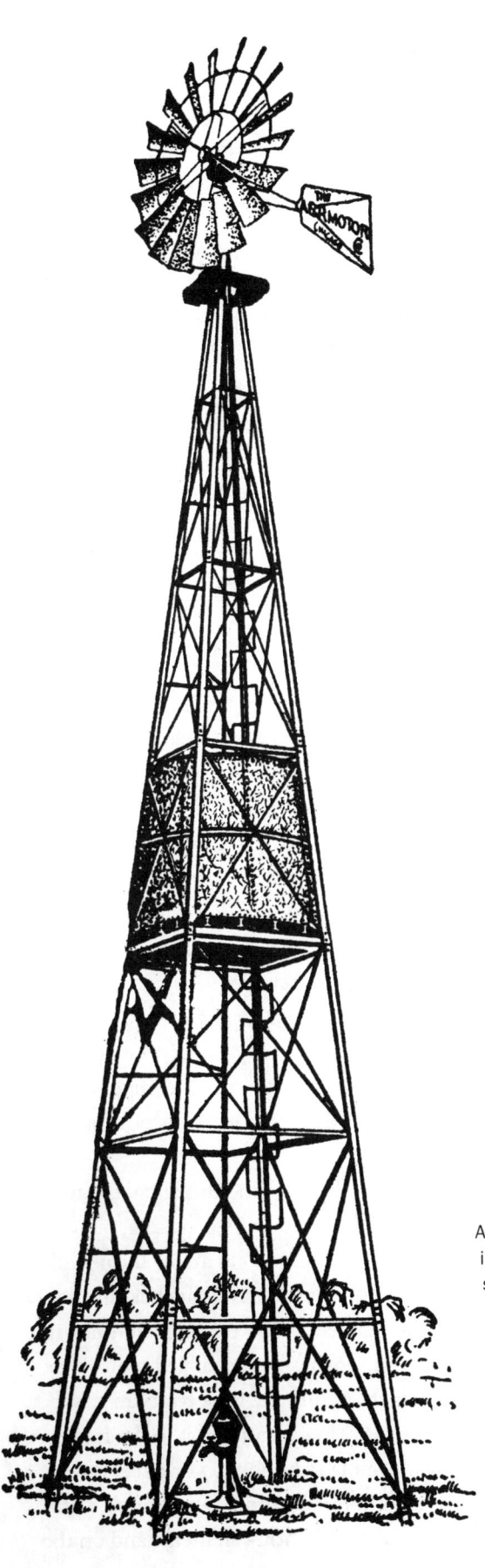

Aermotor suburban outfits
included either steel or wood tanks
suspended within the tower legs
and used hydrostatic pressure
to deliver water for household
or agricultural purposes. From
Aermotor Company, *Price List*[,]
*Aermotor Gasoline Engines for
Pumping and Power* (Chicago,
1920), n.p.

pipe and lumber. "It will carry large loads of castings, or parts of machines which are to be moved from one department of the factory to another. A box of any desired size can be easily fitted onto the body of the truck for handling or storing bolts, nuts or other small parts. There is almost no material handled around the Aermotor factory which is not moved on these trucks. We couldn't do business without them," the company said.[35]

Aermotor's business was not all about water-pumping windmills. Early on, it successfully developed and sold a "power" mill for mechanical applications around the farm. Like its pumping counterpart, the Power Aermotor used the same combination of all-metal wind wheel, poured babbitt bearing, and shaft; however, that is where the similarities ended. The power mill deployed a spur gear to drive a pinion gear and a pair of bevel gears at the top of the tower to spin a vertical steel shaft instead of producing the up-and-down motion used for a lighter pump rod.[36] Where Aermotor, like most of its competitors, used gears with a fractional ratio for its pumping mills, often referred to as a "hunting tooth," the company's power mills employed gearing with an even ratio.[37] Perry and Noyes discovered that to operate a power mill efficiently, its design had to overcome the friction and extreme torque experienced when a fifty- to eighty-foot shaft meshed with gears to transmit rotary power to other machines. Other windmill manufacturers with power mills similarly struggled with these problems of "creep." Aermotor overcame these mechanical challenges first by gearing the windmill to three times the speed of most other mills, using a lighter three-quarter-inch-thick steel shaft, and by arranging its vane to turn the mill into the wind.[38]

Power mills were generally placed on small towers attached to the tops of barns or shed roofs. Aermotor recommended to its agents and customers placing the wheel at least fifteen feet higher than the top of the roof to avoid the eddying effect of the wind glancing off the roof and to receive an unobstructed wind flow. In 1891 the company's first power mills were sold in two wheel sizes—twelve-foot and sixteen-foot—at a price of $100 and $200 respectively, plus the cost of the towers. It dropped the prices by half within a year, further increasing its appeal among farmers and raising the stakes against other power mill competitors.[39] The company also designed its power mill to be as simple to install as possible:

> To put up a 12-foot Power Aermotor on a barn or other building, frame the mast together in the driveway of the barn, put in the shafting, put on the short steel stub tower and the assembled motor, cut a hole through the roof of the barn where the mill is to stand, about 16 by 24 inches, attach the tackle or hay rope to the mast about 6 feet below the mill and draw the mill up through the hole in the roof. As soon as the assembled motor protrudes through the roof, put on the tailbone and the vane (you can stand on the roof to do this), then raise the mast, brace and bolt it to the framework of the barn. Then build a small cupola to be about 7 or 8 feet down from the center of the wheel. Matched flooring sprung into an arch, with a 4 to 6-inch raise, will make the best covering. You can then stand on the

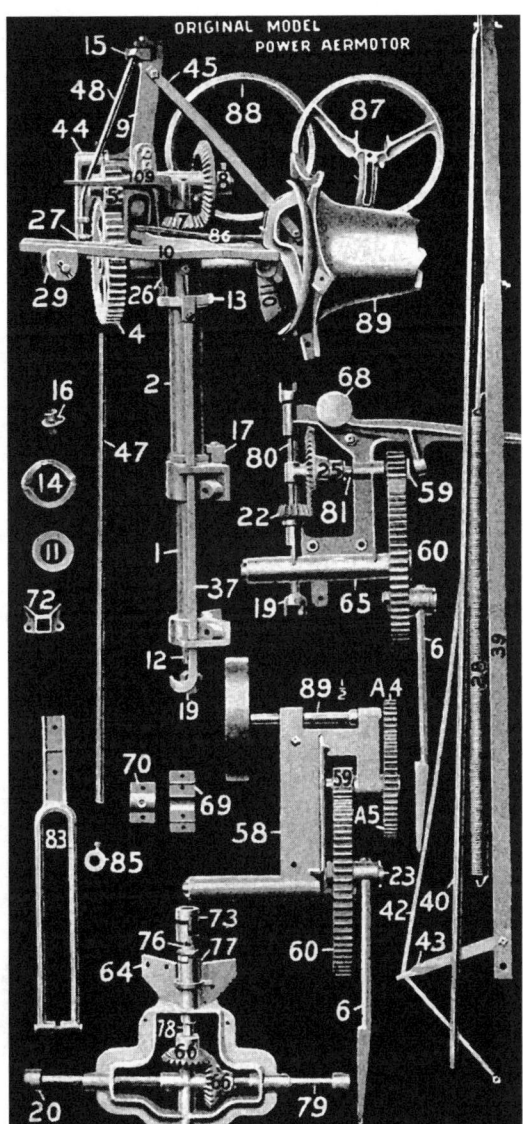

Parts of the original Aermotor power windmill. From Aermotor Company, *Twentieth Annual Aermotor Repair List* (Chicago, 1908), 17.

roof of this cupola and put on the arms and sails of the wheel without any additional scaffolding. In a well framed barn, if the mast passes through or near the ridge pole, no additional braces will be needed above the barn; the light bracing of the cupola will be sufficient. The braces immediately below the barn roof, however, should extend 6 or 8 feet in every direction, in order to give the mast a firm attachment and prevent springing any of the timbers or rafters.[40]

At the bottom of the shaft was a foot gear with safety clutch about four and half feet off the barn floor. In the mid-1890s Aermotor added the fourteen-foot wind wheel to its power mill line and manufactured and sold a host of mechanical devices to go along with the mills, including a circular buzz saw, corn sheller, and feed grinders and choppers. The foot gear was further refined and made versatile enough to attach multiple grinders, if necessary.[41] Farmers also found ways to use

Power windmill, ca. 1905. Power windmills were generally constructed over barns that housed power equipment. Author's collection.

these windmills to power their own homemade equipment. Aermotor playfully remarked in 1900 that its power mill "does the work of two farm hands and a team of horses, and that, too, without bed or board, without coaxing or driving, without shirking or grumbling."[42] There was even the ability to use a power mill to pump water from nearby wells. A pump jack could be placed immediately over a well with a belt connected to the foot gear at the bottom of the vertical shaft, or for distances over twenty feet to the water pump, a gear might be used to operate a pair of triangles.[43] Many farmers used their power mills to not only manufacture their own feed and flour, avoiding the cost of transporting and paying a milling operation to grind their grain, but offered similar services to neighboring farmers, feed suppliers, and grocers, effectively allowing them to generate additional revenue. Aermotor estimated that its twelve-foot power mill would grind through the year on average seventy-five to eighty bushels of meal per day. The company advised that a power mill of this size performed best in wind speeds of between

La Verne W. Noyes at fifty. By this time, he was a multimillionaire, respected by Chicago's industrialists, and active in progressive Republican politics. Courtesy Special Collections Department, Iowa State University Library, Ames.

twenty-five and thirty miles per hour. "With a 30-mile wind one man can scarcely shovel corn into a two-hole self-feed sheller fast enough to keep the hopper full," Aermotor declared.[44]

From the outset, the Aermotor Company was closely controlled. With the exception of a handful of shareholders for incorporation purposes, Noyes owned 99 percent of the company. Those who held the first shares in the company included his wife, Ida; Fred E. Smith, his brother-in-law, who became one of his trusted managers; and Myrtle and Ira Giffen of his sister Francis Adelia's family. The company had listed $100,000 in capital stock in May 1890 and was bumped up to a total of $300,000 by June 1893.[45] Other family members, business associates, and senior managers would receive shares in the years ahead, but the majority stayed in the hands of La Verne Noyes.[46] Ironically, Thomas Perry was not issued any capital stock in the company, perhaps due to his status as a consulting engineer, which gave him the flexibility to continue pursuing windmill-related and other patents without being beholden to Aermotor. Between 1891 and 1893, Perry received a number of patents under his own name that focused on windmill-governing mechanisms.[47]

Noyes made it clear to his employees that he was firmly in charge of Aermotor and implanted his no-nonsense personality upon the office staff and factory workers. He was not beyond confronting employees who did not live up to his expectations. As he described in his own words in a company newsletter in December 1893:

> I do most of the disagreeable things about the office and factory. If any
> clerk leaves paper on his desk at night, I try to be disagreeable about it. If
> a letter is left unanswered over one day, there is another opportunity to be
> disagreeable, and I never miss the opportunity. I try to make it unpleas-
> ant when they are slow about shipping. There is a rule that if an order is

Aermotor office workers, mid-1890s. La Verne W. Noyes valued women, as much as men, for administrative roles within the company. Courtesy Mark S. Welch, Fort Worth, Texas.

unfilled for three days, it shall come to me; but alas, this summer when others have been complaining about want of business, we have seldom been able to ship promptly.

One of the things I especially dislike and inveigh against and try to prevent is Sunday work and overtime work. A man once said, 'you do not appreciate a good man who is willing to work nights and Sundays, and at all times.' I said, 'No, but we do appreciate a man who gets here on time, does a day's work in a day, and goes home and minds his business.'

Among about sixty employees of the office, there were but three tardinesses in October; all were in their places with books out and ready to begin work at 7:30 and 12:30. I set up the pins to secure this result.[48]

Noyes also appreciated and valued hard-working women in his office. He declared, "We believe in 'Women's Rights.' We give them a chance, and many of the best men in the office, and certainly the pleasantest ones, are young women. If any of you have ever been in the claim department and have been held up for cash with order because you let a draft be returned, and have felt the firm, strong hand of the claim department, you will be pleased to know that that chastisement was administered by a young lady."[49]

Much like he did with his dictionary stands, Noyes used the enormous number of agriculture trade publications and newspapers to get the word out about

Open back-geared Aermotor on steel tower at Vaughn, Washington, along the Puget Sound, ca. 1905. Courtesy *Windmillers' Gazette*, Rio Vista, Texas.

his new windmills to individual buyers, hardware stores, and other agricultural equipment dealers. This effort was in addition to producing and distributing handbills, wall hangers, trade cards, and brochures. He encountered stiff competition from other makers doing the same. In an 1889 postcard mailing, Noyes offered newspapers $2 in payment for any pumping windmill sold to a reader of the newspaper provided the publication had been regularly printing free advertisements for these windmills. He drew on his sales experience with his dictionary stands, stating on the postcard, "Since I, as the Dictionary Holder Man, have during the past ten years personally had dealings with at least 5,000 newspapers, and since those dealings have been of such a pleasant character as to make any of those 5,000 papers a good reference, it may add to confidence in this plan of advertising if I personally guarantee the fulfillment of all that is herein proposed, which I cheerfully do."[50] To stand out in the crowd, Noyes developed cleverly worded and abundantly illustrated advertisements.[51] The *Farm Implement News* in an article about Noyes in August 1890 praised his advertising prowess, declaring, "He who has failed to see one of those large red posters portraying the merits of the Aermotor and the Tilting Tower has missed one of the most striking and successful advertising efforts of the decade."[52] Many early Aermotor ads employed healthy doses of customer testimonials about the operational efficiency of these windmills over others. For example, the company published a letter from G. W. Lamb of Randolph, Iowa, in one of its 1890s advertisements, which stated, "There are four 8-ft. Aermotors in this neighborhood, all giving the best satisfaction. They will run when the 10-ft. wooden mills in this locality can't turn a wheel. The Aermotors throw themselves in and out of gear, according to the pressure of the wind. I have pumped 16 bbls. of water in 10 minutes. To any one wishing to purchase a windmill I can highly recommend the Aermotor, as I think it the best mill I ever saw." Another buyer in the same ad, John Colier of Ganavanza, California, boasted, "The 8-ft Aermotor runs and pumps when my neighbor's 12-ft wooden mills are idle. It runs in the lightest wind, and regulates splendidly in storms, and handles the pump easily and with no noise."[53] Noyes could also credit some of his marketing success to the inevitably catchy, memorable name of Aërmotor itself (within a few years, the company dropped the use of the umlaut over the *e* to become just "Aermotor," and it was unseen in company documentation after 1895). Noyes never deviated from this brand on his windmills, unlike other firms of the day that attempted to capture customer attention by using more whimsical names, such as Dandy, OK, X-ray, and Gem, which were disconnected from their parent companies.

Also, unlike some of its competitors, Aermotor in the early years used no third-party agents to handle its sales and in 1893 had only one traveling salesman. The company transacted nearly all of its sales in its first five years via the mail, which was managed from the Chicago headquarters by a group of fifteen "territory men," who were assigned certain states as their territory. "In the rear of the office the desk of telegraph operator is prominent. All messages over the Western Union lines are received and transmitted over our own private wire. This together with local telephone connection, affords the best facilities for rapid communication with all parts of the country," the company said.[54] The Aermotor Chicago office also contained a sample room, which displayed examples of the company's latest products to visitors.[55]

To be in closer contact with its windmill buyers and installers throughout the country, Aermotor had begun opening branch houses in major geographic regions by 1891. These branch houses operated as storage and distribution facilities for Aermotor windmills and parts. Thus it was important for them to be located along tracks to receive railcars of supplies and cover significant but defined geographic sales regions. Since these branch houses sold products exclusively from the Chicago factory, the prices paid by agents for the windmills and related supplies were only slightly higher, mostly to cover the freight transportation costs.[56] Aermotor's first branch house was opened in San Francisco. By the end of 1893, it had similar operations set up at 65 Park Place in downtown New York to serve the New England states, New Jersey, and the southeast of New York State; Buffalo, New York, for the state; Kansas City to cover Kansas, Indian Territory, Oklahoma, Arkansas, Colorado, and western Missouri; Minneapolis to support business in Minnesota, North and South Dakota, and Montana; and Lincoln, Nebraska, to look after the trade in the state.[57] The number of branch houses increased to eight in 1894 and to a whopping twenty throughout the United States by 1895.[58]

To keep its sales agents motivated, Aermotor used a tactic that included both incentives and peer pressure. In an October 1895 letter, the company told its agent in Minco, a town in Indian Territory (now Oklahoma), J. E. Bonebrake, to spend the autumn and early winter months pushing sales of Aermotor "geared mills for power purposes." "We are well aware that it takes more work to sell these geared outfits, but when the trade is once well started you will find it just as easy to sell the power mill as the pumping ones, and the profit on the geared mill should be twice as large," the company said.[59] Agents were also rewarded for their aggressive windmill sales. From August 12 to 24, 1895, Aermotor offered a free eight-foot windmill shipped from the Chicago plant, a $25 value, to the agent who ordered windmills and steel towers at the greatest list value during the two-week period.[60] Big scores were often highlighted in company literature. For example, L. G. Collins of San Diego, Texas, was praised in 1892 for selling the world's largest cattle ranch, operated by Mrs. Richard King, on the Aermotor's capabilities. The ranch, which covered an area of southern Texas the size of Rhode Island, had a herd of nearly 100,000, and wooden windmills of nearly every make scattered across 700,000 acres supplied water to the livestock. In 1890 Collins offered to erect three Aermotors on a trial basis at the ranch to prove the windmill's resilience. Before the end of that year, a dozen more Aermotors were erected, and by 1892 a full two railcars of windmills and towers had been shipped to the ranch.[61] By 1895 the King Ranch had about one hundred Aermotors in operation and was actively phasing out its remaining wooden windmills.[62]

Yet the company encouraged collegiality and corporate pride in the cities in which it set up shop. The Chicago *Daily Inter Ocean* reported on November 1, 1896, amidst the excitement on the eve of the presidential election between Republican William McKinley and Populist Democrat William Jennings Bryan, how the Aermotor Company had erected a 150-foot-tall flagpole over its factory in celebration of Flag Day and how "twenty young women of the office drew 'Old Glory' to its summit amid the cheers of several hundred employees." The newspaper further stated, "It was regarded as fitting that this task should be left to the young women, since there are many of them in the office force, and one of them [Julia M. Flanigan] has been secretary and treasurer of the company since its organization."[63] In the late 1890s Aermotor's Miller Brothers agent in Sacramento, California, fielded a local baseball team.[64]

However, another, and perhaps more important, reason for Aermotor's rapid rise in early sales was the general interest throughout the country, not just for ranches and farms, in access to reliable flows of fresh ground water, and its windmill offered that capability quickly and easily. In the Northeast, windmills for large estates, hotels, and general municipal purposes were often placed on top of elaborately built enclosed wooden towers that could also be climbed and enjoyed for observation purposes. Windmills for use on the Great Plains in the early 1890s, on the other hand, were acquired for practical purposes, and in many cases out of the emergency need to counter a multiyear drought that gripped this region of the country. "The years 1891–1896 might be called the experimental years for the use of the windmill in irrigation," wrote Nebraska historian A. Bower Sageser in 1967.[65] Garden City, Kansas, during this period became a media focal point for its abundance of windmill-filled irrigation ponds.[66] It was also a period of significant research by hydrographers such as Edward Charles Murphy, professor of civil engineering at the University of Kansas in Lawrence. In 1897 Murphy published an extensive, multiyear analysis into the efficiency of windmills used for irrigation at Garden City. The five Aermotors—one eight-foot, three twelve-foot, and one sixteen-foot—in a pool of twenty-five windmills produced by many of the top manufacturers of the day significantly outperformed the competing models when measured by the number of gallons pumped per hour in wind velocities of eight, twelve, sixteen, twenty, twenty-five, and thirty miles per hour.[67] About 1900 Aermotor at its plant set in motion what it claimed to be the largest piston pump ever and attached it to a sixteen-foot windmill, with a cylinder diameter of thirty-six inches, sixteen-inch stroke, and a capacity of seventy-one gallons per stroke.[68]

Just as important as keeping a windmill efficiently pumping water was the ability to shut it down. Without this, tanks might overflow, causing a muddy mess. Aermotor, like many competitors, developed mechanical devices to take their windmills out of service either manually or automatically.[69] One of the simplest was a wooden cutoff lever attached at the base of the tower that used a steel cable or wire mechanism to essentially pull the windmill's vane parallel to the wheel, thus taking it out of the wind. Throw the lever in the opposite direction and the vane would unfurl, allowing the wind wheel to resume spinning. Preferable to many busy farm families were the automatic regulators either attached to the windmill tower or inside the pit of the well. When the pumped water reached a predetermined level in an adjacent tank, a wire connected to the float triggered the regulator to turn off the windmill. As the water in the tank decreased, the float would activate the regulator to unfurl the windmill vane and restart the pumping process. Aermotor warned its customers that it should be protected from freezing in the winter, because the regulator would not work when the surface of the water was frozen.[70]

Aermotor in its startup years did not rely initially on large orders from industrial users, like the railroads, as a number of the windmill industry's founding members did. Those opportunities were largely over by the time Aermotor entered the market. The company did, however, make some inroads into the rail market. It sold windmills, for example, to the Chicago, Milwaukee & St. Paul Railway at Saukville, Wisconsin. Onward Bates, engineer and superintendent of bridges and buildings for the railroad, told Noyes in a December 20, 1894, letter that the Aermotor "so far as I know it is the best of windmills," noting that he didn't pay these types of compliments lightly.[71] However, many railroads had become dissatisfied with the performance of some windmills and their ability to hold up, especially in the harsher weather of the upper Great Plains. They sought stronger, all-metal windmills to replace their decrepit wooden mills.

One of the more colorful examples of Noyes demonstrating his competitiveness—some might have said arrogance—was at the windmill exhibit during the 1893 World's Columbian Exposition in Chicago. Aermotor literally stood out in the crowd of several windmill displays. The exhibit consisted of a fifty-five-foot-tall Dutch-style windmill on which was mounted an eighty-seven-foot-tall galvanized steel tower for a total height of one hundred fifty feet. Visitors were attracted to the Dutch windmill's wood balcony, from where they could look up at the Aermotor windmill and view the rest of the exhibition area. Inside the Dutch windmill's housing were examples of the company's other wheels and its array of feed grinders and cutters and other attachments for its power mills. Next to this windmill was a seventy-foot galvanized steel tilting tower supporting one

Windmills on display at the World's Columbian Exposition in Chicago, 1893. The exhibit included the sixteen-foot Aermotor mounted on an eighty-seven-foot tower and placed upon a Dutch-style windmill for a total height of one hundred and fifty feet (far left). From H. C. Ives, *The Dream City: A Portfolio of Photographic Views of the World's Columbian Exposition* (St. Louis: N. D. Thompson Publishing Co., 1893), n.p.

of Aermotor's twelve-foot wind wheels.[72] The overall exhibit was described by an agricultural publication as "the finest and fullest collection of wind mills, including fixtures, towers, tanks, pumps and machinery of various kinds for attachment, the world ever saw. . . . The numerous tall towers, like masts of ships grouped closely in harbor with colors flying, surmounted by wind wheels of various forms and sizes whirling in the bright sunlight, and throwing off sparkling rays of many colors, afford from the distance a unique, lively and brilliant spectacle."[73] In total, there were sixteen windmill manufacturers displaying from two to four different windmill products. Water was easily reached by driving pipes eight feet into the ground, allowing the manufacturers to show their windmills in action.[74]

In late August 1893, unbeknownst to the other windmill exhibitors, Aermotor started construction of a forty-foot steel tower with a twelve-foot power mill atop one of the stock barns south of the fair's amphitheater. The other companies attempted to stop Aermotor by going to W. I. Buchanan, chief of the Department of Agriculture for the exhibition. Buchanan halted the windmill's construction for two days on a Friday, during which time it was alleged that a group of men from the other windmill companies used a rope to pull down Aermotor's unfinished tower while its work crew was away at supper. Buchanan came back on the following Monday with an approval for the windmill to be erected.[75] The *Farm*

Implement News, a prominent Chicago-based agricultural periodical covering the event, published an article siding with the offended windmill competitors, referring to Aermotor as "an aggressive wind mill concern which does not have much consideration for others in its line of business, or respect for unwritten laws that govern exhibitors in their dealing with one another."[76] Noyes, while undaunted by the negative publicity, claimed he offered the Aermotor power mill as a tool to provide readily accessible ground grain to the exhibit's livestock, but the other windmill manufacturers saw it as a new exhibit.[77] The Aermotor president, when confronted by his competitors, even offered up to $1,000 to help them erect their own power mills. There were no takers.[78] For Aermotor, as the sole exhibitor of a power mill at the livestock barn, the move paid off by attracting numerous onlookers as the mill "cut feed as rapidly as two men could get it to the feed-cutter and ground at the rate of fifteen to twenty bushels an hour," resulting in an abundance of new customer orders and inquiries from those eager to become the company's sales agents, both in the United States and abroad.[79]

Yet Noyes also viewed the importance of collective action by the industry when necessary. On September 20, 1895, he became one of the founding members of the National Association of Wind Mill Manufacturers, which included, in addition to Aermotor, fourteen of the country's largest windmill makers. Noyes was named vice-president of the trade group and an executive committee member.[80] While no details about the association's by-laws and mission have been found, many industry-specific trade groups of the day were focused on combating unjust commercial regulations and controls as well as freight railroad transportation abuses. Shortly after its founding, the press accused the windmill association of attempting to regulate and control prices, which members vehemently denied.[81] It is uncertain how long the windmill association remained in existence, but it appeared to meet in Chicago at least twice a year and on June 15–16, 1899, congregated to "discuss advancing cost of materials and necessity of advancing prices of finished goods," the *Farm Implement News* reported.[82] In the late 1890s Noyes became active in industry groups such as the Illinois Manufacturers' Association and the National Association of Manufacturers, often attending the meetings, speaking out against industry abuses, and even holding various officer positions.[83] He was made president of the Illinois Manufacturers' Association and was praised for standing up to the railroads for unfair freight transportation rate practices. Noyes was equally zealous against public officials taking bribes and, as president of the Civic Federation from 1900 to 1901, proposed an amendment to Illinois state law to make it easier to convict public officials who accepted bribes.[84] He also succeeded in lobbying for a bill to abolish township organization in Cook County and to ensure that power remained in the hands of city and county governments. He was also at the forefront of Chicago's municipal court reform, a system that at the time was notoriously inefficient and run by judges who were routinely influenced by politics in their decision making. Noyes, considered by his peers to be a progressive Republican, was even encouraged to run for state senate to represent his district, but he declined.

As the company rapidly expanded its presence throughout the United States and the world, Noyes's ability to control his business empire was occasionally challenged by potential cheating within his own ranks and competitors attempting to infringe upon his trademark and patents. In 1895 two former managers of Aermotor's San Francisco office, W. E. Hampton and August Holtgen, were arrested and charged with felony embezzlement. Both men were employed with competitor Pacific Tank Company in San Francisco at the time of their arrest. The *San Francisco Call* reported on November 21, 1899:

> Hampton came here from Chicago in 1892 [although historic records show Hampton and the San Francisco branch office were in place by 1890] as manager for the Aermotor Company, who manufacture windmills. He continued as manager for two years, with Holtgen as his assistant, and then left the company and started the tank company. Holtgen was promoted to manager of the Aermotor Company, and the allegation was made that Hampton and Holtgen conspired to divide the agents' discounts on sales of windmills and tanks, Holtgen sending false returns to the head office in Chicago of the sales having been made to agents, instead of to customers direct. The aggregate amount obtained in this way was said to have been about $5000, and the Aermotor Company got judgment in Superior Court against Hampton and Holtgen for that amount.[85]

Under Noyes's leadership, Aermotor quickly became a substantial exporter and within six years of starting operations had its windmills pumping water across every continent but Antarctica. Aermotor was certainly not the first US windmill maker to export; the industry's earliest players, such as the U.S. Wind Engine and Pump Company, A. J. Corcoran, and the Eclipse Wind Engine Company, had been shipping their windmills abroad by the thousands since the early 1870s.[86] In the late 1860s the US government imposed high tariffs on imports to protect domestic industry and help it rebuild after the Civil War, which further encouraged US windmill companies to export. The closest markets for US windmill exports were Canada and Mexico, which by the late nineteenth century were easily accessible by train direct from the factory. One of the most prolific uses of windmills in North America occurred in the late 1880s and 1890s at Yucatán, Mexico. Although dry on the surface, the windy area had an abundance of water twenty-five to fifty feet below the surface. Edward H. Thompson, consul for the United States in Progeso, Mexico, recounted in his 1903 report to the US Department of Commerce how he and a friend imported the first two US-manufactured galvanized metal windmills to Mérida in the late 1880s to provide water for homes. By 1903, Mérida had an estimated 1,200 windmills, of which 95 percent were of galvanized metal construction and 75 percent came from a single US firm.[87] Although Thompson didn't name the predominant windmill's manufacturer, it was Aermotor, based on numerous photographs showing scores of its products in the region during this period.

By the end of the nineteenth century, Aermotor's pumping windmills dominated the skyline of Mérida in Mexico's Yucatán. Courtesy *Windmillers' Gazette*, Rio Vista, Texas.

The biggest overseas markets for US windmills were the Pampas of Argentina and Uruguay, northern and southern Africa, India, and Australia. Like the western United States, these areas had ample subterranean water supplies but required pumps like windmills to bring them to the surface. Nevertheless, regions with surface water and wetter climates, including Europe, Central America, southern Asia, and the island nations of the Caribbean and South Pacific, also imported their share of US windmills in the late 1800s and early 1900s. Aermotor claimed that one of its first windmills erected in China attracted some undue attention from local villagers for its automatic water-pumping ability. "They discussed it constantly," said an Aermotor catalogue of the villagers, "and soon, to the horror and surprise of the missionaries, numbers of them were falling down before it every morning and worshipping it as the 'God of the Fields.'"[88] Not everyone overseas had the same reactions or experiences with US windmills. In India tests were conducted in the early 1900s on behalf of the Madras provincial government by the School of Arts and Crafts to demonstrate the effectiveness of a sixteen-foot Aermotor windmill contrasted with traditional draft animals to supply water to rice fields. After a year of testing, the school, which complained about the pump's mechanical performance, reported that the Aermotor achieved only 53 percent efficiency of maximal power in a consistent 10 mph breeze. However, a French reviewer in the December 1903 publication *Le Mois Scientifique et Industriel* questioned the Madras government's agenda in presenting these results.[89]

While showing a physical presence overseas via trade fairs was an important way to introduce US windmills to new markets, it required significant expenditures of money, time, and logistics on behalf of the manufacturers; consequently, only the larger players could afford to engage in these activities. Aermotor, for instance, made a prominent showing at the Paris exposition in July 1900. The *Farm Implement News* reported that the exhibit included an Aermotor on a three-post, eighty-foot tower with sixteen-foot wind wheel; one three-post, seventy-foot tower with fourteen-foot wind wheel; one three-post, sixty-foot tower with twelve-foot wind wheel; and another on a three-post, fifty-foot tower with ten-foot wind wheel. "The most striking feature of the exhibit is the enormous irrigating pump, which the sixteen-foot mill operates. The cylinder of this pump is thirty-six inches in diameter, twelve inches greater than any heretofore built. At each stroke of the piston it throws two barrels of water. The pump lifts the water to a ten-foot tank, allowing it to flow back directly into the reservoir. The other mills are geared to operate saws and feed grinders," the publication said.[90] A less expensive way for US windmill companies to access overseas markets was to translate their advertising and sales literature into other languages, such as French, Spanish, Portuguese, German, Italian, and Afrikaans.[91]

The most successful way for many of these US companies to bridge cultures was to employ local sales agents. One of Aermotor's biggest and most successful overseas importers/agents in the early 1890s was Agar, Cross and Company in Buenos Aires, Argentina. At the height of the US export trade, this company ordered one to three carloads of windmills a week from Aermotor's Chicago plant.[92] By mid-1893, C. Cadle of Wellington Quay, Dublin, was named Aermotor's sole agent for the United Kingdom and, while exhibiting at the Gloucester and Chester shows of the Bath and West of England Society and the Royal Agricultural Society, respectively, that year generated much attention with his Aermotor tilting tower display. The windmill was noted in the English trade press at the time for its favorable "price, handling and efficiency."[93] On the European continent, one of Aermotor's most active agents was Remi Van Sante-Baetens of Wetteren, Belgium. Starting in 1894, he both sold and erected these windmills for decades, mostly across Belgium and northern France, and even invented special pump attachments to connect with Aermotors.[94] Pierre Leclercq was Aermotor's agent in Tunisia, and he supervised the sale of many of these windmills across central North Africa, including erecting a power mill that could pump water, saw wood, and grind grain for experiments at the Agriculture School in Tunis.[95] In Australia, a handful of agents scattered around the country sold thousands of Aermotor windmills in the late 1890s and early 1900s. Among the most active of these Australian agents were W. D. Moore & Company of Fremantle, James Martin Company Ltd. of Sydney, Robison Brothers and Company Pty Ltd. in South Melbourne, Smellie & Co. Ltd. in Brisbane, and A. G. Webster & Sons Ltd. in Tasmania.[96] Aermotor allowed some overseas agents to paint their own names and logos on the windmill vanes, perhaps in an attempt to boost sales through local name recognition or appeal.[97] This is interesting since the company and other large US windmill makers zealously guarded their trademarks and patents in overseas markets against copycats. For example, Noyes was issued a patent from the Australian government in May 1908 for the company's windmill improvements.[98] However, it didn't stop overseas companies from attempting to copy the Aermotor's name or design characteristics. Belgian manufacturer Jules Dutrieu patented an all-metal water-pumping windmill in 1910 with the name "Aéromoteur Detrieu," which with the exception of a split in the sheet metal at the wide end of the tail vane looked strikingly similar to the Chicago-built Aermotor.[99] Plissonnier (or Société Lyonnaise de Construction de Machines Agricoles), a farm implement manufacturer that was located in Lyon, France, built and sold an Aermotor-style windmill from 1900 to 1914 without an agreement with the Aermotor Company.[100]

An understated but significant benefit of US windmills, when it came to shipping, was their ability to be easily broken down, securely crated for transport, and erected with the most basic hand tools at destination. Noyes had always considered efficient shipping a top priority at Aermotor, for without it, the business would wither. Overseas purchasers counted on Aermotor to protect its windmills against damages as the crated components moved from factory dock to boxcar and were unloaded in a rail yard and stowed in the hold of a cargo ship. In developed countries, the process would essentially be reversed once the ship arrived in port. The arrival in underdeveloped countries, however, increased the risk of delivering a damaged or incomplete windmill. Aermotor highlighted how one of its power windmills was transported on the shoulders of men for 350 miles from the coast to a missionary station at Kamundongo (in modern-day Angola) and successfully installed on top of a fifty-four-foot-tall sycamore tree trunk. A witness to this windmill, Dr. Charles F. Clowe, wrote to Aermotor in May 1895 that the villagers used the tree because it was "impossible to procure sawn or heart lumber suitable for a tower, and it would be too expensive to import it."[101]

Aermotor was in a state of constant innovation with its windmills and related pumping and power equipment, generating more than fifty patents between December 1887 and February 1899 alone.[102] In its first ten years in business, the company continued to make changes to the way its windmills were balanced on the towers, how wheel shaft bearings were lubricated, and how the wind wheel arms were attached to the hubs.[103] The company made the most significant visible change to its windmill design in 1898 by introducing a new hub and bearing to its pumping and power mills. In the so-called 1898 Model, Aermotor abandoned the former bell-shaped hub, to which V-shaped wind wheel arms were attached, for a cylindrical hub. It also introduced short-lived caged roller bearings rather than the graphite-impregnated babbit bearings used in older mills.[104] The new 1898 Model hub, which was designed to include an internal oil pocket, allowed oil to be carried twice with every revolution of the bearing shaft and spread it evenly to the windward end of the shaft. "The 8-foot wheel, which weighs 115 pounds and has an inch shaft, is carried by this bearing, and, in a fair wind, makes 120 revolutions per minute, or 7,200 revolutions an hour, or 172,000 revolutions a day. . . . Nothing should, therefore, be left undone to reduce the friction, to lessen the wear, or to increase its longevity."[105] In its 1899 Model, Aermotor included additional changes to the vane stem and furling device and added brass, screwed-on oiling cups, all in an effort to reduce wear and tear on its windmills. Also in the 1899 Model, the company begrudgingly returned to the use of babbit bearings in its hubs due to failure and overall customer dissatisfaction with the 1898 Model's roller bearings.[106]

Bearing wear was a constant battle for windmill manufacturers, and Aermotor was no exception. Noyes and his engineers were determined to find a way to get more working life from the Aermotor babbitt bearings, which were destined to wear out over time and to require replacing. Even before it secured the patents, the company in 1903 began selling "removable arms" for supporting the shafts of the Aermotor's wind wheel. "The main casting was simply made in three pieces instead of one so that the shaft arms could be turned in their sockets. By this means three perfect bearing surfaces are provided for each arm. Whenever a bearing becomes worn in one place, through overloading or neglect, it is only necessary to loosen a nut and give the arm one-third of a turn to secure a new and perfect bearing," the company explained.[107] The arms of the R.A. Model, as it was dubbed, prolonged the life of the windmill "six to eight times," according to Aermotor.[108] "No single improvement in the Aermotor has ever done so much to increase the popularity as the removable arms. They are just the right thing in just the right place," the company proclaimed.[109] Shortly after, Aermotor came out with the 1903 Model. The only difference between the R.A. Model and this windmill was that the R.A. Model's mainframe was made of three parts, whereas the 1903 Model was one solid piece.[110]

Throughout this development, the windmill company kept a standard numbering system for identifying its components. Parts used in the eight-foot mill, for instance, were numbered the same as in all its other mills. To distinguish these parts in the repair list, the eight-foot pumping mill was called the "A" mill; the ten-foot, the "B"; the twelve-foot, the "D"; the fourteen-foot, the "J"; and the sixteen-foot, the "F." The twelve-foot power mill was called the "G," while the fourteen-foot power mill was the "L," and the sixteen-foot power mill, the "H." However, what became the predominant identifying number for Aermotor's windmills at the turn of the century and going forward was the main casting number. For example, "A-402" refers to the main casting of the eight-foot 1899 Model; "B-402," the main casting for the ten-foot 1899 model mill; and so forth.[111]

In the spring of 1900 the company announced that it was moving from its location at Rockwell and West Twelfth Streets, where it had been for nearly twelve years, to a twenty-acre tract at Chicago Heights, between Wentworth Avenue and State Street, to build an upgraded factory. The site was two and half times larger than its existing facilities, and the area already hosted some of the city's other large manufacturers, such as Sargent Steel Casting, Sheldon-Foster glass works, and the Hamilton Organ Company. The cost of Aermotor's new plant—to be made of brick, stone, and steel—was about $850,000 and included electric power, lighting, and heating systems for the latest manufacturing and labor-saving techniques of the day. The plant would include the employment of 1,000 highly skilled workers. Another benefit of the new location was its proximity to the railroads, including the Chicago Terminal; Elgin, Joliet & Eastern Belt Line; Michigan Central; Chicago & Eastern Illinois; Lake Shore & Eastern; and Chicago Heights Terminal railroads, which agreed to build spur tracks to the Aermotor plant and would allow the company to eliminate teaming with freight wagons.

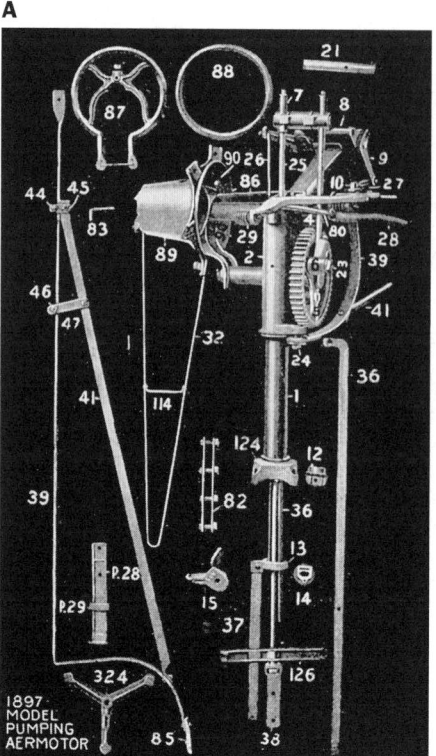

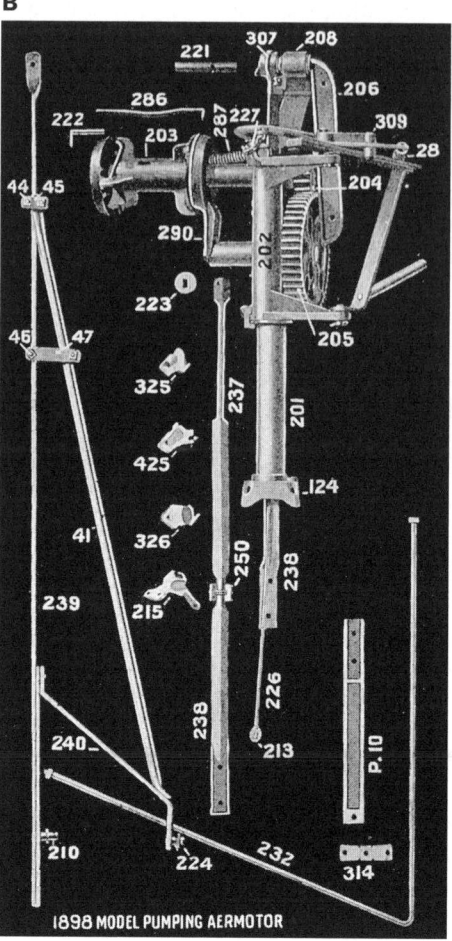

Above and opposite: Parts diagrams of Aermotor pumping windmill models available to customers by the early twentieth century. From Aermotor Company, *Twentieth Annual Aermotor Repair List* (Chicago, 1908), 1–4.

The company broke ground on the factory site in mid-May 1900 and planned to become operational shortly thereafter.[112] However, the company's plant expansion was stopped in its tracks, likely due to the effects on its business from the 1901 stock market crash and subsequent recession that lasted from September 1902 to August 1904. The company called a special directors' meeting, led by Noyes, on April 20, 1905, authorizing the sale of the Chicago Heights property for $349,847.94, with these proceeds being divided up among the Aermotor stockholders, which included Noyes, his wife Ida, Myrtle Giffen Johnes, Iva Giffen Gill, Frederick E. Smith, Lewis C. Walker, Frederick L. Dole, A. J. Berger, G. A. Williams, Joseph J. Fraser, A. F. Chapman, W. P. Gooding, and Julia M. Flanigan. The company also acknowledged at the time that "it seems likely that it will take a very considerable amount of time to dispose of a large part of the machinery included in the Chicago Heights plant."[113] Meanwhile, Aermotor continued to make additions, adjustments, and upgrades to its existing Rockwell and West Twelfth factory location, including expanding its warehouse and shipping room, modernizing the galvanizing plant, and constructing a new forging room that encompassed an acre of ground. The company claimed it was just keeping up with orders for its windmills, towers, pumps, and tanks, noting in October 1906 that

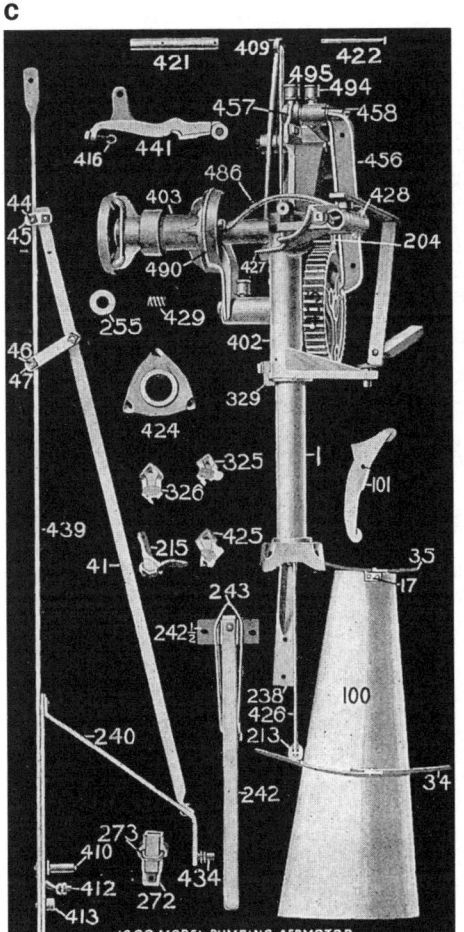

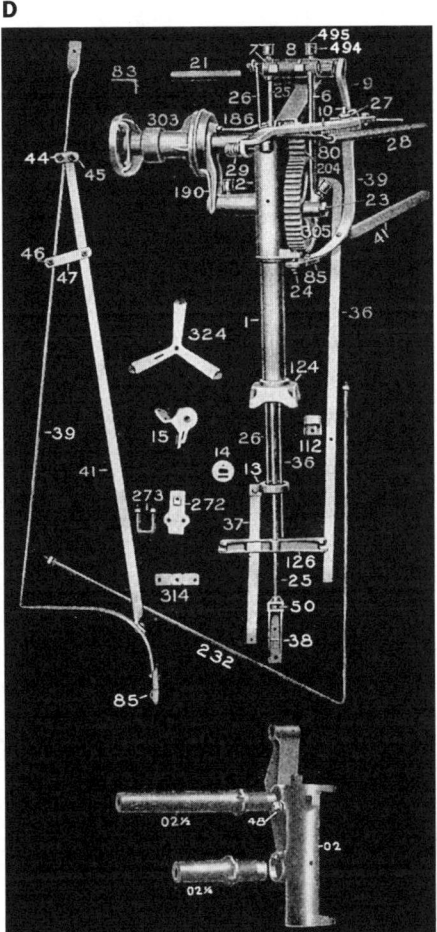

1899 MODEL PUMPING AERMOTOR

"by running our factory day and night, we are able to handle the heavy fall trade with our usual promptness."[114]

By 1899 Aermotor had cornered a large share of the windmill market in North America and established a comfortable presence abroad. "With the ability to direct a new enterprise in an unexplored field, ripened and broadened in the hard school of experience, and with the competitors already in the field sleeping the Rip Van Winkle sleep of twenty years, the Aermotor Company rushed forward and had half the world's windmill business before its drowsy predecessors ceased rubbing their eyes with wonder at having their long, delightful sleep thus ruthlessly disturbed," the company wrote of its competition at the end of the nineteenth century.[115] The company also jabbed at competitors that sold a variety of wind wheels under various brand names:

> A maker who offers you a variety of wheels knows that every wheel is wrong. The only reason for variety is the fact that it offers a refuge. You select your own wheel; and, when it fails, the maker falls back on the claim that the fault was yours in the choosing. He has another style to suggest to you.

And some makers go even further. They operate under different names. When one name comes into disrepute in a community, they start in under another.[116]

Aware of its strong name—becoming synonymous for "windmill" in the minds of many people—Aermotor rigorously guarded its trademark against any potential encroachment by competitors. "The trade-mark is usually displayed by painting the same upon some portion of the windmill, by printing the same on tags attached thereto, or by marking it upon boxes containing the several parts of the windmill," the company wrote in its 1906 trademark registration renewal statement.[117]

At the end of the 1800s, Aermotor predicted new uses for its windmills, including electric power generation for lighting and heating of homes and barns and compressed air for propelling farm equipment in the coming century. "Early in the twentieth century, the ox will cease to be beast of burden on the farm, and the horse will be only a toy, or a pet, or relic of historic interest," Aermotor declared. The company provided artists' renderings of windmills attached to farm equipment in its March 1899 descriptive catalogue, but a quick study of the drawings easily raises questions of the practicality of using windmills this way.[118] Both Perry (who had left Aermotor by 1897) and Noyes, however, had a deep, early interest in the potential mechanical power of compressed air and filed a handful of patents between them for ways to compress air with windmills.[119] By 1899 Aermotor was offering an "air compressor outfit" that attached to its windmills to pump water by force noiselessly from shallow wells or cisterns.[120] Within the next ten years, its compressed air systems became increasingly sophisticated and more efficient. The company touted that its pneumatic pump "affords the advantages of city water works without any of the objectionable features." Its three-way pump of the early 1900s had "an air in the pressure tank when required. It also has a hydraulic regulating cylinder for controlling the windmill. When sufficient water has been pumped into the tank to give the desired pressure the regulating device automatically furls the windmill and holds it out of the wind until enough water has been drawn from the tank to reduce the pressure several pounds. Then the mill goes into the wind again and continues pumping until the maximum pressure had been reached." The system provided steady water flows to kitchens, bathrooms, stationary washstands, lawns, barns, and gardens. "Buggy washing can be made a picnic for the boys instead of drudgery for the 'old man.' Just attach the hose to the hydrant at the barn and let the water squirt. In a few minutes the job will be done and you will feel lots of satisfaction in riding to town in a clean buggy."[121]

US windmill manufacturers in the early 1900s were also undergoing a change as they faced down increased competition from steam- and gas-powered engines that also pumped water and performed mechanical tasks. Some of the railroads during this time began to supplement their windmill pumps or replace them altogether with engines to provide sufficient water to their engine steam boilers. In the height of the steam-powered train era, the nation's railroads used about 600 million gallons of water a year. Passenger trains used between 70 and 120 gallons a mile, while freight trains sucked up from 150 to 350 gallons a mile, depending on the load weight, number of cars, and terrain.[122] "As long as traffic was light and the demand for water low, windmills were an economical solution, but once traffic picked up, a pumping device dependent upon the vagaries of the wind was too uncertain. . . . In 1905, for instance, the Great Northern was using both a windmill and a steam pump at Penn, North Dakota. The tower was nearly sixty feet high with a forty-foot platform, above which the axle of the eighteen-foot wheel turned. The steam pump worked almost constantly, except in a strong wind or when the boiler was being cleaned; then the mill went back into service," railroad historian Harold Cook wrote.[123]

Farmers, as early as the 1890s, began to explore the option of using small internal-combustion gas engines to pump larger volumes of water at faster rates for their crops and orchards. Engine manufacturers cleverly advertised the benefits of their products over conventional windmills, such as operational consistency when needed, versus idle or fickle wind power, and mobility, particularly for mechanical tasks on the farm, including sawing wood, shelling corn, and grinding grain.[124] They often used cartoon-like images of storm-damaged windmills. For example, Madison, Wisconsin–based Fuller & Johnson Manufacturing Company in an early twentieth century advertisement showed Uncle Sam giving a windmill with a human face a black eye, stating: "Think of it! Thousands of windmill owners forced to pump all the water for stock and house use by hand! It takes an 8-MILE-AN-HOUR wind to start a windmill. And it requires a 15-MILE-AN-HOUR wind to make windmills do good work. That's WHY farmers everywhere are doing away with windmills and installing Farm Pump Engines."[125] It was also not difficult to persuade some farmers to purchase a gasoline engine pump just to avoid climbing a windmill tower to oil the mechanism.

The traditional windmill makers did not take this competition lying down and fired back with their own marketing volley against the gasoline engines. Unlike the freely blowing wind, farmers had to pay for fuel and battery materials to power their engines, and many engines would become difficult to start in cold weather and when their ignition chambers became choked with dirt. Kalamazoo Tank and Silo Company, a maker of the Manvel and Kalamazoo windmills, proclaimed in the early 1900s: "It's really no matter for wonder that farmers, after having tried 'em out—and quickly worn 'em out—have gone straight back to their old standby, the Windmill." There was no clear winner in the competition between windmill and gas engine manufacturers for customers. Windmill historian T. Lindsay Baker observed, "The battle between windmills and gasoline engines continued well into the twentieth century, and it might best be considered to have been a 'stand off.' The gas engines did indeed find their places on American farms and ranches, but so did the windmills."[126]

Some windmill manufacturers such as Aermotor, Fairbanks-Morse, and Baker Manufacturing Company took the step of building their own lines of gasoline engines, knowing they would compete with their windmill sales. By 1908

Aermotor had successfully made and sold a small single-cylinder engine adapted to both its eight-foot windmill and hand pumps. These engines generally sold for $27.00 to $37.50 apiece depending on the basic attachments and support beams.[127] The firm also developed a heavy "back-geared" gasoline pumping engine that sat directly over a well and could be used for pumping larger volumes of water into elevated tanks for institutional purposes.[128] Aermotor claimed to have thoroughly tested each of its engines, running them for several hours under actual pumping conditions, before shipping them to customers.[129] The company said its engines could be "used to run a washing machine, grindstone, corn-sheller, churn, cream separator, ice cream freezer or other light machinery which is ordinarily operated by hand. . . . Many of our customers are using these engines on spraying outfits. Others are running small concrete mixers with them. The extent to which they may be used for supplanting hand labor is limited only by the skill and ingenuity of the purchaser."[130]

Aermotor also introduced general-purpose 2- and 4-horsepower engines with twenty-seven-inch and thirty-four-inch flywheels, respectively, for attaching pulleys to drive a variety of household, farm, and shop machinery. The shipping weights for the engines were 475 pounds for the 2-horsepower unit and 750 pounds for the 4-horsepower engine. Several years later the company stepped up the power by introducing a distinctive fluted sheet-steel hopper engine, which came in 2.5- and 5-horsepower and included heavier cast iron flywheels. These engines cost $115 for the 600-pound 2.5-horsepower unit and $165 for the larger model, which weighed about 1,000 pounds.[131] Aermotor continued to sell gas engines into the mid-1930s.

In addition, Aermotor gained early notice from customers requiring specialized steel towers and construction for purposes other than pumping water. In the late 1890s the company was already selling various heavy-duty towers to support large municipal water tanks. A few years into the twentieth century, Aermotor was approached by a Mexican company to develop steel towers to support heavy cables for carrying electricity from a hydropower plant to mines more than a hundred miles away. These towers—1,200 in total—were four-posters, reminiscent of its windmill towers except they were topped with three-inch wrought crossed iron pipes to attach and support the electric cables. By erecting the towers 500 feet apart, which it proved through testing at the Chicago plant, the company could ensure that the steel structures withstood the immense weight of the cables. Aermotor noted it even saved the Mexican customer freight shipping costs, stating, "These towers . . . made over 75 carloads. No other concern could have furnished towers anything like the same strength with less than 100 cars of materials."[132] Capitalizing on this success, Aermotor patented several electric transmission tower designs in 1906 and 1909.[133]

Aermotor's ability to build unique and sturdy towers even caught the attention of the federal government. The US Forest Service, which was formed in 1905, prioritized erecting lookout towers throughout the country's forests to spot and contain fires quickly before they could spread. The agency relied primarily on

Aermotor made a range of gasoline pumping engines, ranging from the smallest No. 1 plant (left), which was "capable of doing the work of an eight foot Aermotor," to its 5-horsepower general-purpose engine, which was topped by a unique fluted cooler. From Aermotor Company, *Aermotor Gasoline Pumping Engine* (Chicago, ca. 1920), n.p., and *Price List*[:] *Aermotor Gasoline Engines for Pumping and Power* (Des Moines, IA, 1920), 8.

Aermotor to build galvanized steel towers, which it selected for the "strength, rigidity, mechanical perfection, and safety" of the design and because "its long life is believed to make it the cheapest in the long run." The first towers were accessed with a ladder straight to the platform, which supported an enclosed structure measuring about six feet by six feet squared.[134] Factory prices for the towers in 1914 ranged from $160 for the fifty-footer to $215 for the seventy-footer.[135] These first towers, sold from the mid-teens to the mid-1920s, had small cabs because the towers were based on Aermotor's windmill tower designs. Aermotor, which sold thousands of these towers to the Forest Service in the decades ahead, continued to make them taller, added easier-to-climb steps, and increased the sizes of observers' accommodations. Aermotor produced the LS-50, LS-25, LS-40, and MC-39 lookout towers which could be configured to different heights of about twenty-foot intervals. There were also some variations that included a catwalk with railings around the outside of the observer's cabin and mid-point decks. The company competed with about a half-dozen other manufacturers of these towers, but its main competitor for this business was International Derrick.[136] Aermotor also made fifty-foot to one-hundred-foot-tall flagpoles, with the base resembling a windmill tower; and various bell towers for schools, churches, and fire departments.[137]

The pressure on Aermotor to deliver on these new and existing products meant long, grueling days on the manufacturing line and shipping department for its workforce. While US manufacturers began to improve workplace conditions and safety by the early 1900s, factories like Aermotor subjected employees to the hazards of moving mechanical parts, chemicals, molten metals, gases, and dust, any of which could easily maim, incapacitate, or kill a person. At the turn of the twentieth century, factories operated six days a week, and individuals could work as much as sixty hours weekly, leading to exhaustion. Unions started to protest poor working conditions and demanded better wages for their members. Aermotor had dealt with occasional work stoppages during protests by the National Association of Machinists, of which some of its workers were members.[138] The windmill manufacturer's attitude toward union demands was no different from that of many industrial enterprises at the start of the twentieth century, holding the line unless forced to change by law. On May 1, 1912, Illinois enacted its first workers' compensation legislation in an attempt to hold companies in certain industries, such as farm implement manufacturing, accountable for supporting workers injured in their workplaces. However, the act was flimsy and allowed companies to opt out of supporting it, which many did, including Aermotor. The company's board, consisting of La Verne Noyes, L. C. Walker, and Frederick E. Smith, on January 16, 1912, declared "[t]he Aermotor Co. elects not to provide and pay the compensation to any employee who has elected to accept the provisions of 'an Act to promote the general welfare of the People of the State by providing compensation for accidental injuries or death suffered in the course of employment' . . . and elects not to be bound in any way, whatsoever by the provisions of the said Act."[139]

Noyes's ambition combined with the reputable quality of his windmills generated enormous profits, making the onetime farmer's son a wealthy man, and with this money came a newfound respect among Chicago's elite. Having no children, Noyes indulged his wife Ida and her passions. She was considered by many acquaintances and friends to be a skilled writer and poet, in addition to an artist and photographer.[140] She studied painting at the Chicago Art Institute and later at several Parisian studios. Ida, as was customary of the day, traveled for months at a time to both Europe and Asia, often while her husband stayed home to tend to the business. Letters written to her husband from her travels abroad revealed her playfulness and charm. In a letter from Heidelberg, Germany, on December 19, 1886, Ida wrote, "How blessed we are in loving each other truly, trusting each other fully, and having no secrets one from another."[141] She often shared his curiosity in how things worked. Ida was also involved in the Chicago chapter of the Daughters of the American Revolution, serving two terms as vice-president-general of the national society. Through the D.A.R., she taught many foreign-born citizens about US history and government. In addition, she was active in other civic and art groups in the city, including the North Side Art Club, the Twentieth Century Club, Chicago Woman's Club, and Woman's Athletic Club.[142] In late 1895 Noyes bought a vacant lot on the northwest corner of Lake Shore Drive and Elm Street, measuring 49 feet by 152.6 feet, for $65,000 to build a spacious home for Ida and himself. The *Chicago Daily Tribune* commented that the property was "one of the few choice vacant lots on the Lake Shore drive and the price paid for it is in excess of any ever paid for vacant residence property in Chicago."[143] Noyes undoubtedly appreciated the benefits that came along with wealth. "What would I do if I were

Ida E. Noyes, wife of La Verne W. Noyes, was known for her appreciation of the arts and women's education. The Ida Noyes Hall (below) was built on the University of Chicago campus shortly after her death. Author's collection.

to start life over again? I would undertake to make a half million dollars in the first ten years after getting through college—not for the sake of money, but for the sake of that peace of mind which comes from knowing that the future is well provided for," he said in an address during the twenty-fifth annual reunion of the Iowa State College class of 1872. He even believed young women college graduates should do the same, adding, "Her chances for making the half million may not be quite so good; but, were I in her place, I would make a bold start for it, and if some gentleman tried to dissuade me from my undertaking, I should tell him to go to——."[144] However, Noyes remained vocal about the benefits of farm work as a young man and how the experience contributed to his drive to be a successful inventor and businessman. "Had I a family of a half dozen boys, they would go to the farm when they were two years old, and would never return to the city or to the city school until they were fourteen. I would in that way give them an opportunity for future usefulness and future greatness that money cannot buy," he once wrote.[145]

Personal tragedy, however, struck on December 5, 1912, when Ida died of an illness at age fifty-nine. It was noted that for the last year of her life she was invalid, "but she was always cheerful."[146] After her death, she was cremated and, for nearly two years, Noyes kept the airtight copper urn containing her ashes and other mementos at their home until he had her interned at Chicago's Graceland Cemetery under a large granite block simply inscribed with both their names. He donated her jewelry to the Chicago Art Institute for public display and gave their cottage at Midlothian, Illinois, where he enjoyed playing golf, to a friend of Ida's, saying that it "lost interest for me after [her] departure."[147] He also donated generously to the community in her name, including constructing a cloister at the Fourth Presbyterian Church on Lake Shore Drive, which they attended, connecting the parsonage and church edifice. He funded the construction of the Ida Noyes Cottage at the Illinois Industrial School for Girls in Park Ridge, Illinois, which housed twenty girls. However, Noyes's most outstanding monument to his wife was a $490,000 gift to the University of Chicago to build the Ida Noyes Hall, an 82,000-square-foot, three-story Gothic-style building to house a gymnasium and social center for the university's female students.[148] On April 17, 1915, a large ceremony was held on the university grounds to lay the cornerstone of the new student hall, inside of which was placed a reflective letter written by Noyes to his wife. Noyes also commissioned the artist Louis Betts to paint individual portraits of Ida and him to hang in the hall.[149] The hall was officially opened in June 1916. He even generously donated to his alma mater, Iowa State College. In 1914 he offered to fund the services of renowned landscape architect O. C. Simonds to design a small lake on campus at a cost of $10,000. The lake was completed in 1915.[150]

On a national front, the market that was once so favorable to windmill manufacturers, allowing them to sell an abundance of steel-based products at reasonable prices, was also turned upside down by the escalating clash of armies across the European continent in early 1915. The war that everyone thought in 1914 would be over quickly was now mired in a bloody stalemate along the Western and Eastern Fronts and required enormous amounts of metals to fuel weapons production. One of those metals facing restrictions was zinc spelter, which many windmill manufacturers by this time used to coat their steel products against rust. Before the fighting broke out, Germany, Austria and Belgium supplied about two-thirds of the world's spelter. US industry, which had been a net importer of this material before the war, was now exporting large quantities of it, among other metals, to Europe for the war effort. Windmill manufacturers gloomily informed their agents in the spring of 1915 about the rising costs of zinc and steel.[151] One windmill company wrote to a customer in May 1915 that before the war the cost to galvanize light steel angles was $19 per ton and now was up to $35 per ton.[152] Another lamented at the end of 1915 that steel prices had risen by 50 percent, zinc by 190

Aermotor's general offices at 2500 Roosevelt Road, Chicago, ca. 1910. Author's collection.

percent, and galvanized sheet metal by 50 percent. The company's windmill prices had increased by 15 percent and its towers 20 percent.[153] Aermotor was no exception in feeling the impact of scarce raw materials, acknowledging in 1915 that zinc is "many times as high as it was before the war" and by this time was "practically impossible to get." And like many of its counterparts, Aermotor encouraged its customers during the war years to buy painted goods. "Today if you ride across the continent you will see many painted Aermotors and towers. They have all been up from 23 to 27 years. Others continued to paint them for many years," the company said.[154]

By 1915 Aermotor had spent the past twenty-seven years since its inception basking in the success of its all-metal, open-gear windmills, becoming one of the biggest and most recognized windmill brands in the country and, for that matter, the world. During the war years, census records indicate that thirty-one companies in the United States were engaged in windmill manufacturing.[155] However, before the end of the decade, Aermotor's well-being, as well as that of others in the industry, would be put to the test with the onset of mechanical changes to water-pumping technology, an increased need for product diversification, the spread of electricity and advent of small electric pumps, and management changes to the longtime closely held organization. In some cases, Aermotor would continue to prove itself on the leading edge of technology, while at other times it would be playing catch-up with others in what continued to be a fiercely competitive market.

CHAPTER FOUR

Scholes's Resolve

A **windmill in need of lubrication** will generally let its owner know through an audible "squeak or grind," which I. W. Dickerson deftly described in his 1918 article, "Care of Windmills," as a "cry for oil."[1] It was not a sound that the rancher, farmer, or homeowner wanted to hear because it meant scaling the windmill tower to apply lubricant to silence it. Climbing windmill towers with a bucket or tin of grease or oil was dangerous work, and especially unpleasant for those with a fear of heights. A fall from a height of a dozen feet or more could easily result in crippling injury or death. However, the ignored windmill could eventually lock up as the metal-on-metal friction intensified and gears and other moving parts ground to a halt, leaving the structure vulnerable to being blown to pieces by strong winds. Even Dickerson commented in the early 1900s how the countryside was already littered with wrecked windmills, many of which were casualties of failure to properly and routinely lubricate.[2] Because the earliest of these geared mechanisms were constantly exposed to the weather and airborne abrasives, such as dirt, lubricating a windmill could easily become a near-weekly event for some owners.

As early as the 1880s windmill manufacturers struggled with how to minimize the maintenance of their machines, especially in terms of reducing friction between moving parts. There were perhaps more promises than true breakthroughs from these companies during this period. As long as the metal parts were exposed to the weather, simply put, routine applications of lubricant were required. Thomas O. Perry, who is credited with engineering the Aermotor Company's first all-metal windmill design, understood this dilemma. When the company started operations in Chicago in 1888, it offered customers a tilting tower to lower its windmills to the ground for lubricating as an alternative to climbing a fixed tower. Another windmill builder at the time, Star Manufacturing Company of Carpentersville, Illinois, also sold a tilting tower. While Aermotor sold thousands of these tilting towers over two decades, the design lost favor due to its higher cost and greater susceptibility to high-wind damage than fixed towers.[3] (See chapter 3 for more details.)

The emergence of US windmills coincided with that of the budding petroleum industry. Led by John D. Rockefeller in the late nineteenth century, oilmen learned how to raise petroleum from beneath the surface of the earth, refine it, and use its lubricating and illumination properties for a variety of industrial purposes. The most common form of windmill lubricant in the late 1800s and early 1900s was known as "cup grease."[4] With the goopy consistency of heavy axle grease, this lubricant could be smeared on a windmill's gears with a stick or daubed into iron or brass grease cups. Gravity and a little time would allow the substance to sink into and between bearings and joints. The problem with cup grease was its response to hot and cold temperatures. In the summertime heat, it had the tendency to become runnier and not adhere long to metal parts, whereas in the winter cold it would congeal and become harder to apply or flow to bearings. The manufactured lubricant could be purchased in a variety of quantities from either the windmill companies or third-party providers, ranging from quart-size tins to upward of fifty-gallon barrels.[5] More resourceful farmers might be inclined to make their own lubricants from animal fat mixed with coal oil for thickener. However, its ability to hold up to nature's elements was much shorter-lived than its petroleum-based counterpart.

In the 1890s Aermotor sold its own windmill oil, heavily promoting its ability to stay fluid even at temperatures nearing 0° Fahrenheit. "Why is it that some dealers need repairs in winter and not in the summer? It is because their machinery is oiled with an inferior quality of oil that hardens up before it reaches the bearing," the company declared.[6] Aermotor sold its oil for about sixty-five cents a gallon, almost twenty cents more than many competitors, based on its all-temperature fluidity claims. The oil could be purchased in one-gallon, two-gallon, five-gallon, and ten-gallon cans and in barrel lots of fifty gallons.[7] Aermotor's oil came in cylindrical cans, which had a funnel-shaped spout for easier pouring, and were capped with a wood or cork stopper. A handle made of wire and a wooden grip allowed for easier and more comfortable carrying. Most of the can's exterior was covered with a wooden jacket, stamped with the words "Aermotor Oil," to protect the inside tin against dents and punctures. The company even touted the reuse of its cans, stating, "The jacket can in itself is an excellent thing and very useful for holding kerosene and other oils."[8]

During the 1890s windmill makers attempted a variety of mechanical ways to regulate the flow and distribution of lubricant from grease cups to main shafts and bearings, especially as they transitioned from the use of heavy greases to more fluid oils. Some companies simply recommended that their customers stuff bits of old cloth or waste wool into the cups to slow the flow of the oils.[9] This did not necessarily eliminate frequent trips up the towers to administer the lubricant. Aermotor developed clear glass receptacles for its ten-foot 1897 Model windmill that allowed users to see from the ground when their oil cups were empty and needed refilling. The company claimed the glass oilers held enough oil for two months at a time.[10] However, Aermotor didn't produce the glass receptacles beyond that year.[11] The reason may be explained in the company's 1899 descriptive catalogue,

Aermotor, like many windmill manufacturers, sold its own brand of oil in tin cans wrapped in wooden jackets to prevent puncturing. From Aermotor Company, *Aermotor Price List* (Chicago, 1897), 65.

which states, "If an air-tight glass oil cup is used in the hot sun it will vaporize enough oil to force the remainder of the unvaporized oil out of the cup in a few hours."[12]

In 1899 Aermotor introduced a new oiler that consisted of a brass cup with a screw-on cap. The center of the cup projects upward to prevent oil from running directly into the bearing but is "carried up and over by the capillary action of a twisted wire." The company explained that "[i]f one of these cups is placed where it can be observed, it will be found that it carries over a drop of oil about every three hours."[13] Also by the end of the nineteenth century, Aermotor's larger hub bearing included a number of bronze and friction washers, in addition to an oil pocket. "Across the oil pocket are partitions which, twice in every revolution, carry some of the oil toward the windward end of the shaft, and some of it up to deposit on these friction rings. This affords constant and automatic lubrication, but an oil hole also permits them to be oiled directly. The Aermotor throughout is self-oiling," the company explained.[14] Even so, a trip up the tower was required periodically to add oil.

In the early 1890s some windmill manufacturers believed they could eliminate the use of oil altogether by infusing bearings with graphite. The silver-black, carbon-based substance can be found in large underground pockets and cheaply mined. While it feels greasy to the touch, it is a dry lubricant. Windmill manufacturers, such as Challenge Wind Mill and Feed Mill Company and the Perkins Wind Mill Company, pressed graphite into their bronze or brass bearing rings. They heavily marketed the graphite bearings, with Challenge claiming its Dandy steel windmill "will run twenty-five years without oil" and "therefore needs no attention," once set in operation.[15] Perhaps ahead of their time (graphite bearings are used in many mechanical applications today), the windmill bearings were generally a failure, creating more friction than bearings using oil or grease lubricants. Aermotor openly chastised the graphite bearing movement within the windmill industry, calling it a "fraud" and further stating in 1893 that there were many instances in which the bearings "squeak and scream before they have run four weeks" and "make the windmill run very hard, wear out very quickly, and detract much from its power."[16] However, even with Aermotor's technical sophistication, it had not overcome the routine climb up the windmill tower.

At the turn of the twentieth century, inventor and businessman Stephen E. Burke of Edon, Ohio, developed a mechanism that allowed for more prolonged oiling of windmills from the ground. The device resembled a tin oil canister and was mounted above the windmill's gearing. The oil was channeled through a series of delivery pipes running from the bottom of the canister to the various bearings of the windmill. When the mill needed oil, a person on the ground tugged on a wire cable that ran through the pump-rod housing and triggered a spring-loaded valve sleeve to open and release a certain quantity of oil.[17] The oiler, which Burke began to sell in 1901 through his Burke-Bollmeyer Oiler Company, was praised by trade publications for its simplicity and labor-saving qualities. The *Farm Implement News* wrote, "It not only does away with climbing to the top of the mill in all weathers, but it enables the mill to be oiled regularly—a duty that is often neglected, when the farmer knows that every time he does it he risks his life." The publication further reported that "years of life can be added to the mill" and the device "is worth investigating."[18] The company, which contended that its device could be attached to many windmills on the market, attempted to explain the benefits of the windmill oiler with a combination of mathematics and fear:

> If owners of wind mills are going to oil their mills twice a week for twenty years, as they should be, and the mill will wear ½ longer if they do, they would have to climb to the top of the mill two thousand and forty times more than with the Burke-Bollmeyer Oiler, which costs $6.75, or ⅓ of a cent a trip! And it don't seem to us that any sane man would do it, or send his boy or hired man to do this work for the price, as it is a dangerous job and people get killed doing this work.
>
> If it takes twenty minutes to oil a wind mill twice a week it means forty minutes a week; three and a half days a year; seventy days in twenty years; seventy days work at less than ten cents a day for climbing to the top of the windmill tower, as against the price of Burke-Bollmeyer Oiler, and two thousand and forty less chances of getting killed.[19]

The company continued to market its oiler heavily in the farm press.[20] One of the large windmill manufacturers that promoted the Burke-Bollmeyer Oiler was the Flint and Walling Manufacturing Company of Kendallville, Indiana, which in 1902 showed the oiler attached to its Steel Star direct-stroke steel windmill. The oilers could also be mounted on top of the open back-geared Aermotor windmills.[21] However, it's unclear how successful the product was on the market. People still had to climb the windmill tower eventually to refill the Burke-Bollmeyer Oiler.

Behind the scenes, Stephen Burke and his business partners were making plans to enter the windmill manufacturing business themselves. In March 1902 Stephen Burke received a patent for a windmill that was designed to hold a certain amount of oil that would continuously lubricate several bearings with the motion of the windmill itself. His design also protected the oil bath from the weather.[22] In August 1902 the company filed articles of incorporation with a capital stock of $100,000 to manufacture windmills under the brand Red King and changed its corporate name to the Burke-Bollmeyer Manufacturing Company.[23] The theme of the company's initial windmill marketing centered on proper oiling and how competitors paid a disservice to their customers by not addressing this fundamental mechanical issue. It even took aim at Aermotor's tilting tower, which had disappeared from its catalogue by 1899, for its instability and failure to stand up to strong storms.[24] With the exception of the oiler, Red King direct-stroke and back-geared windmills with eight- and ten-foot wheels were not much different in appearance than other windmills on the market.[25] Burke continued to refine his oiler's design and operation, but in late 1904 he left Burke-Bollmeyer Manufacturing Company, which was renamed Red King Manufacturing Company a year later

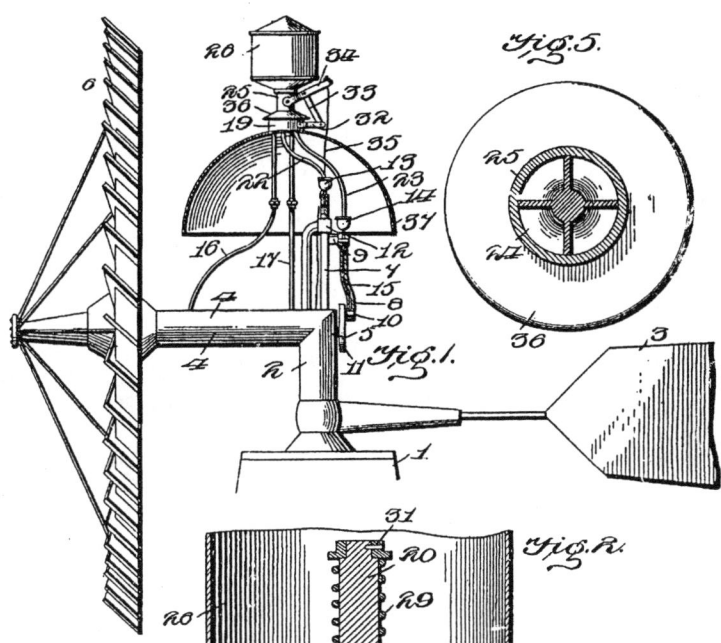

Stephen E. Burke of Eaton, Ohio, patented this device to allow oiling of windmills from the ground. Courtesy US Patent and Trademark Office, Washington, DC.

and saw an expansion of its product line to include other types of water pumps and tanks.[26]

Another firm, the Red Cross Manufacturing Company of Bluffton, Indiana, in the spring of 1903 introduced a patented oiler to its windmills, which from a distance looked similar to the Burke-Bollmeyer Oiler and used a wire from the ground to dispense the oil. The company offered the oiler for an additional price on its entire windmill line, including its back-geared steel mills with four- to sixteen-foot wheels and its direct-stroke steel mills with wheels ranging from eight feet to sixteen feet.[27] Its windmills and oilers were also sold by Sears, Roebuck and Company under the name Kenwood. Self-oiling attachments for windmills could be purchased from the catalogue retailer in 1906 from $3.89 for the six-foot windmill up to $7.98 for the sixteen-foot windmill.[28] Red Cross continued its investigations into improving the oiler's design and operation in the first decade of the twentieth century.[29]

Stephen Burke also sought to improve further the windmill oiler as late as 1913.[30] A new lubricating advancement was about to sweep all of this aside—the "oil-bath" windmills. This innovation of continuously bathing windmill components in a set amount of oil in a contained reservoir had been emerging as early as 1905. The Elgin Wind Power and Pump Company of Elgin, Illinois, in 1906 introduced to the market its Little Giant, a partially covered twin back-geared windmill. While its gears, pitman arms, and rocker arms were exposed to the weather, the bearings of the windmill's main shaft and crankshaft were encased along with an oil reservoir. As the shaft turned, oil was continuously lifted from the reservoir to flow over the bearings. The company claimed that just one gallon of oil would

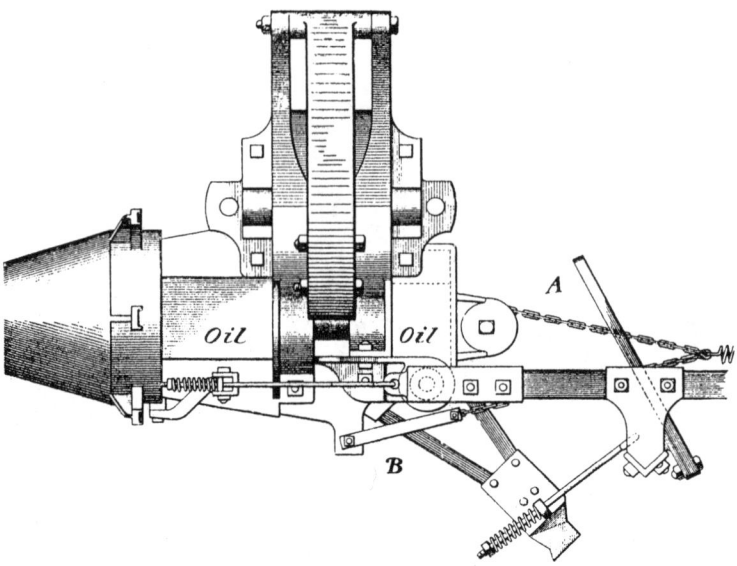

The Elgin Wind Power and Pump Company introduced its Little Giant windmill with contained oil pockets to keep gears continuously lubricated for up to a year. From Elgin Wind Power and Pump Company, *1882 Up-to-Date Windmills 1906* (Elgin, IL, 1906), 11.

last twelve to eighteen months before requiring a refill of fresh oil.[31] By 1912 Elgin had delivered to the market a completely enclosed oil-bath windmill called the Wonder.[32] As historian T. Lindsay Baker described the mechanical breakthrough of the Wonder in his 1985 book, *A Field Guide to American Windmills*, the working parts of this windmill were protected and bathed in oil, much like the parts inside an automobile's crankcase. In this case, the two crank gears—when in motion—lifted oil continuously from the reservoir to the main shaft and pinion bearings, while wooden bearings sliding on steel guides moved oil to the crosshead and upper pitman bearings. The oil then flowed back into the reservoir where the process started again.[33]

Aermotor found itself under intense pressure after 1912 to come up with its own oil-bath windmill after witnessing the meteoric success and popularity of Elgin's Wonder. Most windmill users would prefer to climb their tower once a year to change the oil rather than weekly to refill grease cups or every few months to top off an oiler. La Verne Noyes, founder and president of Aermotor, turned to his engineering department, in particular to a young man named Daniel Ransom Scholes, to come up with a practical oil-bath windmill design quickly to counter the competition.

Scholes was born in Chicago on August 11, 1882, to Thomas Jefferson Scholes, a Civil War veteran and educator, and Lydia Ransom Scholes, a homemaker who practiced homeopathic medicine. He was the middle child of seven, two of whom died before reaching two years of age. While in high school, Daniel began working for Noyes as an office boy. Noyes was so impressed by the young man's aptitude that he loaned him money to attend Cornell University in Ithaca, New York, where he earned a mechanical engineering degree in 1904.[34] He immediately

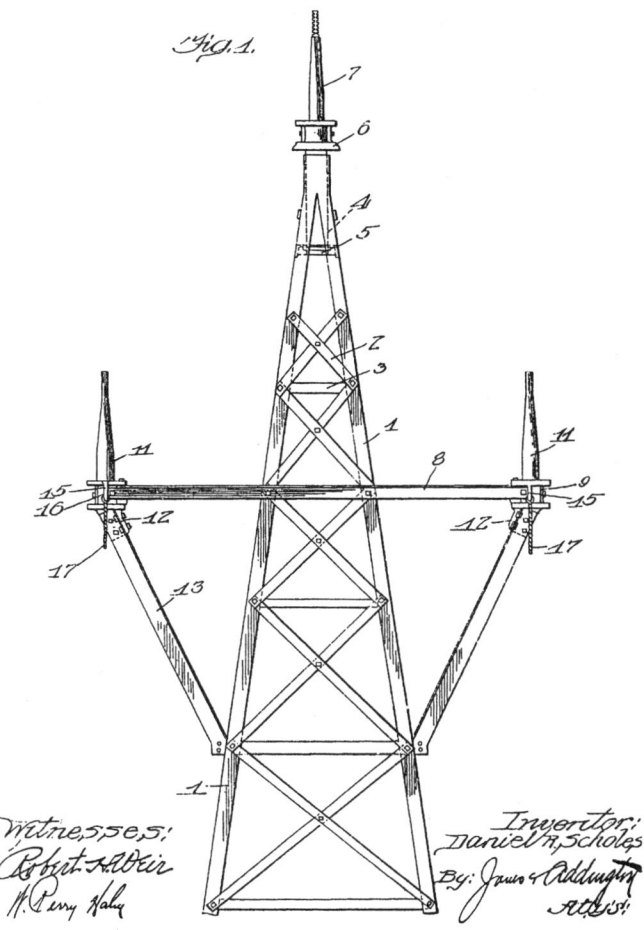

Daniel R. Scholes started work at Aermotor as an office boy in the late 1890s and with financial assistance from La Verne W. Noyes earned a mechanical engineering degree from Cornell University in 1904. Courtesy Division of Rare and Manuscript Collections, Carl A. Kroch Library, Cornell University, Ithaca, New York.

Scholes received his first patent in 1906 for this electrical power transmission tower. Courtesy US Patent and Trademark Office, Washington, DC.

returned to Chicago upon graduation to work for Aermotor.[35] When he arrived back at the company, he did not start with windmills. Noyes instead assigned him to the development of electric power transmission towers for which Aermotor was building an increasingly larger customer base. Scholes patented a number of tower designs, including special anchor posts, between 1906 and 1909. The towers had to be strong enough to support the enormous lateral strain imposed upon them by the power lines. In addition, Scholes had to incorporate suitable insulators to prevent contact between the power lines and the steel towers. Some of his tower designs drew upon structural characteristics long used in Aermotor's windmill towers, including dovetailed apexes and ladders for climbing.[36]

In 1913 Scholes, under the watchful eye of Noyes, set out to develop a new windmill bearing lubrication mechanism, which was no easy task since it required some fundamental changes to Aermotor's traditional design. By the summer of 1914 Scholes had received a patent for a mechanism "for effecting the continuous circulation of lubricating oil from a receptacle through the bearings to be lubricated thence back to the receptacle."[37] He also designed a new double-gear mechanism to drive the pump rod.[38] Noyes had a hand in the new auto-oiled

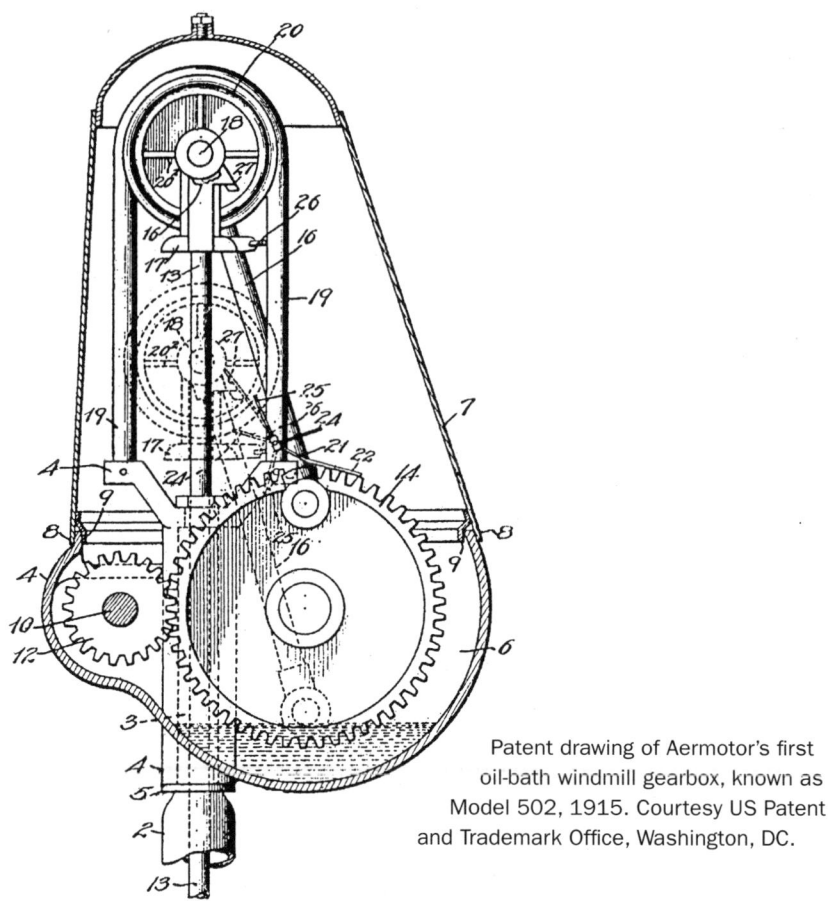

Patent drawing of Aermotor's first oil-bath windmill gearbox, known as Model 502, 1915. Courtesy US Patent and Trademark Office, Washington, DC.

structure as well, securing a number of related patents under his name, including for the design of the new gear casing and for a sheet-metal hood to cover and protect the power transmission unit.[39] One of the critical pieces of Aermotor's auto-oil mechanism was the oscillating lubricant carrier, consisting of a thin piece of wire that doubled back on itself and was tightly curled at one end. As the large gears inside the casing churned up oil, the wire picked off some of it without ever touching the gear teeth. "At every stroke of the pump it throws a splash of oil onto the yoke. From this it drips into the guide wheel and runs along the cross shaft, oiling all of the upper bearings," the company explained.[40] Another critical design change involved the introduction of a new furling mechanism for the windmill tail vane and vane assembly buffer.[41] The company claimed to have one of its employees turn one of these windmills on and off 10,000 times in a single day to test the mechanism for ease of use and durability. The company said, "A small child can easily furl this windmill or an automatic regulator can take care of it," yet to do so the windmill required a thirty-pound pull on a wire.[42] The new windmill's tail vane also had concentric corrugations pressed into sheet metal to make it stand out from its earlier models. Realizing that the new design would attract competitors' attention, the company was quick to seek patent protection overseas.[43]

In addition, Aermotor introduced new Easy-To-Build-Up towers to go along with its auto-oiled windmills. Earlier fixed steel towers were designed for ease of

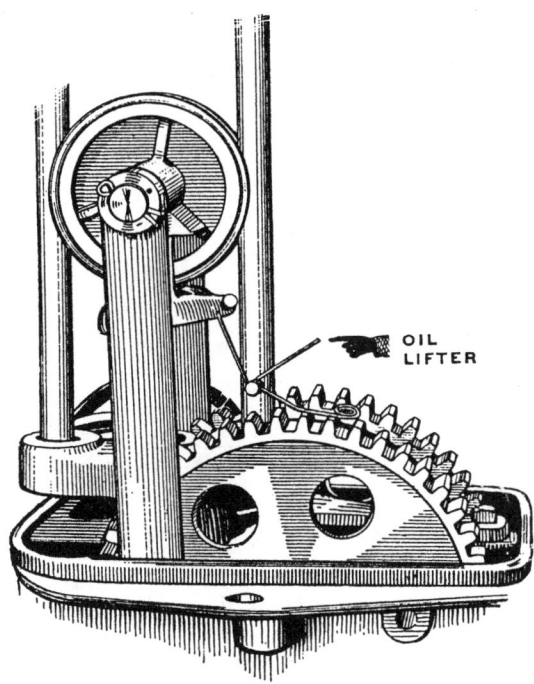

OIL
LIFTER

Model 502 had a spring-like wire oil lifter that with each stroke of the pump threw a splash of oil onto the yoke to keep the upper bearings lubricated. From Aermotor Company, *The Oil Lifter* (Chicago, ca. 1915), n.p.

assembly lying on their sides on the ground, requiring that they be raised as one single unit. The revised Easy-To-Build-Up towers could be assembled this traditional way or alternatively could be built up from the ground one piece at a time. Every aspect of the tower was given consideration by the company to ensure that two men with a wrench and rope could "quickly, easily and safely" erect it.[44]

> The girts all come below the splice in the corner posts. A man standing on a plank at any set of girts can easily reach to build up the next section of the tower. The assembling is done from a staging on the inside, which gives a feeling of perfect security no matter how high the tower.
> The steps are put on before the erecting is started, so that there is a ladder to climb the tower as fast as it is built up.
> Two can men can easily put on an 8-foot mill without any tackle. For the larger sizes a short gin pole and light tackle are needed to lift the motor and vane into position.[45]

A distinctive feature of this new tower was the corner posts made in seven-foot lengths, which the company believed was a comfortable length for an average man to handle. For the preceding twenty-five years, Aermotor's standard steel towers had consisted of ten-foot sections divided by horizontal girts. Then in 1912 the company revised its girt heights to not exceed six feet, seven inches, which it said increased the strength of its towers.[46] The Easy-To-Build-Up towers took the 1912 design a step further by allowing an open tower base so that anyone could access the pump beneath the windmill without obstruction. Aermotor also noted that the Easy-To-Build-Up tower braces were interchangeable for three heights of towers, including the twenty-seven-foot, thirty-three-foot, and forty-foot towers, while another set of braces fit its forty-seven-foot, fifty-three-foot, and sixty-foot

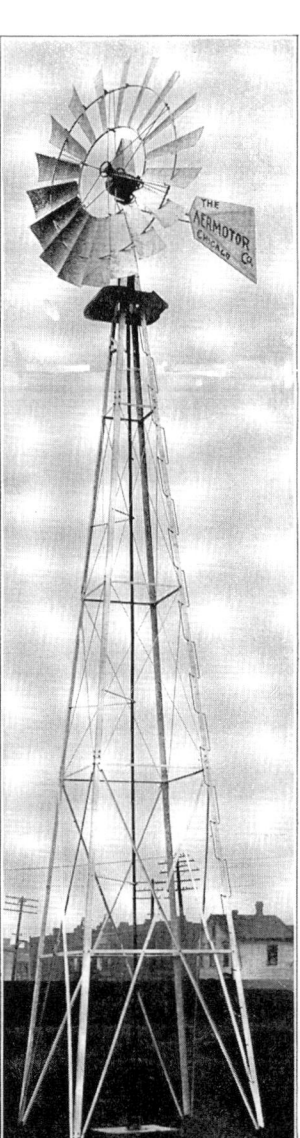

Aermotor claimed that all that was needed to erect one of its new Easy-To-Build-Up towers, which ranged in size from twenty-seven feet to eighty feet tall, was two men, a wrench, and a rope. From Aermotor Company, *Auto-Oiled Aermotor with Duplicate Gears Running in Oil and the Easy-to-Build-Up Tower* (Chicago, ca. 1915), n.p.

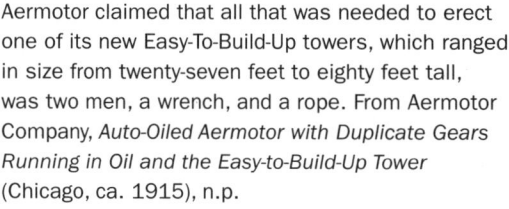

towers. "This means that the dealer who carries a 40-ft. tower in stock can put up on either a 27-ft., 33-ft. or 40-ft. tower from the material on hand without ordering any special parts. The same is true of the 60-ft. tower and the two shorter heights," the company said.[47] These towers could even be built to seventy-three-foot and eighty-foot heights.

Aermotor began a major marketing campaign at the start of 1915 to introduce its new line of oil-bath windmills to farmers and ranchers. To windmill installers and well men, the company said, "There are hundreds of windmills in your vicinity which are badly worn, which need repairs frequently, which have to be oiled every few days. Every one of these windmills, old or new, should be replaced by this auto-oiled mill, which runs a year or more with one oiling."[48] The company encouraged its dealers to offer their windmill purchasers packages that included annual inspections and oil changes.[49] An eight-foot Aermotor windmill gear case, for example, required two quarts of oil to be replaced each year.

Most customers were uninterested in tearing down their old towers just to erect an entirely new windmill. Aermotor, like other early oil-bath windmill manufacturers, realized this and found ways to work with these frugal customers. The company advertised, for example, the new Auto-Oiled Aermotor with steel stub tower section that could be bolted onto a sawed-off wood or steel tower of any make. Aermotor even offered to sell the customer—for just fifty cents—a hacksaw to make the tower cuts and a punch that with any hammer could make the one-eighth-inch holes in the old tower angles to bolt on the new stub tower. For older Aermotor towers, Auto-Oiled Aermotor buyers could request an extra piece and punch to hang a new furl lever.[50] In some cases, all the farmer or rancher had to buy was the new gearbox and tailbone to switch out the wheel and vane sheet from his open-gear Aermotor with the new Auto-Oiled unit, but this required the existence of a steel tower and a wind wheel with round arms and a long regulating spring.[51]

At the time the first Auto-Oiled Aermotor, also referred to as the "502" based on the part number given to its prominent cast gearbox, was introduced to the market, World War I had begun in Europe and metals such as zinc—a key ingredient for making galvanizing dips for steel—were in short supply for commercial purposes, as mentioned earlier. During the war Aermotor encouraged customers to consider buying painted products, such as windmills, tank and bell towers, and gearbox hoods, instead of galvanized ones. The company boasted that there were many of its early painted towers still in use around the country, emphasizing their durability. Although by 1915 zinc was "practically impossible to get," Aermotor with its dwindling stock still offered customers galvanized components but at prices from 5 to 25 percent higher than painted products.[52]

Aermotor's rush to market of the 502 proved disastrous. It appeared that in the company's haste, little if any vetting was done with its customers prior to introducing the new windmill to the market. Worse yet, Noyes and his management might have arrogantly assumed the machine would quickly dominate the market

A Canadian dealer for Aermotor with a display of new oil-bath back-geared Model 502 windmills, ca. 1915. Courtesy *Windmillers' Gazette*, Rio Vista, Texas.

like its open-gear models had done in the early 1890s. The 502's flaws—particularly with the upper-end and bearing oiling mechanisms and vane stem and buffer design—must have become immediately apparent in the field, as a year later Aermotor introduced its so-called "1916 Model," often referred to as the "602." Interestingly, Aermotor's marketing literature for the 602 is nearly identical to that of the 502 in terms of describing the erecting and operational benefits of the windmill, except that it was now pictured with a flat sheet-metal vane sheet instead of the corrugated version.[53] The 1916 Model came in eight-foot, ten-foot, twelve-foot, fourteen-foot, and sixteen-foot sizes and retailed from $37.50 for the eight-foot windmill (excluding tower) to $210 for the sixteen-foot machine.[54] While Aermotor pressed forward with its improved Auto-Oiled windmill, the relationship between Scholes and his longtime mentor Noyes for reasons unknown soured, and for a time he left the company.[55]

By now in his late sixties, Noyes during the 1910s became active in his philanthropic activities. He financially supported the beautification of the Iowa State College campus, his alma mater, and in November 1914 hired well-known Chicago landscape architect O. C. Simonds to design a reservoir, including adding the appropriate trees and plants. The university board accepted Noyes's offer and in May 1916 further agreed to name the body of water Lake La Verne.[56] The college then bestowed on him an honorary doctorate of engineering in 1915. Noyes also became deeply concerned about the effects of World War I, especially on the US soldiers and sailors sent over in late 1917 and 1918 to fight on Europe's battlefields. While the fighting raged, Noyes thought about how he might assist the returning veterans. To the surprise of many friends and business associates, he announced on July 25, 1918, that he would donate $2.5 million to start the La Verne Noyes Foundation, which would be administered by the University of Chicago and help pay tuition for honorably discharged servicemen or their descendants. The gift was supported by the sale of various properties owned by Noyes throughout the city, including his Lake Shore Drive home, the La Verne and Pickwick buildings on South Michigan Avenue, the Shops Building on North Wabash Avenue, the Chemical Building on North Dearborn Street, a quarter interest in the Metropolitan Building on South State Street, a vacant lot on Dearborn Parkway, and the 9.75 acres of the Aermotor plant at Campbell Avenue and Twelfth Street. Twenty percent of the total was to be used for paying salaries to staff engaged in teaching US history or political science. The scholarships were to be made available to "deserving students without regard to differences in sex, race, religion, or political party, who shall be citizens of the United States."[57]

Noyes meanwhile maintained a firm grip on the business activities at his company. He oversaw the annual stockholder meetings, which also included brother-in-law Frederick E. Smith, who served as vice president, and Lewis C. Walker, treasurer and secretary. Together these men decided the overall direction of Aermotor, selection of corporate officers, and dispersal of funds and dividends. Of the 3,000 certificates of Aermotor capital stock valued at nearly $340,000, Noyes in 1916 controlled 2,935 shares, with Smith at 25 and Walker 40.[58] Noyes would oversee his last stockholders and directors meeting on Jan. 21, 1919, and the last special directors meeting that declared a 25 percent dividend on May 29, 1919.[59]

On June 4, 1919, La Verne Noyes was taken to Chicago's Presbyterian Hospital with the measles. Shortly thereafter bronchopneumonia developed in both of his lungs, severely weakening his heart, and he remained gravely ill. After seven weeks, his physicians, Drs. B. W. Skippy and Ralph Brown, had all but given up hope on Noyes. Shortly after midnight on July 24 he appeared to rally, but only briefly. At 1:30 a.m., Noyes passed away. His niece, Mrs. John C. Horning, her husband, and Lewis Walker, secretary and treasurer of Aermotor, were by his side. He was seventy years old. For the next several days, newspapers in Chicago and across the country reported the death of Noyes, the "great philanthropist" and "friend of veterans." The *Chicago Journal* noted in an editorial:

> Only the man who in his boyhood pumped water by hand for a herd of thirsty cattle knows the debt of gratitude which the farmers of America owe to La Verne W. Noyes. There were windmills before his time, to be sure, but he made them practicable and cheap, set them up on a steel framing that defied the winds, added contrivances to make them oil themselves and made them as near accident proof and fool proof as human inventions often get.
>
> His inventions brought him a large fortune—and who ever grudged him a dollar of it? It was payment for proved service, payment on the percentage plan at that. A water supply is the first requisite of comfort, health and prosperity. Noyes's devices were and are the source of water supply in hundreds of thousands of homes. And the money made in service has largely been spent in service. In the last year 568 young men from the army and navy have been enrolled at the University of Chicago, thanks to the endowment of La Verne W. Noyes.

La Verne W. Noyes shortly before his death in 1919. Author's collection.

A school teacher by training, an inventor by instinct, a champion and supporter of education, a philanthropist, a patron of art; he made substantial additions to the happiness and the wisdom of the world, and the community of which he was so long a part mourns deeply at his passing.[60]

The *Chicago Evening Post* said that "he took the windmill and standardized and universalized it as Ford did the automobile. And it will . . . be his best monument."[61]

While known widely in business and politics, Noyes requested a private funeral service at his Lake Shore Drive home, which was held on July 26. Attending were his sister, Adelia F. Giffen, and her children, Mrs. William F. Johnes of New York; Mrs. Percy C. Gill of Salt Lake City; Mrs. John C. Hornung of Glencoe, Illinois; Ernest N. Giffen of Ely, Nebraska; and W. Herbert Giffen of Marshalltown, Iowa, together with close friends and business associates.[62] Noyes was buried next to his wife Ida in Chicago's Graceland Cemetery. Rev. Samuel M. Gibson, assistant pastor of the Fourth Presbyterian Church, preached, and Rev. Thomas W. Goodspeed, Noyes's friend and biographer, gave the eulogy. Gibson closed with the same prayer read at Ida Noyes's funeral on December 5, 1912. The pallbearers included Frederick L. Dole, Joseph J. Fraser, G. A. Williams, G. A. Oliver, W. P. Gooding, and N. R. Stephen.[63]

However, the most anticipated news was the instruction in Noyes's will for dispensing of his estate and the future of Aermotor. Two days after his funeral, the details started leaking out in the Chicago press. It was speculated that his fortune was valued at close to $4 million; yet during the last seven years Noyes had disposed of at least half his holdings with the money turned over to charitable causes.[64] Details of the will were made public on July 28, 1919.[65] It was noted that Lewis Walker, Joseph Fraser, and Frederick Dole were named executors. The will provided his only sister, Adelia Giffen, in Central City, Iowa, with $500 a year for the rest of her life. "My said sister is already well provided for and I therefore leave her this small annuity merely as an expression of my affection for her," Noyes wrote. Other relatives were given between $500 and $5,000 annually from his estate.[66] United Charities was given $5,000 a year through December 31, 1934. An estimated $1 million remaining from the estate was also designated for university and college scholarships at the discretion of the trustees.[67] Noyes offered the University of Chicago first pick from his paintings, furniture, rugs, and other household items for use in the Ida Noyes Hall, with the remainder to be donated to other schools, societies, and institutions.[68] To Aermotor employees with five or more years at the company, he authorized his executors to divide among them $25,000 as a gift for their service.[69] As for the overall company, Noyes stated in his will: "My Trustees may retain the ownership of the stock in the Aermotor Company . . . and continue to operate said Company for a period of twenty-one (21) years after my death. At the end of such period they shall sell or otherwise dispose of (and they may do so at any time previously when they deem it for the best interests of the Trust Estate) all of said stock at such price or prices and on such terms as to them may seem best."[70]

The Des Moines branch as it appeared in the early 1930s. Courtesy Stanley A. Anderson, Skiatook, Oklahoma.

In the months following Noyes's death, Aermotor's new leadership was put in place. Prior to that, the management structure had remained virtually unchanged, with Noyes serving as president for thirty-one years. At the September 15, 1919, special directors' meeting, Frederick Smith resigned as director and vice president of the company. Noyes, however, had seen to it in his will that Smith and his wife would each receive $5,000 a year for the rest of their lives, in addition to the twenty-five corporate shares that Smith retained. Frederick Dole was named Smith's successor and became one of the three directors and vice president. Lewis Walker, also a director, filled the role of president, and Joseph Fraser was elected director, secretary, and treasurer of Aermotor. The three men also set their respective annual salaries at $14,000 for Walker, $12,000 for Dole, and $10,000 for Fraser.[71] On September 17, 1919, Daniel Scholes was welcomed back to the company and named second vice president, starting with a salary of $800 a month; his annual salary would be $11,000, starting January 1, 1920, which was raised again to $12,000 in 1921.[72] By the start of 1920, new corporate letterhead was printed with the names of the four company officers. The company was identified as "Aermotor Company (Founded 1888 by La Verne Noyes)[,] Manufacturers of Windmills, Engines, Pumps, Water-Supply Goods[,] Steel Structures for Electric Transmission Lines." In the early 1900s Aermotor had reduced its number of branches to Des Moines, Iowa; Kansas City, Missouri; Minneapolis, Minnesota; and Oakland, California.[73]

In August 1921, the University of Chicago sold Noyes's former Lake Shore Drive mansion, vacant since his death, to Charles S. Peterson, president of the Regan Printing House and the Peterson Linotyping Company, for an undisclosed amount of money.[74] One Chicago newspaper conservatively speculated that the

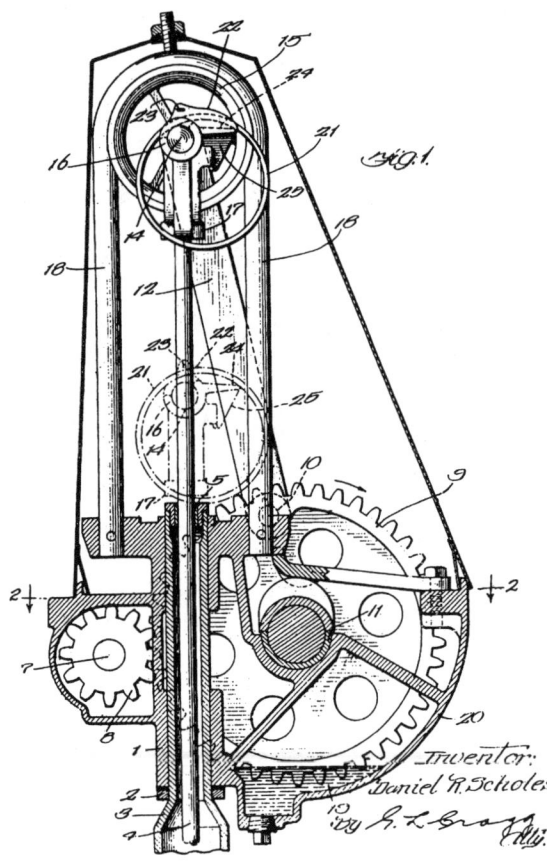

Patent drawing of Aermotor's Model 602 with oil ring for lubricating the upper bearings, 1925. Courtesy US Patent and Trademark Office, Washington, DC.

home sold for just over $200,000, although the university had offered the property for $250,000.[75] The money was more valuable to the university than the property itself and would further support Noyes's scholarship program, which by 1923 had increased to as many as 650 former military veterans compared with 286 in 1920–21.[76]

At the start of the Roaring Twenties, Aermotor, like many US manufacturers, watched its fortunes increase as people spent their rising wages on automobiles and other modern amenities, including new oil-bath water-pumping windmills for farms and rural households. The press continued to track the success of Aermotor in the early 1920s, which translated into additional scholarships. Essentially after paying its bills and employees, the company's profits were put toward additional scholarships in the name of La Verne Noyes.[77] In early 1923 the Noyes estate trustees announced 400 new scholarships for former World War I veterans and their direct descendants to twenty-two universities and colleges. For example, that year 100 scholarships were provided to Northwestern University and 40 to the Lewis Institute of Chicago.[78] Also in 1923 female war veterans received a batch of Noyes scholarships, including 20 to the Teachers' College of Columbia University; 5 to the George Peabody College for Teachers in Nashville, Tennessee; and 5 to the University of California at Berkeley.[79]

Aermotor recognized its workforce on December 20, 1921, by releasing an additional $32,262 in compensation to be divided among the employees. This amount

An artist's rendering of Aermotor's popular Model 602 windmill. From Aermotor Company, *Aermotor Co. Price List* (Kansas City, MO, 1930), n.p.

was further increased to $46,932 at the end of 1922, $66,642 in 1923, $111,652 in 1924, and $119,297 in 1925, dropping to $118,495 in 1926 and $95,949 in 1927, and then rebounding to a peak of $128,465 by 1928.[80] In addition to bonuses, the senior executive management raised their annual salaries to $17,000 for Walker, $15,000 for both Dole and Scholes, and $14,000 for Fraser.[81]

A significant change to the corporation occurred in mid-April 1925, when the University of Chicago agreed to sell the plant property back to Aermotor for $625,000. Noyes had willed the windmill factory's 9.75-acre property to the university before his death, and the company continued operations under a lease agreement. To cover the cost, Aermotor took out a $200,000 bank loan, which it had to pay back at an interest rate of 4.5 percent.[82] However, the move helped solidify the company's control over its operations and reduced the risk of closure or being sold to another firm. An Aermotor official told the *Chicago Tribune* that "future expansion of the company warranted its owning the land rather than being merely a lessee."[83] Work on a new building was mired on Christmas Eve in 1927 when a bomb made from dynamite was exploded in a window, causing an estimated $500 in damage and cracking windows in nearby blocks. Aermotor president Walker blamed the attack on disgruntled unionists because the building was constructed under the Landis award committee's open-shop plan, a citywide effort to break corrupt building trades unions.[84]

Meanwhile, Daniel Scholes wasted no time after his return to the company to further refine Aermotor's Auto-Oiled windmill. In December 1919 he applied for a patent on a redesign of the lubrication mechanism to convey oil from the basin to the upper ends of the pitman arms, which connected to the pump rod. The change was simple. Instead of the wire oil lifter, Scholes replaced it with a metal ring: "As the pump rod approaches the lower end of the stroke, the Ring comes in contact with the large gear, over which it is suspended, and is revolved about one-fourth of its circumference. In its contact with the gear it is loaded with oil, which by the revolution of the Ring is carried to the upper shaft. By this means, all of the upper bearings and the guide wheel are kept constantly dripping with oil."[85] The ring oiler avoided the problem of the wire oil lifter being bent when the wheel during variable winds sometimes feathered backward; it also proved more durable than its former wire counterpart during shipping.[86] Scholes also sought to improve the Auto-Oiled windmill's circulation of oil to the wheel bearing.[87] Over the next five years, he refined the mainframe bearing for the tower mount and vane control mechanism so that the windmill responded more efficiently in the wind.[88] The company introduced adjustable-stroke pitman arms to give the user the option of either a longer or shorter stroke. The short stroke position increased the windmill's pumping depth by one-third but decreased the pumping capacity by twenty-five percent.[89] Scholes even altered the lip to the base of the sheet-metal helmet for a snugger and more weather-resistant fit with the gearbox.[90] Lastly, he improved the way the windmill's wheel spokes were attached to the hub for increased strength and durability.[91] Each of Scholes designs was also tested in a large wind tunnel machine which the company housed in one of its buildings. The machine produced controlled breezes on up to hurricane-force winds.[92] The result was a simpler, sturdier, and more efficient 602, marketed by the company as the Improved Auto-Oiled Aermotor in 1928.[93]

Since its inception in 1888, Aermotor stood by its claim that the ideal size for a wind wheel was eight feet in diameter, and for the next thirty-seven years it remained the company's smallest windmill. The largest-size wheel it carried by this time was sixteen feet in diameter. Then in late 1925 Aermotor unveiled its six-foot Auto-Oiled windmill to "meet the demand for a small windmill for shallow wells."[94] The long stroke of the six-foot Aermotor was five-and-a-half inches. The windmill was designed to fit on the same four- or three-post towers as its eight-foot models. "Where there are no wind obstructions, and a cheap outfit is wanted, we recommend the 21-ft. Ranch Tower. . . It has a spread of 6 feet at the top of the anchors. The angle braces are arranged so as to have a clear passage to the pump. It is made on the same plan as our Easy-To-Build-Up towers which have gained such wonderful popularity," the company wrote.[95] Aermotor, which retailed its six-foot oil-bath windmills for $33.50 in 1925, made them available to dealers from its branches in Des Moines, Kansas City, and Minneapolis.[96]

The rapid and continuous changes made to the 602 during the 1920s tested the company's production line as well as its dealers and installers in the field. (It's not uncommon today for windmill collectors and rebuilders to find 602 windmills with slight mechanical and physical variations.) One of the more curious oil-bath Aermotors to appear briefly in the mid-1920s was the so-called Model 612, based on the number on the gearbox, with the standard eight-foot wind wheel.[97] Its design differences center on the main shaft's oil return, with a separate oil travel cavity having been added to the forward end of the case. As a result, the barrel of the hub is enlarged to accommodate the increased size of the snoot area.

Aermotor constructed this wind tunnel in the mid-1920s and used it to test windmills in various wind conditions through the 1930s. From Aermotor Company, *Be Thrifty—Pump with Wind* (Chicago, ca. 1940), n.p.

Other than that, it replicates the 602.[98] Although some windmill historians and rebuilders have anecdotally referenced the 612 as a windmill sold to the US military, no evidence has been found by this author to substantiate the claim. Only small quantities were made and have been primarily found in Illinois, Indiana, and Michigan.[99] Aermotor's prices in the early 1920s for its windmill products also increased due to a combination of market conditions for raw materials, such as iron, steel, and spelter, and changes in manufacturing costs related to its new oil-bath windmills.[100] The standard eight-foot oil-bath Aermotor in 1927 retailed for $44, while the largest sixteen-foot windmills sold for $300 apiece.[101]

During this period most US windmill manufacturers had begun to offer oil-bath machines and, like their open-gear predecessors, brought about a new level of competition and marketing schemes to attract customers. In 1923 an Aermotor advertisement rationalized:

> If the wheel of an Aermotor should roll along the surface of the ground at the same speed that it makes when pumping water it would encircle the world in 90 days, or would go four times around in a year. It would travel on an average 275 times per day or about 30 miles per hour for 9 hours each day. An automobile which keeps that pace day after day needs a thorough oiling at least once a week. Isn't it marvelous, then, that a windmill has been made which will go 50 times as long as the best automobile with one oiling?[102]

According to the 1919 US census of manufactures, there were thirty-one windmill companies employing 1,932 workers. Illinois had the most windmill manufacturers at eight, followed by Wisconsin with five, four each in Indiana and Nebraska, three in Kansas, two in Michigan, and one each for Iowa, New Jersey, North Dakota, Ohio, and Texas.[103] These same windmill manufacturers also in this intense period

Restored Aermotor Model 612 hub at the Windmills at Riverside Farm park in Poland, Indiana. Courtesy Neal Yerian, Westfield, Indiana.

of technology change risked tripping over each other's work, resulting in occasional legal actions taken for alleged patent violations. Aermotor, for example, brought a suit in the US District Court at Lincoln, Nebraska, in 1921, charging Dempster Mill Manufacturing Company of Beatrice, Nebraska, with infringing on a 1906 windmill bearing lubrication patent, which it had acquired from William P. Brett of Decatur, Illinois. Aermotor had incorporated part of Brett's invention into its Auto-Oiled windmill and discovered that Dempster was attempting to do the same. The companies settled the lawsuit by requiring Dempster to pay Aermotor $6,000 as the amount to recover from the infringement. Dempster also took a license for the remainder of the term of the patent, which was seventeen years under US patent law at the time, and was limited to use the invention solely on its direct-stroke windmills.[104] In another case, the Challenge Company of Batavia, Illinois, avoided a patent infringement showdown with Aermotor in 1928 by modifying the mechanism that moved its Challenge 27 mill pump rod up and down from a single-roller crosshead guide, which Aermotor had patented for its oil-bath mill, to a crosshead guide with two smaller wheels moving along the outside of two vertical crosshead guide rods.[105]

While oil-bath windmills substantially reduced the routine need to climb towers, they still required attention. It was now easier than before to perhaps neglect oiling a windmill for two to three years. Aermotor recommended its customers in the fall to thoroughly wash out the gear case with kerosene, in addition to removing the plug in the oil pocket of the hub and squirting enough kerosene into the collector to completely wash out the oil return passage and pocket. The old oil in the gear case could then be emptied into a bucket by unscrewing the plug at its base. The level of the new oil in the gear case was to be no higher than a quarter inch below the shaft that carried the small gears. Like most windmill manufacturers, Aermotor recommended its own line of "zero-flowing" oil for its mills. While up on the tower, Aermotor users were also urged to make sure all bolts and nuts were tight, the oil ring was in place and properly working, oil was applied to the turntable through the hole in the main casting near the top of the pipe, axle grease was heavily applied to the furl rungs, and the windmill's helmet was properly secured to the gear case to keep out airborne dirt and debris, which could harm the internal gears.[106]

While windmills proved themselves to be vital for pumping life-sustaining water, interest was building in the years leading up to World War I among farmers and small town residents to use the power of the wind to generate electricity. Rural people were aware, through newspapers, magazines, and hearsay, that electric power already lighted the country's cities and provided their inhabitants with newfound labor-saving devices and luxuries, such as lighting at night, heating and cooling appliances, washing machines and clothes irons, and radios for news and entertainment. For most people living in rural areas of the United States in the early 1920s, however, receiving electricity from large central power plants through copper lines at their homes, farms, and businesses was nonexistent.[107]

Since the early 1880s, inventors and engineers had studied the possibilities of using the wind to generate electricity. This concept was first proposed in a speech delivered by Sir William Thomson, also known as Lord Kelvin, in 1881 to the Mathematical and Physical Science Section of the British Association.[108] But harnessing the wind to create electricity by using existing power and water-pumping windmill technology proved easier said than done. Besides the unpredictable winds, the technology to connect a windmill with a dynamo and some form of apparatus for regulating the electric power, storing it in batteries, and releasing it for use was not quite there, leaving many to consider it impractical. That is not to say that there were not examples of wind-electric plants at this time. In 1887 Charles F. Brush, considered a founder of the US electrical industry, constructed a windmill on his property in Cleveland, Ohio, to supply electricity to his home. The Brush windmill had a rectangular, sixty-foot-tall, 80,000-pound tower to which was attached a fifty-eight-foot-diameter wind wheel with 144 wooden blades and a large vane tail to turn the wheel into or out of the wind. The wind wheel's twenty-foot-long, 6.5-inch shaft drove a series of shafts, pulleys, and belts connected to the 12-kilowatt dynamo. Brush operated the windmill until 1909.[109] Another inventor, Poul la Cour of Denmark, developed a successful windmill in the 1890s to generate electric power. The windmill, which had a forty-foot-diameter wind wheel with four sails reminiscent of the European windmills for grinding grains into flour, turned a dynamo capable of generating 160 volts and 120 amperes. This power was stored in a battery bank with a capacity of up to 66,000 watt-hours.[110]

While Brush and la Cour would be remembered for their groundbreaking work on turning wind into electricity, there were numerous other tinkerers and experimenters throughout the world at the time attempting to do the same. Occasionally these individuals would gain the attention of a publication and their stories would be told.[111] In North America some early twentieth-century farmers attempted to generate electricity by connecting traditional water-pumping or power windmills to dynamos.[112] One of those individuals, J. F. Forest of Poynette, Wisconsin, attracted the interest of electrical engineer Putnam A. Bates, who wrote an article about his electricity-generating windmill in a 1912 *Scientific American* article. Forest told Bates after observing the mechanical work of his twelve-foot windmill, he surmised that it must be possible to use it to also drive the armature of a dynamo to produce electricity. At an estimated $250 investment in parts and time, he connected to the windmill's vertical shaft a series of pulleys, bevel gears, and a set of grinder rings to turn the dynamo on the second floor of his barn, which also housed a fourteen-cell storage battery unit. Forest's windmill generated enough electricity to power fourteen tungsten lamps throughout the house and farm buildings. He was also able to use the windmill to turn his drill press, corn sheller and feed grinder, saw and grindstone, washing machine, and grain elevator.[113]

Oliver P. Fritchle, who had been manufacturing battery-powered automobiles since 1904, began offering for sale in 1917 an electricity-generating windmill under the name Fritchle Wind-Electric Plants. The wooden wind wheel and tail, along with the steel tower, from a distance looked like a standard water-pumping windmill, but upon closer inspection it lacked a pump rod. The generator sat on top of the tower behind the wind wheel, and the electric power was transferred by wires down the tower to storage batteries on the ground. Although intriguing, the windmill was not a commercial success, and the Woodmanse Manufacturing Company, which produced them for Fritchle, sold only modest numbers of the machines, mostly to customers on the Great Plains and some as far away as Brazil and Argentina. Fritchle discontinued the windmill in 1923.[114]

Other US windmill manufacturers also made attempts to enter the electric-power generation market. La Verne Noyes of Aermotor was a steadfast believer in the prospect of using windmills to produce electricity. As early as 1895, Aermotor used a windmill on top of its building at Park Place in New York City to charge batteries for office lighting. He was also aware that some farmers were using power Aermotors to do the same. In 1911 Noyes filed a patent application for a windmill that consisted of six wind wheels, three each connected to vertical supports that attached to the ends of a horizontal bar that was balanced on the tower turntable and directed into the wind with one large tail vane. Each wheel would turn a generator for charging batteries. The windmill was never built for commercial use.[115] During those early years Noyes was unconvinced that the generator and battery technology was at a point where his company could successfully produce a reliable electricity-generating windmill for the market.[116] It wasn't until the start of the 1920s that Aermotor's senior management instructed chief engineer Daniel Scholes to design a commercial variant.

Scholes, who was already versed in the emerging electric power industry, filed a patent for an electricity-generating windmill on December 1, 1921.[117] His design included an electric generator located behind the wind wheel. Gearing transferred the wind wheel shaft's rotational motion to that of the generator's armature. With the exception of the cylindrical generator mounted on top of the mainframe, the design from a distance did not appear much different from Aermotor's water-pumping windmill setup.[118] The Electric Aermotor, sold only with a twelve-foot wind wheel, could be ordered in two sizes of direct current generators—110 volts (with a fifty-six-cell battery unit) and 32 volts (with a sixteen-cell battery unit). The wind machine would begin to generate power in eight- to ten-mile-per-hour winds and increase its output in stronger winds until it reached the capacity of the generator, which in the case of the 110-volt unit was 2 horsepower. While the smaller 32-volt Electric Aermotor would provide enough power for lighting a small country home, the company recommended the 110-volt unit for "where the points at which light or power are to be used are widely separated, or when it is desirable to erect the windmill at some distance from the place where the power is to be used."[119] The electric current from the windmill generator was channeled down the tower through two wires to a switchboard consisting of volt and amp meters and switches for the incoming and outgoing lines. The company claimed that once its 110-volt Electric Aermotor fully charged the batteries, up to ten 25-watt lamps could operate for forty hours and ten 4-watt lamps for twenty-five hours without the generation of additional current. Aermotor also said the fully charged battery bank had the energy to individually power a 1-horsepower motor continuously for 12 hours, pump 35,000 gallons of water to an elevation of fifty feet, and operate a flatiron for 20 hours, a twelve-inch electric fan for 150 hours, or a washing machine for 50 hours.[120]

Much like it did in 1888, Aermotor presented its new electricity-generating windmill to an already crowded market. But this time was different. Not only was Aermotor's competition coming from several water-pumping windmill manufacturers, such as Perkins Corporation of Mishawaka, Indiana, and Stover Manufacturing and Engine Company in Freeport, Illinois, in the early 1920s there were also numerous gas-powered farm electric light plant manufacturers, including the likes of the Dayton Engineering Laboratories Company (Delco) in Dayton, Ohio; Fairbanks, Morse & Co. of Chicago; and Westinghouse Electric & Manufacturing Company in East Pittsburgh, Pennsylvania, which sold hundreds of thousands of their machines to homeowners, farms, and businesses throughout the country and the world.[121] The multibladed windmills, like the Electric Aermotor, also faced significant technical hurdles, including high-torque wind wheels, which were good at operating at low-wind speeds and suitable for pumping water, but not for electric-power generation, and the fact that too much of the useful power from the wind got absorbed by the blades themselves, a phenomenon known as "wind congestion."[122] A new breed of ambitious young windmill developers understood these drawbacks and opted for sleeker propeller-style, lightweight wooden blades, inspired by early airplanes. The propeller permitted the wind to easily pass through it, giving an airplane lift. These types of propellers, used in the context of wind-based electric-power generation, took off quickly in winds

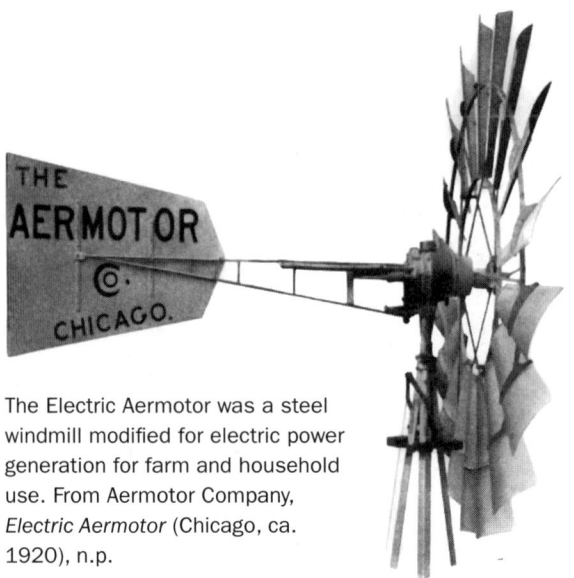

The Electric Aermotor was a steel windmill modified for electric power generation for farm and household use. From Aermotor Company, *Electric Aermotor* (Chicago, ca. 1920), n.p.

ranging from eight to twenty-five miles per hour, and their shape allowed them to use the lift from the wind to turn the generator quickly and more consistently. By the end of the 1920s, names like Aerodyne Company and Jacobs Wind Electric Company, both of Minneapolis, Minnesota; Herbert E. Bucklen Corporation in Elkhart, Indiana; Parris-Dunn of Clarinda, Iowa; Wincharger Corporation of Sioux City, Iowa; and Wind-Power Manufacturing Company of Newton, Iowa, dominated the electric wind power business.[123] Consequently, Aermotor sold only a limited number of its electricity-generating windmills.

Outside its popular water-pumping windmills, Aermotor had more success designing and manufacturing specialized steel towers and structures. By the 1920s, the company supplied a number of different tower types to the electric power transmission industry. Competition for this business was also fierce. On February 15, 1924, Scholes was called into a meeting with Aermotor president Walker and corporate officers Dole and Fraser to address claims from Foster Milliken of New York City alleging an infringement on his patent for a transmission tower and initiating legal proceedings against Aermotor and one of its customers, Consumers Power Company, in Michigan. After reviewing the matter with Scholes, the company decided it had not infringed on Milliken's patent and "would defend it to the limit of our ability."[124] (No further mention of this lawsuit was found in the record.) Other types of specialized Aermotor towers available at the time included those for bells, sirens, beacons, and antennas. Still, these towers often employed elements of its standard windmill towers.[125]

In addition to the sophisticated observation towers developed by Aermotor for the US Forest Service, the company received $12,500 in July 1924 from the US Navy to supply twenty steel galvanized towers, likely for beacons, in addition to twenty spare sets of bolts for the towers and twenty sleeve nuts for the round braces.[126] One of the more interesting government contracts for Aermotor came from the US Coast and Geodetic Survey (C&GS) in late 1926. Jasper S.

Restored Electric Aermotor in the private collection of
Roger Bailey in McCool Junction, Nebraska. Courtesy
Mike Brigolin, Columbus, Michigan.

Bilby, chief signalman for the agency, developed a new tower design from which
surveyors could measure the topography of the land surface. Earlier C&GS towers had been generally hand-built wooden or steel-pipe structures that took five
to six days to erect. Bilby had experience with different tower types during his
more than forty years with the agency and was impressed with Aermotor's all-
steel windmill tower configurations. After receiving approval from his superiors,
Bilby approached Aermotor design engineers in Chicago with his structural concept, which essentially consisted of a tower within a tower. The tri-post inner
tower held the sensitive survey instruments, and the four-legged outer tower supported both the observer and recorder. The towers were assembled so as not to
touch each other, which eliminated any vibration to the inner tower when surveyors scaled or came down the outer structure. The outer tower had many of
the same characteristics as Aermotor windmill towers, including the looped ladder rungs. Other requirements for the tower, according to Bilby, were that they
must be easily erected and taken down by a four-man crew with basic hand tools
in a day, and the components had to be light enough to transport by a one-and-a-
half-ton truck from location to location. Inner tower heights ranged from 24 feet
to as high as 129 feet, with the outer tower heights at 10 feet taller, so that the tall-
est Bilby towers were 159 feet tall. The first Aermotor-built Bilby towers entered
service with C&GS in the eastern United States in 1927, and their popularity as a
survey platform took off.[127] Other US government agencies used them, including
the Army Corps of Engineers, Air Force, and Geological Survey, and addition-

ally they were deployed internationally by the Inter-American Geodetic Survey in Central and South America, the Canadian Geodetic Survey, and the British Ordnance Survey.[128] Bilby towers afforded C&GS a quick payback on the investment by ensuring a 75 percent cost savings, or as much as $3,072,000 between June 1927 and June 1932, over erecting traditional wooden towers. Jasper Bilby was even congratulated in September 1927 for his tower engineering and cost savings to the agency by the then US commerce secretary Herbert Hoover.[129]

While Scholes increasingly put his stamp on the future products and direction of Aermotor, the last of the duo who started the company, Thomas O. Perry, died at his Oak Park, Illinois, home on January 25, 1927, after a two-week illness. He was seventy-nine years old.[130] After Perry left Aermotor in 1897, he continued to pursue windmill-related inventions, including further refining the design of the wind wheel itself to have six broader swept blades to "increase the simplicity and decrease the cost of manufacture by requiring fewer parts of simpler construction and less weight for necessary strength than any form of wind-wheel heretofore known."[131] One of his patents included a hub mechanism in which the six windmill blades could mechanically pivot, or adjust, to the varying conditions of wind speed, a concept that was decades ahead of its time and would become an important component to modern electricity-generating wind turbines.[132] He also attempted to refine his tilting tower design, which was already falling out of favor with windmill users.[133]

However, Perry's most practical inventions during this period were his air compressor pumps, which he developed as a consulting engineer for United Pump and Power Company of Chicago starting in 1906. During this time, homeowners without access to mainline public water hookups were looking for alternatives to water storage tanks on towers or in attics, which used gravity to feed water through the pipes. These aboveground tanks risked freezing in the winter and developing damaging leaks. Some pump makers offered systems that used compressed air to pump water stored in buried tanks or basements. The problem with these systems was that the water risked becoming stagnant. Perry's device, marketed by United Pump as the Perry Pneumatic Pump, voided the step of water storage altogether to source water directly from a well. The system used a small electric motor or other source of power, such as a windmill or gasoline engine, to produce and store compressed air in a tank that in turn drove the submerged pneumatic pump in the well whenever a faucet was opened. The pumps came in a number of sizes, ranging from a small half-horsepower motor to supply up to 500 gallons of water an hour, which was sufficient for a rural home or farm at the time, to a 10-horsepower pump to provide up to 300,000 gallons daily for a factory or small community.[134]

In the later years of his life, Perry's scientific work looked to the skies and even the far reaches of the solar system. One of the proudest moments of his life occurred on August 26, 1911, and then again five days later when he spent a collective hour of flying above the Oak Park area in a plane operated by Calbraith Perry Rodgers, who became the first pilot to fly a plane across North America from New

Thomas O. Perry shortly before his death. Perry spent the last years of his life developing designs for vertical-lift aircraft and calculating the size of the known universe. Courtesy Bentley Historical Library, University of Michigan, Ann Arbor.

York to California later that same year. Perry wrote that during the second flight the airplane "circled three to four hundred feet over the roof of my own home and later reached an altitude of more than 2,000 feet."[135] Interestingly, Rodgers had only completed his flying lessons at the Wright School at Simms Station in Dayton, Ohio, in June 1911. A newspaper reported on August 27, 1911, that Perry became the oldest man, at sixty-four years old, to fly in an airplane in the United States.[136] This experience, compounded with his extensive knowledge and experimentation with blades and aerodynamics, led Perry in 1916 to design and patent a vertical flying machine, making him one of the fathers of helicopter technology.[137] One of his experimental aircraft models was donated to the Smithsonian National Air and Space Museum in Washington.[138] Perry similarly devoted a great deal of thought and time to understanding the size of the solar system, meticulously charting the known planets, with Neptune being the farthest away from earth. His scale for drawing the chart measured 92,897,000 miles to the inch.[139] Before his death Perry completed a manuscript for a book titled *Life in the Universe*, in which he gave scientific consideration to natural events, organic and inorganic properties, and humans' relationship to the universe. The book was published posthumously by Perry's nephew, Frank Perry Keeney.[140] Although Perry was not a religious man, Albert W. Palmer, pastor of the First Congregational Church in Oak Park, said at his funeral: "He faced the universe with an honest and inquiring mind and worked out a philosophy of life which helped to steady and comfort him. He accepted the universe and trusted it, and that is the essential basis of all religion."[141]

Despite all his work and success post-Aermotor, Perry was bothered by the inventive credit often given to his former employer, La Verne Noyes, for the popular

Aermotor windmill display at Chicago's Museum of Science and Industry in the mid-1930s. Courtesy Museum of Science and Industry, Chicago.

all-metal windmill. He chastised the editor of the *Chicago Evening Post* in an August 1, 1919, letter for a misleading article about Noyes and the Aermotor windmill, stating:

> If the article in your paper involved only omissions of proper credit which belongs to others, that might be excusable in an obituary notice. One hundred years from now it will probably make no difference to anyone whether windmill history is correctly presented or not. But there are several still living who are sufficiently conversant with the history of Aermotor, or who know some of its history by hearsay, to make it unpleasant for the living to have their versions apparently called in question by what you have published. Besides it is inconceivable to me that Mr. Noyes could wish to have claims made for him, after his death which can be so easily disproved, since he freely gave due credit to me while he lived. If I should leave this matter unnoticed, it would be interpreted as agreed to by myself, and I would be a contributor to the promulgation of such loose statements as lead Napoleon to assert that "History is fable agreed to."[142]

Perhaps the fiercest defender of Perry's record was his wife, Mary E. Keeney-Perry. On March 5, 1927, she wrote a letter to Julius Rosenwald, chairman of Sears, Roebuck and Company, who a year earlier pledged $3 million to build a museum celebrating Chicago's scientific and industrial achievements, offering what she called "undisputable proof" of her late husband's work on the all-metal windmill.[143] Rosenwald's secretary responded to Mary Perry on March 8, stating that her letter would be passed to Leo Wormser, who was "looking after the museum project for Mr. Rosenwald." After a telephone conversation with Wormser on March 10, she mailed a letter the same day containing a list of models that she was willing to donate to the planned museum. These included an Aermotor windmill model, a four-foot model of the wooden wheel and dynamometer that Thomas Perry used for his various windmill experiments, and a glass model of the pneumatic pump that he designed for United Pump and Power Company after he left Aermotor.[144] Perhaps Mary Perry was eager to place these pieces since she had sold her home in Oak Park and had to vacate the property in mid-May 1927.[145] The pieces remained in storage at the home of her nephew Frank Perry Keeney at 1900 S. Prairie Avenue until July 1930 when the museum rented warehouse space for artifact storage. Keeney, by that time, was president of Domestic Engineering Publications.[146] According to museum records, only one five-piece apparatus was accepted, which was a model of the Aermotor windmill.[147] It's uncertain what happened to the remainder of the models, with the exception of the Perry helicopter model, which ended up in the possession of the Smithsonian Institution. The Museum of Science and Industry in Chicago officially opened its doors on June 19, 1933, with the Perry-designed Aermotor windmill on display. Aermotor Auto-Oiled windmill models were also added. While it is unclear how long the display ran, it was likely removed and placed in storage in the early 1940s, when Major Lenox Lohr took over as president and changed the appearance of the museum exhibits.[148] Mary Perry died in Oak Park Village, Illinois, on July 28, 1946, at the age of ninety.[149]

Yet inside the administrative office of Aermotor Company in Chicago, the portraits of both La Verne Noyes and Thomas Perry hung on the wall as a constant reminder of the men who set in motion one of the most prominently produced and recognized windmills in US history. Their images perhaps inspired Scholes and other design engineers of the company to persevere through the evolving agricultural and consumer market and withstand significant changes in competition and manufacturing of the post–World War I years in the United States and worldwide.

CHAPTER FIVE

New Heights

The economic outlook in the United States, as well as the rest of the industrialized world, during the early 1930s was unequivocally bleak. Initiated by the stock market crash in New York on October 29, 1929, many people once able to get by on their incomes were suddenly cast into poverty as numerous banks, businesses, and farms failed during what would become known as the Great Depression. It was estimated that at the start of the decade, about fifteen million Americans, or a quarter of the nation's workforce, had lost their jobs. Many farmers, however, had already spent most of the 1920s attempting to recover from the post–World War I collapse in farm commodity prices. During the conflict, farmers were encouraged by the government to increase their crop production as exports of grain and other commodities increased to feed Allied armies in Europe and Russia. They took on debt to buy new equipment and expand useful acreage. By 1920, Europe and Russia had significantly curtailed grain imports as their farmers recovered after the war and were subsequently protected by their own government tariffs. US crop production remained robust during the two years following the war, creating a rapid surplus, and prices in 1921 abruptly fell to half their wartime values. For example, wheat prices, which during the war reached $2.50 a bushel, were selling for about a dollar a bushel in 1921, and eggs, which at the start of 1920 were about forty-four cents a dozen, fell to about twenty cents per dozen by 1921.[1] Dwindling profits meant many farmers were unable to pay back their bank loans and lost their properties. Windmill prices dropped as farmers had less money to spend for machinery. The Aermotor Company's popular eight-foot windmill in 1917, for example, retailed for $37.50 without the tower, but the price had fallen to $29.00 by 1921. The firm's sixteen-foot mill dropped in price from $210.00 to $175.00 over the same period.[2] A recovery of agricultural commodity prices began gradually between 1923 and 1924, and Aermotor responded by raising its eight-foot windmill prices to between $44.00 and $45.00 between 1923 and 1927.[3] However, for many farmers and agricultural equipment providers, hopes for prosperity were dashed by the stock market crash of 1929 and the ruinous Dust Bowl, which resulted from a persistent drought combined with effects of decades of poor plowing and planting techniques and overgrazing by cattle. The Great Plains region, in particular, became littered with abandoned homesteads as farmers migrated to California and other states in search of work. Those individuals, businesses, and farms that survived found ways to economize and eliminate redundancy and waste.

Yet the water-pumping US windmill, with its simple arrangement of mechanical parts, relative ease of repair and maintenance, and use of free-flowing wind, stood as an enduring figure during these hard times. With less income, however, fewer people were able to buy new windmills in the early 1930s. As industry historian T. Lindsay Baker described, the Great Depression ushered in the era of "bargain windmills" or "New Deal specials."[4] Many rural Americans generally sought to repair their existing windmills first, or even buy them secondhand, before considering the purchase of a new one. This economic reality placed an enormous strain on the windmill manufacturers of the day, who watched their sales plummet and workforces and manufacturing equipment become increasingly idle. Numerous tactics were taken at the start of the 1930s to keep their windmill lines alive, including significantly reducing their unit prices to dealers and customers and trimming manufacturing costs by decreasing the number of parts used to operate their machines. Even the export market offered no relief, as many countries overseas were experiencing the same economic hardships as the United States. Congressional passage of the Smoot-Hawley Tariff Act in 1930, which raised tariffs on numerous imports in an effort to protect domestic industries from foreign competition, inspired other countries to react in kind, making it increasingly difficult to sell US-made windmills abroad. With these dismal prospects, some firms simply shifted away from windmill manufacturing, if they were fortunate enough to have diversified product lines, or went out of business altogether. By 1929, eleven years after the end of World War I, the US windmill industry had shrunk from thirty-one firms to fifteen.[5]

Aermotor, one of the most sophisticated windmill manufacturers of the day, did not escape the ill effects of the Great Depression. The company dropped the retail price on its eight-foot windmill from $45.00 in 1933 to $30.00 by 1937. Over the same period, its sixteen-foot windmill fell in price from $310.00 to $197.50.[6] Nevertheless, the company's tightly knit senior management, which included Lewis C. Walker, president; Frederick L. Dole, first vice-president; Daniel R. Scholes, second vice-president; and Joseph J. Fraser, secretary and treasurer, appeared to be unfazed by the economic downturn at the end of 1930. They offered the La Verne W. Noyes estate trustees, as well as themselves, healthy 25 percent dividends on the corporate stock and an end-of-year bonus of $87,362 to be divided among the company's officers and employees.[7] This situation had changed by late 1931, when Aermotor slashed the dividends on its corporate stock to 10 percent and cut the annual bonus amount to officers and employees to $24,980.[8] No bonus was given to employees at the end of 1932, and in April 1933 Aermotor officers even reduced their pay and that of their salaried employees. The magnitude of the reduction in annual compensation is unclear, but it was likely in the range of 10 to 15 percent.

The stockholders in 1933 approved the following salaries for the immediate future: Walker, $14,000; Scholes, $13,625; Dole, $12,500; and Fraser, $11,500.[9]

In November 1932, Democratic candidate Franklin Delano Roosevelt won the presidential election with promises to lift the country out of the severe economic slump and put its vast numbers of unemployed back to work. During Roosevelt's first hundred days in office he and the Congress instituted a number of legislative measures to correct what they perceived to be wrong with unbridled capitalism. One of those measures to impact US businesses was the National Industrial Recovery Act, which was enacted into law in June 1933. The act essentially asked companies to collectively develop fair competition practices, with the goal to "raise wages, create employment, and thus increase purchasing power and restore business."[10] The National Recovery Administration sent instructions to all US companies on July 27, 1933, that outlined the requirements for compliance. Aermotor, as a member of the National Association of Farm Equipment Manufacturers, carefully studied the document and, along with the association, accepted its provisions. Since the company was located in a major city, it had to pay its factory employees no less than forty cents an hour and work these individuals no more than forty hours a week. Farm machinery manufacturers were permitted under the agreement to pay female employees five cents less per hour than their male counterparts under the National Recovery Administration minimum wage scale. The larger goal of the National Recovery Administration was to reduce the volume of production in order to stimulate market demand for industrial products. Part of the goal was sought by stabilizing prices through the "codes."[11] By signing the industry-wide document, Aermotor president Walker on August 15, 1933, enabled his firm to display in advertising the National Recovery Administration's distinctive membership logo, the blue eagle clutching a gear in one talon and lightning bolts in the other. Bold red letters above stated "NRA member" and below read "WE DO OUR PART."[12] However, the New Deal program was largely considered a failure and opposed by most businesses at the time. In May 1935, the US Supreme Court ruled that the National Industrial Recovery Act was unconstitutional and the program dissolved.[13]

While the Depression wrecked many businesses by 1934, Aermotor still offered a token annual bonus of $3,400 in December that year. This money was distributed among its 346 employees in the Chicago factory as well as among its nineteen employees in its branch houses in Dallas, Texas; Des Moines, Iowa; Kansas City, Missouri; Minneapolis, Minnesota; and Oakland, California.[14] Interestingly, in January 1935, the corporate officers voted to adjust their annual salaries upward, with Walker receiving $15,000; Dole, $13,500; Scholes, $15,000; and Fraser, $12,500.[15] With labor unrest brewing throughout the country, Aermotor corporate officers had to keep this increase confidential. The plant and branch workers again had to divide among themselves a total of only $3,375 in 1935 as their end-of-the-year bonus.[16]

Aermotor persevered through the depths of the Great Depression, cutting from its catalogue products with diminished or lackluster sales, such as gas engines and windmill-based electric generators, and focusing primarily on water-pumping machinery and steel towers. Before the start of the decade, the company promoted sales of its Model 602 windmill with a giant, twenty-foot wind wheel. Wheels of this size were nothing new to the market. There were numerous examples of wooden windmill manufacturers from sixty years earlier offering wind wheels of double or even triple the size. Aermotor slowly introduced different sizes of wind wheels, firmly believing its eight-foot wheel was still the ideal size. Its twenty-foot wind wheel was likely developed and offered after repeated customer and dealer requests seeking a self-lubricating windmill to pump deeper wells or to raise larger amounts of water from shallow wells. The twenty-foot wind wheel had eighteen galvanized steel blades, each measuring six feet long and thirty-two inches at the outer edge and weighing about forty-five pounds. A windmill of this size and weight required some changes to Aermotor's traditional four-post tower structure. (The company recommended a minimum-size tower of forty-feet tall for the twenty-foot windmill.) To withstand the enormous pressure of the wind against the windmill and tower, the company endorsed use of its "basket type" anchors, which had a rectangular base of five feet by three feet eight inches and were placed six feet deep in the ground. Set at the lower edge of the wind wheel was a six-foot-diameter platform accessed by a loop-rung ladder that branched off vertically from a single corner post about two-thirds up the tower. To reach the massive gearbox behind the wind wheel, Aermotor provided a ten-rung, galvanized steel ladder that mounted permanently onto the vane stem. Standing on the stem to remove the nearly man-size galvanized sheet-metal hood to change the oil and inspect the gears surely took skill and nerve. Aermotor noted that its twenty-foot windmill could be furled by a worm-geared winch attached to a single corner post at the bottom of the tower. "A small boy can easily operate it," the company claimed. With a maximum pump cylinder of fourteen inches in diameter and thirty inches long and the wind wheel turning in a fair wind of eleven to seventeen miles per hour, the twenty-foot Aermotor windmill could pump upwards of 9,000 gallons per hour from a twenty-foot-deep well.[17]

By the late 1920s and into the early 1930s, Daniel Scholes and his staff were also busy making design improvements to the Model 602, with a focus on making the windmill more profitable without sacrificing its mechanical simplicity and ease of maintenance.[18] Aermotor's so-called "improved" windmill, the 702 model, which was officially introduced to the market in the summer of 1933, included an enlarged wheel shaft to withstand better the stresses placed on the wind wheel. The company added wheel arms that screwed into the hub for better balance and arm crosses, which supported the inner band of the wheel, forged together in a manner so that the "careless erector cannot possibly get the wheel together wrong." The bearings for the main shaft and large gears were designed to be more easily removed and replaced, if necessary. These interchangeable bearings, like the older poured-in-place bearings on the 602 mills, were made of babbitt metal and were constantly bathed in oil. Aermotor gave its two cut pinions longer hubs, which met in the center of the bearing for better support and quieter rotation and

Twenty-foot oil-bath back-geared Aermotor windmill in an unidentified tropical setting in 1939. Courtesy *Windmillers' Gazette*, Rio Vista, Texas.

were secured together by a long key to keep them from working loose. The company adopted a regulation system that allowed the windmill to pump at slower speeds. This could be done by adjusting the position of the governor spring on the vane stem. The company made it easier for the adjustable brake band tension to be increased by the user and added a waterproof oil chamber to the pole swivel to keep the windmill turning more efficiently into the wind while reducing potential breakage of the wooden pump poles and guides.[19]

Aermotor aggressively marketed its new Model 702 in numerous agricultural and well-drilling publications around the country and commented to its dealers in late September 1933 that "it is making a big hit."[20] The company first rolled out the 702 in the six-, eight-, and ten-foot mills that year, and by October was taking orders for the twelve-foot windmill, with the fourteen- and sixteen-foot mills appearing at the Chicago factory before the end of the year.[21] In late 1933 the assembled 100-pound, six-foot 702 windmill motors, excluding towers, retailed for $20.50 each and up to $175 for the 1,220-pound, sixteen-foot mill.[22] The company

An Aermotor windmill and electric pump display at an unidentified trade fair along the Gulf of Mexico in the 1930s. Photograph by Robert R. Miller, Dallas, Texas. Author's collection.

also offered the option to purchase for a third of the price the factory-assembled 702 motors, allowing customers to use their 602 wind wheels and tail vanes, or for half the price, sold "stripped" motors, which included just the 702 main frame, wheel hub, small gears, pitman guide, guide wheel, and removable bearing for the large gears. If in good condition, Aermotor's Auto-Oiled windmill parts, including the V-shaped wheel arms, tail bone, large gears, and other Model 602 parts could be incorporated to complete the stripped motors.[23]

One of the outstanding sales coups for Aermotor's new Model 702 was its selection by Pan-American Airways in early 1933 to supply water on the remote islands of Midway and Wake. There the airline housed crews and provided stop-over services for its Clipper passenger seaplanes transiting the Pacific over 8,000 miles from Alameda, California, to Canton, China. Pan-American's New York office worked with Aermotor in Chicago to have the ten-foot windmills with standard forty-foot galvanized tank towers delivered to the islands for setup.

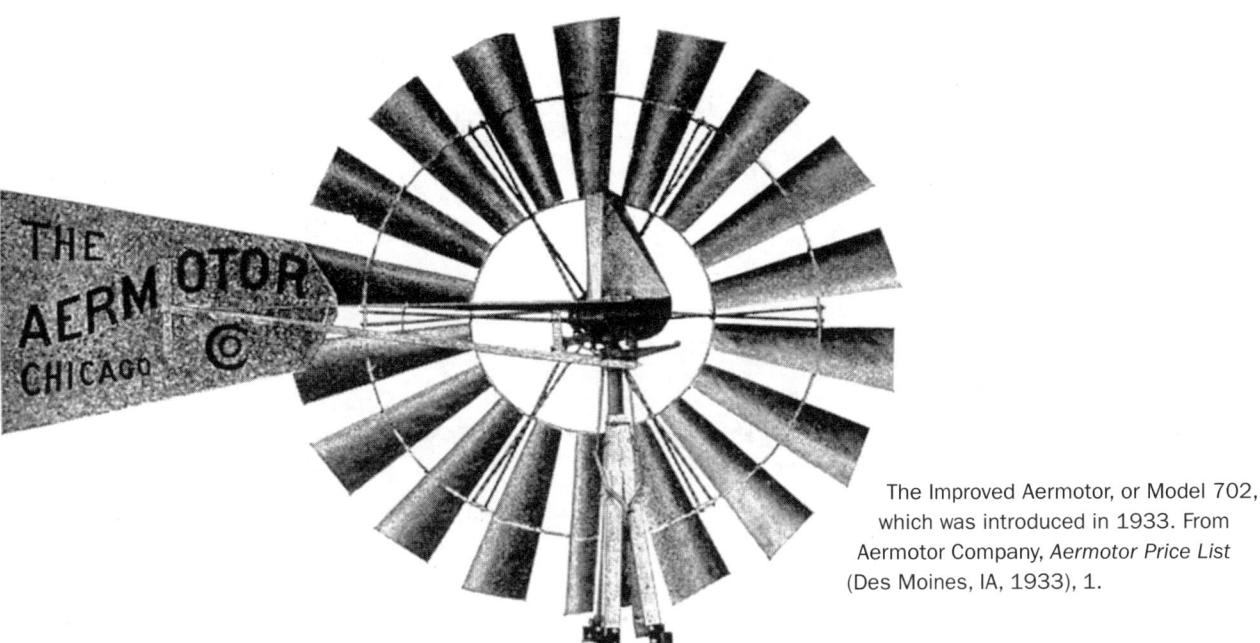

The Improved Aermotor, or Model 702, which was introduced in 1933. From Aermotor Company, *Aermotor Price List* (Des Moines, IA, 1933), 1.

Management and employees of Aermotor's Dallas branch show a crated Model 702 windmill. Author's collection.

The wooden tanks, also manufactured by Aermotor, measured ten feet high and twelve feet in diameter and had a capacity of 7,000 gallons of water each. Using gravity from the elevated towers, fresh water was piped throughout the airline's operations bases. Pan-American engineers also found that the windmills pumped water nearly twenty-four hours a day due to the constant winds sweeping across the islands.[24] In a follow-up to the use of its windmills by Pan-American, Aermotor boasted how traditional and new technologies had come together:

> This is a strong endorsement of the reputation of Aermotors for reliability as these windmills will be depended upon to supply all water for airports, radio stations, power houses, staff living quarters, and passengers on arid

PAN AMERICAN AIRWAYS *Insure* PACIFIC WATER SUPPLY

with **IMPROVED AERMOTORS**

THE WORLDS MOST DEPENDABLE WINDMILLS

Aermotor's Model 702 windmill was selected by Pan-American Airways in the early 1930s to supply water on the islands of Midway and Wake where its Clipper service made rest stops. Courtesy *Windmillers' Gazette*, Rio Vista, Texas.

islands where the securing of water is the most complex problem of living. And, what is more natural than that the Aermotor and the Airplane should be joined in service. The Aermotor was the first steel windmill and opened the way to greatly improved methods for providing water for domestic and livestock use all over the world. The Airplane is providing swift and sure transportation to the ends of the earth. Both depend upon the free air for their operation.[25]

By the early 1930s, however, Aermotor started witnessing a shift in the application and reliance on mechanical windmills to electrically operated pumps for rural households and farms. President Roosevelt created the Rural Electrification Administration (R.E.A.) by Executive Order 7037 on May 11, 1935. The new federal agency received congressional statutory authorization a year later, on May 29, 1936. It was believed by the Roosevelt Administration and many in Congress at the time that connecting power lines to the nation's farms would help lift the economy out of the Depression. Only 11 percent of the country's 6.8 million farms were served by central power stations. Those households without power line access, if they had electricity at all, obtained their power from small, stand-alone electric light plants and wind generators, which charged banks of batteries with just enough power to burn a few light bulbs and operate a radio or other small appliance. Once the batteries ran down, the process of recharging them had to start all over. On the other hand, direct power access from the grid was predictable and continuous, unlike keeping an engine filled with fuel or waiting for the wind to turn a set of blades. Contrasted with other New Deal programs, the R.E.A. was generally deemed successful. Within two years, the agency had helped set up 350 electric cooperatives in forty-five states, capable of delivering power to 1.5 million farms. Each co-op member paid a monthly service fee, in addition to a per-unit cost of electricity, to the provider.[26]

Access to electricity further raised the desire among Americans to acquire electrical appliances that would reduce their physical workloads. This included an increased demand for indoor plumbing, without the drudgery of bringing water into the house by hand from an outside mechanical pump. Windmills and hand pumps were increasingly viewed as external sources of water for livestock and garden irrigation rather than for human consumption.[27] Aermotor understood this emerging trend and set out to develop its own line of household electric pumps at the start of the 1930s.[28] Interestingly, the design of Aermotor's ground-mounted deep-well electric pumps incorporated a number of mechanical aspects found in its oil-bath windmills, including the large gears and pitman arms with oiling ring.[29] A major difference between this electric pump and those of Aermotor competitors was a patented U-shaped guide for lifting the pump rod. "This new and ingenious engineering achievement has made it possible to build a pumping unit without the usual elongated opening in the front of the pump," the company proclaimed.[30] Like older-style pump jacks and windmills, these initial Aermotor electric devices produced reciprocating strokes that actuated traditional piston-type pumps underground at the water table.

The changes in Aermotor's business now required its salesmen to understand both mechanical and electrical pump equipment, and when and where it was appropriate to push a particular option to a customer. To support this change, the company began publishing a range of separate sales catalogues and brochures to highlight its line of electric pumps. Farmers and ranchers could still buy windmills for their livestock, while people living in towns and emerging suburbs might be swayed to purchase electric pumps to support their residential plumbing needs.[31] To promote sales of its electric pumps, Aermotor encouraged its salesmen to "canvass along new power lines," contact schools and dairies, look out for new suburban home and resort construction, consult with architects, and follow up with old customers who might want to swap out their windmills for electric pump outfits.[32] By the end of the 1930s, Aermotor offered customers a range of deep- and shallow-well electric pumps attached to various-size pressure tanks depending on the average daily amount of water required. The company even offered a windmill connection to its electric pumps, allowing the windmill to "do the heaviest pumping, and to switch to electric operation at a moment's notice" by shifting only a single connecting pin.[33] This permitted the customer to still use the "free power" of the wind via a traditional Aermotor windmill and save money on his or her electric bill.[34]

With the introduction of these new products, Daniel Scholes's prominence in the company increased. He took a major step forward after Frederick L. Dole's death on September 4, 1936. Dole had worked for Aermotor since 1891 and served as a vice-president and board director from 1919. To fill Dole's vacancy, Scholes was elevated from second vice-president to first vice-president, while Frederick Smith, La Verne Noyes's brother-in-law (who had resigned from the company in 1919 during the reorganization but remained an active shareholder) was brought back to serve as second vice-president.[35] In October 1936, Scholes was elected to the board and given a single share of the company's capital stock.[36] Business for Aermotor started looking up that year, as the board voted to give Dole's widow, Harriet, $7,000 and to distribute $42,360 among the company employees as an annual Christmas bonus.[37]

After nearly eighteen years of successfully leading the company's engineering and product development, Scholes was now in a position to demand a more visible role in Aermotor's overall management structure and to set its direction forward. In October 1937, the board chairman, Walker, acknowledged that Scholes "had contributed greatly to the success of the corporation." Perhaps concerned that Scholes might be hired away by the competition, Walker asked him to commit to a five-year contract. Scholes initially refused to enter into the contract unless he was given the option to buy more shares of Aermotor's capital stock. This put the board in a quandary since 2,933 shares of the capital stock were tied up in the Noyes estate trust, and the remaining 67 shares were divided among the corporate officers: Walker 45, Smith 25, and Fraser and Scholes each with one share. The board thus decided to purchase 273 shares from the Noyes trust at a cost of $300,300 and agreed to sell Scholes a minimum of 35 shares for $600 apiece during the duration of his contract, running from 1937 to 1942. The remainder of the 273 shares would be held in the company's treasury where it would be available for purchase. Scholes's contract with the company was finalized on October 28, 1937.[38] Also, after fifty years of service, Smith retired from Aermotor on December 31, 1939, reducing the number of corporate officers back to three—Walker, Scholes, and Fraser.[39]

In 1938, the company's fiftieth year in business, the US economy was still plagued by the effects of the Great Depression. That year through 1941 Aermotor sold off tracts of real estate across Chicago and its suburbs, along with various stocks and bonds of other corporations acquired during Noyes's tenure, to help keep the company afloat.[40] A mere $3,492 was to be divided among the company's more than 300 employees that year for the annual Christmas bonus, which was similarly followed by the issuance of a $3,628.77 end-of-the-year bonus in 1939, $3,389.29 in 1940, and $3,474.22 in 1941.[41] Aermotor did not provide pensions for its managers and factory workers. However, it did reward certain individuals for their many years of service upon retirement. For example, Aermotor gave G. A. Williams, who joined the company in 1899 and served as factory supervisor from

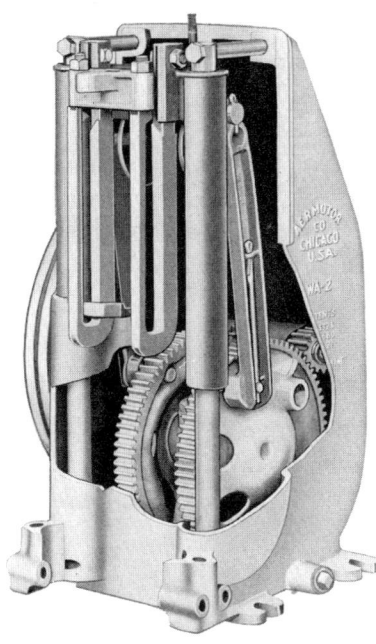

Cutaway of Aermotor pump jack. Aermotor introduced electric pumps at the start of the 1930s. From Aermotor Company, *Aermotor Electric Pumps for Deep or Shallow Wells* (Chicago, 1933), 3.

1904, a $5,000 check "in recognition of the faithful performance of his duties during forty one years of service" when he tendered his resignation on January 15, 1941.[42]

For those windmill companies, like Aermotor, that survived the Great Depression, there would be even greater pressure in the early 1940s. The United States was hastily drawn into World War II after the December 7, 1941, Japanese attack on Pearl Harbor, Hawaii, and soon after came a call by the federal government to convert civilian manufacturing into wartime production. For the next four years, windmill production in the United States practically ceased. At the same time, more cattlemen and crop farmers desired larger outputs of water for their livestock and irrigation requirements, testing the ability of traditional mechanical windmills to keep up and driving a shift toward large gas engines and electric motors as pump power sources. However, the windmill was still highly prized by many rural Americans, who were gradually recovering from the economic devastation of the 1930s, for its simplicity and ruggedness. US Department of Agriculture engineers C. L. Hamilton and Hans G. Jepson wrote in 1940 that "engines and motors pump water faster and furnish a more dependable source of power than do windmills, but they are more expensive to operate." Carl Rohwer and M. R. Lewis, also USDA engineers, concluded in their irrigation research that same year that if the pumping requirement was fifty gallons per minute or less and the lift from an underground well was not more than fifty feet, then "the possibility of using windmills should be investigated."[43]

CHAPTER SIX

Disillusion

Even before the Japanese bombed Pearl Harbor, many Americans believed it was only a matter of time before the United States would be drawn into the fighting in World War II that had already spread across much of Europe, North Africa, and the Far East. US lawmakers had passed the Selective Training and Service Act on September 14, 1940, permitting a peacetime draft of men between the ages of twenty and forty-four years of age. In early spring of 1941, Congress signed the Lend-Lease Act, authorizing the president to provide war materiel to nations for which defense was deemed vital to US interests. This was shortly followed by the creation of the Office of Price Administration to recommend price controls on commodities to avoid possible wartime inflation. The country was just beginning to recover from a decade-long economic depression and did not want to risk allowing soaring prices on consumer goods and commodities to dampen its progress.

In the months after the Japanese attack and subsequent US declarations of war on Japan and Germany, more legislation and executive orders were passed that shaped the way manufacturing and commercial sales would be conducted for the duration of the conflict. These included the president's War Production Board, which was put in charge of all war production and supply; passage of the Emergency Price Control Act to curb rising prices in the face of inflation; Roosevelt's wartime freeze on wages, salaries, and prices; and rationing on everything from meat and sugar to rubber and metals. There was also the need to add women to the workforce as men were either drafted or voluntarily left in droves to join the armed forces.

Upon entering the war, US manufacturers were ordered to contribute to the production of much-needed war materiel. No matter the size, each company had a role to play. The Aermotor Company, which for many years operated a highly integrated factory of about 350 skilled machinists and other workers who were capable of making windmills and other water-pumping equipment from start to finish, was an ideal enterprise for producing military hardware. In the years before the war, the company had supplied various products to the US government, including observation and fire control towers for the Forest Service, Bilby towers to the Geological Survey, and beacon towers for burgeoning airports across the nation.[1] Marvin Isvik, who in September 1941 joined Aermotor as a "sales correspondent" responsible for correspondence and orders in the Midwest territory, recalled the immediate change inside the company following the Pearl Harbor attack:

> I was 28 years old, the perfect age for the draft law. The company wanted to keep me and secured me "war work," which gave me a deferment. I supervised fifteen women off the street who we trained to do machine work on precise parts for Bell and Howell Company cameras for the Army, Navy and Air Corps. These women learned very quickly and were good with their hands.[2]

Aermotor did not stop making windmills during the war, but their numbers and sales were greatly reduced due to mandatory rationing and other government restrictions.[3] To buy a windmill, tower, tank, or electric water pump, a customer had to apply to his or her County Agricultural Conservation Committee, administered by the US Department of Agriculture, for approval of a special ration certificate. Only after the certificate was issued could the consumer then go to a local supplier to purchase the item, and still there was no guarantee it would be immediately available. This bureaucratic process was frustrating for farmers and ranchers who needed quickly to repair a windmill or pump to ensure adequate water supply for their livestock.[4] Exports overseas were essentially nonexistent. During the Aermotor annual meeting of shareholders on January 21, 1943, Lewis C. Walker, president of the company, "spoke of the severe restrictions placed on the manufacture and sale of Farm Machinery and Equipment and attachments, under the United States Government Limitation orders known as Limitation orders No. L-26 and L-170. He pointed out the marked effect of these limitations on the operation of the company, particularly in the last quarter of 1942."[5] It is unknown how many windmills Aermotor actually produced and sold during the war years. By late 1942, the company was unable to ship its popular eight-foot windmills "because we are all out of sheet steel for the wheels," Walker wrote in a letter to Aermotor dealers.[6] However, Aermotor was certainly not alone in the stresses that wartime rationing caused to its windmill business. By the start of 1943, for example, the Butler Company, a windmill manufacturer in Butler, Indiana, decided to withdraw its salesmen from the field that year due to the lack of sales.[7] United States Engine and Pump Company (successor to the U.S. Wind Engine and Pump Company) wrote to a potential Australian customer in 1944 that "owing to our pre-occupation with war work production of windmills has been at least temporarily suspended for the duration of the war."[8]

The end of the war with the announcements of Germany's surrender on May 8, 1945, followed by Japan's on August 8 that year, in addition to the expiration of government rationing mandates, did not result in an immediate revival of the US windmill business.[9] Some of the most notable industry names, including U.S. Wind Engine and the Challenge Company, also of Batavia, shuttered.[10] The Baker

Manufacturing Company of Evansville, Wisconsin, reported in 1948 that its sales of windmills, pump jacks, and other water systems were $4.7 million, double that of 1945. However, Baker Manufacturing's punch presses for making windmill parts were still underutilized, so the company began using the machines also to make television antenna towers.[11]

In June 1945 the Office of Price Administration allowed the windmill manufacturers to raise their prices on new windmills and repair parts by 7 percent above those of April 1, 1942. The government agency in late July 1945 raised the allowable price increase on water pump systems to 10 percent. Aermotor embraced the price increase, asking its dealers in a letter dated July 30, 1945, to "take this increase into consideration in figuring out your remittance on orders sent to us from this date forward."[12] Dealers were also increasingly asked by their customers to either erect new windmills or make repairs to existing machines, escalating order requests to Aermotor. Daniel R. Scholes, vice president, explained in a September 15, 1945, letter to J. W. George of Albany, Texas, who was pleading for a delivery of six- and eight-foot Aermotor windmills and towers, "We certainly wish that we were in position to supply all of our dealers with the full amount of their demands. At the present time, our production is slowly increasing but it still is far from the point whereby we can make immediate shipment of orders."[13]

Aermotor spent the remaining years of the 1940s attempting to build back its windmill and electric water pump business. Marvin Isvik, the newly hired Aermotor sales representative before the war, accepted a position as a field salesman, living in north-central Iowa and overseeing a massive territory that went as far west as the Sand Hills of western Nebraska. In a company car—a 1946 Ford—he drove between 40,000 and 50,000 miles a year meeting with dealers and well drillers. He described his personal experiences from the late 1940s:

> When I went out, I went out to a wide variety of dealers. We sold windmills and pumps to hardware stores, groceries, well drillers, and pump installers. I saw this could not go on. Gradually I eliminated all the dealers who were not in the water supply business.
>
> I got to know basically all the well drillers in the state of Nebraska and know them well. Almost all of them were customers of ours. . . .
>
> Aermotor had a postcard that we sent to dealers to let them know we were coming. It read '_____ will call on you on _____.' I would send these cards out, would write them by hand, and used them really successfully. . . . Dealers would wait on me.
>
> I also made some cold calls when I didn't know them. . . .
>
> When I called I almost had a canned speech why Aermotor was better [than the competition]. I could say with sincerity they would have a protected territory. I would tell them about terms and discounts. . . .
>
> I gave them the straight company line. I had no special terms or discounts to offer. There are other things than price that one has to sell: quality, dependability, and advertising.[14]

Glenn LaVerne Roberts, one of Aermotor's top salesmen, posed at a work site of George Patrick Well Drilling, Albion, Nebraska, in the mid-1950s. Courtesy Roberts Pump and Supply Company, Grand Island, Nebraska.

Throughout the war, Aermotor maintained its branch offices in Dallas, Des Moines, Minneapolis, Kansas City, and Oakland. After nearly five years of field sales, Isvik hoped to relocate to the Des Moines branch, but the company had other plans. He was instructed in 1949 to open a new branch in Omaha, Nebraska, which was attractive for its railroad and highway connections. In addition to Nebraska, the Omaha branch served the eastern half of Iowa.[15] Aermotor's branch locations remained closely controlled by the home office in terms of paying salaries and hiring salesmen. Windmill and pump orders were often telegraphed to Chicago and were relayed using a long-standing specific list of telegraph names, such as "Atlanta" for an Auto-Oiled eight-foot windmill or "Dwight" for the twelve-footer.[16] Product inventory at the branch offices was traditionally controlled by Chicago, but after World War II some branch managers started exercising independence in terms of the ancillary products that they carried. Isvik said he "took the liberty of buying non-Aermotor merchandise for sale. Omaha was the first branch house to do that. There was nothing that competed with the Aermotor Company. We sold pipe, plastic pipe, pressure tanks. Aermotor didn't make any of these, and customers wanted them. We tried to supply complete water systems, not just Aermotor products."[17] It was not uncommon for windmill installers in the Nebraska Sand Hills to build wooden towers and attach the metal Aermotor stub tower to the tops instead of buying the metal Aermotor towers. Sometimes a farmer might just want to buy an Aermotor tower to use with another type of windmill, or just the Aermotor windmill head to go on top of another manufacturer's tower.[18]

It took nearly three years after the war before orders for both windmills and electric water systems at branch houses were accepted on availability and dealer

Marketing Aermotor windmills and pumps, Stanley A. Anderson (left) and Bill Otery display the company's products at the Nebraska State Fair in 1949. Courtesy *Windmillers' Gazette*, Rio Vista, Texas.

quotas based on past purchases.[19] Unlike many competitors, Aermotor branches remained under strict rules as to whom they could sell their products: namely, they could work only with authorized Aermotor dealers, who were mostly well drillers, and make no direct sales to farmers or ranchers.[20] The branches and their salesmen were also careful to stay within geographically defined sales territories. They did occasionally contact each other for assistance if they experienced product shortages. It was also not uncommon for an experienced manager or salesman in one territory to fill in at other branches when short of manpower. Executives from the Chicago headquarters often visited the branches and encouraged their managers to share with them ideas to boost sales or improve product quality.[21] "It took some getting used to Aermotor's system, and for a new employee you learned the system and not what you might have learned in school or college," recalled Stanley Anderson, a longtime salesman who started with the company at the Des Moines branch in 1949.[22]

In the immediate years after the war, Aermotor pressed hard to restore its windmill business and ran a postwar advertising campaign stating, "The wind is

Aermotor Model 702 windmill on a twenty-one-foot, wide-spread steel tower on an unidentified Great Plains ranch, ca. 1950. Courtesy *Windmillers' Gazette*, Rio Vista, Texas.

free—use it!," once again hoping to connect with frugal farmers and ranchers.[23] Even its hand pumps and its own brand of oil were still available for purchase. Census Bureau figures for the number of windmill heads sold peaked for the industry in 1928 at 99,050 units at a value of $3,944,499. Two years later, in 1931 and with the weight of the Great Depression bearing down on the national economy, the number of windmill heads sold fell to about a third at 37,523 with a value of $1,294,992. In 1943, US windmill manufacturers sold 34,523 heads at a value of $1,537,449, reflecting purchases based on necessity and rising prices of metals. By 1947, the number of windmill heads sold had reached 39,076 units at a value of $2,424,457, in line with postwar replacements and inflation.[24] Aermotor's standard eight-foot windmill head, without tower, was priced at $86 in mid-1947, while the sixteen-foot and twenty-foot windmills were $542 and $1,398, respectively.[25] Windmill sales remained strong in the Texas–New Mexico area, where rural electrification was slower to expand, making the company's Dallas office the strongest branch for orders after the war.[26] It even constructed a new branch building at Dallas in 1952:

> From the front door to shipping platform, the building has been designed with you in mind . . . to handle your orders with speed and efficiency . . . to give you quick shipping service.
>
> Aermotors, towers, electric water systems, repair parts and other water supply goods demanded in your territory are here in abundance to make your Aermotor business most profitable to you.
>
> Carload orders and motor truckload orders of not less than 10,000 pounds should be sent to the new Dallas address, but shipments will be direct from the factory, as customary, at Chicago prices.[27]

Yet, by the early 1950s, many Americans were starting to view the windmill differently. Once a symbol of prosperity and modernity, the machine was now increasingly viewed as a relic of the past. In response to a 1953 *Chicago Tribune* article stating that water-pumping windmills were fading away, E. M. Fleming, an Aermotor sales manager, responded that the "windmill business is not dead by a long shot." He told the newspaper the company had seven branch houses in midwestern and western cities and distributors in several others.[28] Aermotor also benefited from postwar exports of windmill and repair parts to countries such as South Africa and Australia. Its best-selling windmills in the postwar years were the six-footer (17 percent), eight-footer (50 percent), and ten-footer (8 percent) for 75 percent of total sales, with the remainder divided up among the twelve-, fourteen-, sixteen-, and twenty-foot windmills. The company did not give out actual sales figures to its salesman in the field nor has a record been found of its annual windmill sales.[29]

To continue making its windmills attractive to customers, Aermotor began offering customers newly designed "wide spread" windmill towers in twenty-seven-, thirty-three-, forty-, forty-seven-, and fifty-three-foot heights, making them easier for installers and repairman to work under.[30] It also made available to dealers tower extensions as "an economical means for increasing the height of an existing tower no longer high enough to provide adequate wind exposure for the windmill." The base extensions were sold in both seven- and fourteen-foot heights.[31] In the early 1950s, an engineer was assigned to review the company's products and recommend improvements. One of those changes came in the form of Aermotor's tower loop steps. For many years, the company used a looped step bolted to a corner post with one raised end pointing upward to help prevent slippage when climbing. The step was redesigned so there were two raised ends per loop, making it simpler and more efficient to install tower steps.[32]

While the Aermotor 702, introduced in 1933, proved a successful and desired windmill, it was not without its occasional problems. In 1950 the company heard complaints from customers about a "clicking noise" coming from inside the gearbox during operation. Aermotor engineers observed that the noise generally occurred at the top of the stroke, when the start of the reversal caused "a sudden movement of the gear so that the contact between the gear teeth shifts from one side of the tooth to the other," often referred to as "backlash" in geared machinery. The company said the backlash occurred because the large gear teeth were cast thicker and might be slightly irregular after being manually filed, unlike the small steel pinion gears that had teeth accurately machined in a special gear hobbing machine. "In assembling the Aermotor, it is intended that this extra thick tooth shall be filed, if necessary, just enough to let it pass. If the assembler should file it too much, there will be more backlash than need be and this would tend to produce a noticeable clicking noise at the top of the stroke when the Aermotor is working under load," the company explained. Aermotor said the clicking noise should not occur if the gears were set tight to their shafts during assembly. "Now, while much effort is expended by the factory management to make sure that these operations are done carefully and correctly, it is always possible that a new assembler, or a careless one, will slight this work," the company warned. It advised its salesmen to check that all gears were tight to their shafts and report instances of loose-fitting gears immediately to the factory.[33]

Aermotor also received complaints in the early 1950s of oil leaks on its six-foot mills. Company employees investigated the problem and found "Excessive runout of an X-705 gear may permit it to touch the X-579 inverted cup washer, transfer oil to its under side and force it over the cast iron ring and down the post." The company corrected the problem by providing its branches with new cup washers with a diameter 1/16-inch smaller and identified the replacement parts by painting them "dark red." It also sent branches a 7 5/32-inch length of 1¼-inch pipe to be used as a "gage" in checking the X-500 pipe. To use the gauge, Aermotor said, the locknut and cup washer must first be removed, then "slip the pipe gage over the small end of the pipe so that it rests against the top turntable washer. If the end of the X-500 pipe is not even with the end of the pipe gage, remove one or two X-576 flat washers in the turntable ring, as necessary, to make them even. Then restore the lock nut and the new red-painted cup washer." The company explained in a memo to its branches that all six-foot windmills manufactured in its plant after March 5, 1953, "have been assembled with special care as to gear runout and other possible leakage. These mills will be marked for a time by the letter 'R' stamped on the end of the shaft."[34] Oil leaks, however, persisted. In 1955 the company found that a small hole normally drilled in a downward position in the pillar to allow hand-oiling of the windmill turntable was prone to leaking during high wind speeds. Aermotor corrected this problem by drilling a three-sixteenth-inch hole horizontally in the rear side of the pillar post, about five-sixteenth inch above the rim of the eight-foot windmill frame and similarly for the other size mills. It was also suggested that the holes in older mills could be plugged with "lead, wood, or other suitable materials."[35]

Across much of the United States, electric water pumps increasingly generated a larger portion of Aermotor equipment sales and profits. Marvin Isvik explained this change in customer preferences:

> When I first went out [as a sales representative in 1946], windmills were still being sold for pumping. REA changed much of this. One of the first things a farmer bought after getting REA power was an electric pump. Sales of windmills in rural areas basically died. If they needed a windmill, they bought one from a neighbor who didn't need one he had.[36]

The company responded to this market change by making a concerted effort as early as the late 1940s to educate and encourage its dealers to think of Aermotor as more than just a windmill company and to embrace the sales opportunities of electric water pumps in their territories.[37] To symbolize this shift, Aermotor in the mid-1950s crafted a new corporate logo for its brochures, letterhead, and other advertising which consisted of a large trapezoid containing the word "AERMOTOR," with a smaller elongated one across the bottom foreground that read, "Water Where You Want It!" To facilitate sales of electric pumps in burgeoning

The Aermotor factory in Chicago as it stood in the mid-1950s. From Aermotor Company, *How to Choose Your Water Pumping System* (Chicago, 1954), 2.

markets, Aermotor opened additional offices in locations such as Amarillo, Texas; Cordele, Georgia; and Harrisburg, Pennsylvania.[38]

Significant changes were also taking place within the Aermotor Chicago corporate office and factory during the late 1940s. At the close of the war, Lewis C. Walker remained president, with Daniel R. Scholes as vice president and Joseph J. Fraser as secretary. Frederick E. Smith, brother-in-law of late founder La Verne W. Noyes and a former vice president, died April 22, 1940, at age seventy-four shortly after retiring with fifty years of service to the company.[39] Smith's widow, Mildred, held twelve and a half shares of Aermotor corporate shares. She was represented by William H. Hamilton, who was named second vice president in January 1945.[40] By mid-1946, Walker's age and health started catching up with him, and he started cutting back his duties after fifty-three consecutive years with the company. He also agreed to reduce his annual salary to $10,000 and subsequently raised the monthly incomes of other up-and-coming managers, including Melvin T. Jensen and Donald H. Anderson.[41] On January 28, 1947, Walker announced he was no longer able to perform his duties, and the job of president was turned over to Daniel Scholes, who by now held 115 corporate shares and earned an annual salary of $15,000. Hamilton, now vice president, was granted a salary of $10,000. Anderson, second vice president, and Jensen, treasurer, each earned $8,400 a year, followed by longtime secretary Fraser at $6,000.[42] Walker remained a board director of the La Verne Noyes Estate, which controlled 2,660 corporate shares for the company's university and college scholarship program, but died shortly after the transition on June 11, 1947. His widow, Geneva, received $10,000 as pension for her husband's service to the company.[43] In October that same year, Fraser resigned, citing health problems, and the role of both treasurer and secretary was filled by Jensen.[44] Anderson resigned from his post as second vice president and left the company on May 19, 1948.[45]

The changes within Aermotor's workforce after World War II were even more dramatic. As war work ended and the military veterans returned to their work stations from the battlefields of Europe and Asia, the company's highest-skilled blue collar positions, such as tool and die making, machining, and foundry work, were mostly dominated by older Europeans, predominantly from Poland, Germany,

Scandinavia, and the Baltics, which reflected the prewar neighborhoods that surrounded the Chicago plant. It was not uncommon within a small manufacturing outfit like Aermotor to find skilled workers who spoke very little English and used hand gestures to explain processes or make a point. Some of this may have been cultural stubbornness or a way to protect one's job within the company, but the variation in languages undoubtedly made things difficult for younger employees who only knew English to learn the company's operations effectively. Increasing numbers of African Americans and Latinos soon joined the company. Before the war, there were very few African Americans employed at Aermotor. "[W]hen I started there [in the mid-1930s], you could count on one hand the amount of black people that worked there," recalled Clarence Kleinke, a foundry supervisor.[46] Most hourly wage earners walked or took the commuter train to work since they lived close by, while their counterparts in management and supervisory roles, who made more money, could afford to live in the suburbs and drove or carpooled to the factory each day.

Working conditions and operations inside the walls of the Aermotor plant, even at the end of World War II, appeared to have changed little since the days when Noyes ran the company. Raw materials arrived at the factory by train or truck. Most of these supplies were handled manually, and even as late as 1949 two men with shovels unloaded railcars of foundry core sand into a wagon pulled by two horses.[47] Metals were cast, shaped by hammers, bending machines, or heavy punch presses, before the pieces were moved to galvanizing, grinding, assembly and painting, and eventually to packaging and shipping for either domestic or export customers. The men on the foundry and factory floor were constantly exposed to dangers, such as filthy air and chemicals, hot metal, whirling overhead pulleys and belts, and bone-crushing machines. The brick walls and wooden floors of the cavernous factory made it bone-chilling in the winter and stifling hot in the summer.

William Crist, who joined Aermotor as an apprentice in 1953 and worked in the Chicago factory until 1961, described a typical day for the men who labored in windmill assembly. Clad in denim coveralls or matched shirts and pants, the men punched in at a time clock and headed to their benches just before the eight o'clock factory whistle blew to start the workday. The men would take their four-wheel Aermotor-made handcarts to pick up windmill parts off racks and out of bins, which had been brought up to the second floor of the factory from the foundry department. Men responsible for the six-, eight-, and ten-foot windmills were expected to assemble between ten and twelve of a single-size unit each shift. Each man spread out his gear cases and related parts across a large steel bench. He would start by installing a single part, such as a counter pin and gears. Once that part had been installed across all the heads along the bench, he would begin the process all over again with the next part. Each man quietly tended to his own work. The only sounds heard were the banging and clanging of hammers against metal. At ten o'clock the whistle blew for a ten-minute break. At noon the men received a half-hour lunch. Most sat at their desks eating packed meals from home.

An Aermotor worker dips sections of wind wheels into a galvanizing bath. Aermotor's Chicago factory workforce had shifted from predominantly northern European descent to African American and Hispanic by the mid-1950s. From Aermotor Company, *The Great Strength of the Aermotor Wheel* (Chicago, ca. 1933), n.p.

They would then wash up and head back to their work stations by 12:30 p.m. The next ten-minute break came at two o'clock in the afternoon. The day was over at 4:30 p.m. If a man finished his work before that time, he could pull stock for the next day or sit quietly at his bench. He was not allowed to socialize with other workers on the floor. When the closing whistle blew, the men would file into their department's washroom. There they splashed themselves with water at long sinks, and many would change into clean clothes before heading home. On Fridays, factory floor employees received their pay in cash envelopes.[48]

Like many factory environments of the day, there tended to be inherent distrust between salaried management handling administration from an office and hourly wage-earning employees on the shop floor. It was alleged that Noyes in his day often strolled the plant floor near quitting time to ensure that his employees kept working until the whistle blew. There was also concern that foremen and supervisors served as spies for upper management, reporting mistakes that could lead to one being fired. The rank-and-file workers at Aermotor had had some form of union representation since the early 1900s. However, the union generally posed little threat to the windmill manufacturer. In 1901 employees who were members of the National Association of Machinists called for nine-hour workdays as opposed to the standard ten-hour day, but unlike their brethren in other factories across the country they did not demand the same level of pay for a nine-hour day as they made for ten hours and avoided walking off the job.[49] The labor unions became more organized and a force to be reckoned by companies after Congress passed the National Industrial Recovery Act in 1933, which guaranteed workers the right to organize and bargain collectively. This was shortly followed in 1935 with congressional passage of the Wagner Act, which said if a majority of employees in a company voted to be represented by a union, then that union became the bargaining agent for all. After World War II, a wave of anticommunism swept across the country, and those unions with former or current socialist political leanings fell out of favor in many industries. In March 1951, Aermotor's employees voted 255 to 10 to oust the "left wing" Farm Equipment Union Local 177 in favor of being represented by the Congress of Industrial Organizations–affiliated United Electrical, Radio and Machine Workers of America Local 1177.[50] The union struck at Aermotor two months later over a wage dispute. One picketer, Frank Sztanga, reportedly collapsed and died during the short-lived strike.[51] Still, most of the company's workers viewed the union as weak and ineffectual in the face of management. "They were afraid they were going to get fired. See, because you really didn't have anybody, the union was not very strong, and if they didn't like the way you looked or you combed your hair or so, they would say, they would can you," Crist said.[52]

The dominating figures among the senior management in the early 1950s were Daniel Scholes and William Hamilton. Scholes pressed the company to continue developing electric pumps to take advantage of the postwar housing boom.[53] Residents of many new suburbs relied on well water for their numerous household fixtures and appliances, including kitchen sinks, bathtubs, toilets, dishwashers, and automatic washing machines, and by the mid-1950s, Aermotor was offering a line of submersible pumps for single-family-home wells. The company's first submersible pump, introduced in 1953, was the brass model "G" line, which had brass bowl diffusers screwed together with a left-hand thread to prevent loosening from motor counter torque. Between 1954 and 1957, the company also sold its Budjet convertible jet pump and the Sumpmatic submersible sump pump. Aermotor highlighted the durability and ease of installation of these devices. Similarly, to keep up with the increasing population, farmers had to increase their crop, dairy, and meat production, which meant installing water pumps with larger and more efficient outputs. Here Aermotor responded with more powerful and bigger deep-well pumps.[54] Hamilton, on the other hand, was more interested in furthering development and sales of the company's line of specialized towers, such as those for observation, power transmission, radio and television antennae, sirens, church bells, and land surveying.[55] Aermotor continued to sell numerous observation towers to federal and state government agencies for forest fire control, but a new customer for these towers emerged in the 1950s—the US Air Force. By the late 1950s, the Air Force, which became an independent military branch in 1947 after more than thirty years of being part of the Army Air Corps, had acquired "hundreds" of these observation towers from Aermotor for its airfields across the country and around the world.[56]

In July 1953, after fifty-five years with the company, Daniel Scholes retired from Aermotor. While he continued as board member and shareholder of the company, the position of president went to William H. Hamilton. Melvin T. Jensen, secretary and treasurer, was also elected first vice president, and James R. Walsh was named second vice president.[57] By the late 1950s, however, it became increasingly apparent to Aermotor's senior management that the factory required an infusion of capital to modernize. Between April 1953 and December 1957, the company began to sell off tens of thousands of common shares and bonds, which it held in some of the biggest US manufacturers of the day, and distributed the

Daniel R. Scholes, president of Aermotor, ca. 1945. Scholes retired in 1953 after fifty-five years with the company. Courtesy Marcia Baldwin, Twin Lakes, Wisconsin.

proceeds to the Aermotor shareholders and the estate of La Verne Noyes for distribution to colleges and universities that were beneficiaries of the estate.[58] At the start of 1958, the primary shareholders of Aermotor were the La Verne Noyes Estate with 2,660 shares, with the rest divided among Daniel Sholes, Barbara Scholes, Hamilton, Jensen, Sylvia Comings, and Mildred Smith, for a total of 2,855. The board of directors for that year consisted of Howard A. Hazelton, Thomas A. Harwood, Elwin T. Jolliffe, Raymond J. Spaeth, and William W. Kerr, in addition to Hamilton and Jensen.[59] As president, Hamilton earned an annual salary of $25,000 in 1958, with Jensen, vice president, earning $23,000 and Walsh, second vice president, $14,000.[60] While its pump and tower products sold well, Aermotor appeared to lack the financial prowess to make the necessary capital investments to modernize its factory equipment and reduce manufacturing costs, and the mood in the company turned desperate. At its January 28, 1958, annual directors' meeting:

> Ways and means were discussed to improve the profit position as soon as possible and it was agreed that some very definite steps should be taken at once, particularly an immediate price increase on windmills and windmill towers; also steps be taken to reduce overhead expenses starting with the factory. The management is to prepare an immediate program for consolidation of departments, etc. to effect a cost savings. Discussions followed regarding advisability of purchasing castings outside thus enabling us to close the foundry. A study of this problem is also to be made immediately.[61]

The company's annual holdover inventory after sales posed a heavy burden. The company calculated that after its 1957 sales it had an excess inventory of $325,229.86 and a two-year holdover of inventory valued at $214,129.31. Aermotor considered closing its foundry and purchasing castings from a third-party supplier. Jensen noted at a directors' meeting in March 1958 that based on 1957 production, Aermotor could have bought castings at about 15.5 cents per pound, or two cents a pound less than the cost of 17.43 cents per pound to make them in house.[62] The company also aimed to sell off all remaining parts in inventory for windmills produced prior to its Model 702.[63] That spring Aermotor sent its export manager, James E. Fetters, to South America to explore the possibility of licensing windmill production in this export territory.[64]

But all of these internal plans came to a halt when in early June 1958 Aermotor announced that it had been acquired by Motor Products Corporation of Detroit for an undisclosed sum. It was speculated that Aermotor sold the business to Motor Products for $1.5 million. The Chicago press reported that Aermotor grossed about $4 million in annual revenue and operated at a profit, but this appeared inconsistent with the company's scramble in 1957 and 1958 to cut operations costs.[65] Aermotor's seven remaining shareholders were bought out, and the La Verne Noyes Trust was separated from the sale with $285,500 to support the university and college scholarship program.[66] Retired president Scholes and his wife, Barbara, had returned by then to Illinois after a few years in Wisconsin renovating a lakeside cottage.[67] Scholes died in La Grange, Illinois, in June 1966 at age eighty-three.[68]

Motor Products had once supplied all the chrome trim for Chrysler automobiles but lost the contract by the late 1950s, requiring it to redefine its business. The company went on an acquisition spree in 1958 and 1959, snapping up truck and tractor winch makers Braden Winch and Arrow Gear, industrial warehouse firm Bond Steel and Storage, electronics research and product developer Trionics, and parking meter manufacturer Duncan, in addition to Aermotor.[69] Although they reported to corporate officers in Detroit, the acquired companies maintained their individual brands and were operated as independent divisions within Motor Products. With the exception of Hamilton, who left the company when Motor Products took over, the existing Aermotor management initially remained intact, with Melvin T. Jensen as vice president and general manager, William T. Milton as vice president of engineering, Robert A. MacDonald as factory manager, Wendell C. Dean, sales manager, James E. Fetters, assistant sales manager and export manager, and Carl Cue, controller.[70]

Motor Products immediately ordered Aermotor to discontinue manufacture and sales of outdated pumps and products and to expand and improve its electric pump manufacturing. By the late 1950s, windmill sales had leveled off, and the business for electric pumps had become highly competitive among manufacturers as more homeowners and small businesses, such as motels, service stations, and restaurants, moved to suburbia. "Anticipating additional increases in sales, we are fully conveyorizing our production lines so that we may handle much greater pump volume," Motor Products said in 1960. The company added, "Our present marketing plans call for intensification of our distribution program for Aermotor pumps by the establishment of additional dealer outlets in our existing marketing areas as well as in territories in which we have not been previously represented.

We are stepping up our market promotion activities to give further support to our dealer efforts and in order to attract new dealers to the Aermotor lines."[71] Included with this initiative was a new product packaging program, "up-to-minute revisions of catalogues," new decals, and the start of a new payment plan to dealers. During the first year under Motor Products, Aermotor launched its MJ multistage jet pumps with two and four stages and with motor ratings of one-half to two horsepower.[72] This was immediately followed by the LJ series of electric pumps, which were designed for shallow-well applications such as small homes in coastal and southern states as well as northern resort areas.[73] However, it was Aermotor's Model S submersible pump, introduced in 1961, that made the company a significant competitor in the domestic pump market. These pumps, which featured plastic parts for the outside walls, were simpler to manufacture and allowed Aermotor to double its submersible pump sales within a year of the Model S introduction.[74] The company showed a 22 percent increase in total electric pump sales from mid-1960 to mid-1961, and during the last quarter of the 1961 fiscal year (ending June 30, 1961), Aermotor sold more electric pumps than in any other quarter in its history.[75] It backed its electric pumps with a five-year unconditional warranty, which it claimed was "a promise no other maker of water pumps has dared to make." The reason for the lengthier warranty was Aermotor's success at resolving the long-running industry problem of sand and other abrasives entering submerged pumps.[76] Aermotor also added to its electric pump sales by distributing purification treatment equipment made by Chicago-based Everpure, Inc. The system removed iron and sulfur and helped prevent acid corrosion in copper plumbing.[77] In fiscal year 1962 (ending June 30, 1962), Aermotor recorded its highest sales during its seventy-four-year history; however, the parent corporation did not break out the division's share of the total operating profit for that year.[78] The Aermotor division began marketing a six-inch submersible pump, purchased from Gold Crown, a manufacturer in Lubbock, Texas.[79]

New windmill sales continued to drop off in the early 1960s. Motor Products (renamed Nautic Corporation in 1960, with its headquarters relocated to New York City in 1961) nevertheless still stated that it considered the windmill replacement market "a secure and profitable source of business in the future."[80] In the early 1960s there were about a dozen firms left manufacturing or selling windmills and their parts. In addition to Aermotor, some of the other notable companies remaining in the industry included Dempster Mill Manufacturing Company of Beatrice, Nebraska; Fairbury Pipe and Supply Company in Fairbury, Nebraska; Baker Manufacturing Company of Evansville, Wisconsin; and Heller-Aller Company in Napoleon, Ohio.[81]

In July 1963, Nautec made the decision to close the Aermotor Chicago plant after nearly seventy-five years of manufacturing. The strategy included integrating windmill, electric pump, and tower production with the Braden Winch Division in Broken Arrow, Oklahoma, a suburb of Tulsa. To accommodate the necessary move of machinery and materials, Nautec built a 90,000-square-foot addition to the existing 100,000-square-foot Braden building. The construction was financed

Well drillers install an Aermotor electric submersible pump. From Braden-Aermotor Corporation, *Annual Report 1968 for the Fiscal Year Ended June 30, 1968* (Broken Arrow, OK, 1968), [6].

The Braden Winch lathe line at Broken Arrow, Oklahoma, ca. 1958. Courtesy PACCAR Winch Division, Broken Arrow, Oklahoma.

by a municipal trust set up by the city of Broken Arrow. The senior management and sales staff in Chicago were first moved temporarily to an office at nearby Elk Grove Village, Illinois. The move, however, did not include Aermotor's Chicago factory workers. Instead, Nautec decided to hire and train about 200 employees from the Tulsa–Broken Arrow area to take over the former Chicago jobs.[82] The Chicago plant property was sold for redevelopment purposes.

The move of Aermotor to Braden's plant in Broken Arrow was nearly complete by the summer of 1964. Nautec claimed, "By integrating their product facilities and sharing much of the machinery—some of it automated—both Divisions expect to make substantial savings and increase profits."[83] Only three Aermotor managers from Chicago were moved to Broken Arrow in May 1965: Wendell C. Dean, who served as Aermotor's marketing director from 1952 to 1965, was named division manager, while Stanley Anderson was put in charge of overseeing branch offices and distributors, and James Fetters became manager of advertising and exports.[84] A film was produced to explain to Braden employees how Aermotor products were made, which conveyed the message that "even a secretary could

A worker checks jet pumps on a conveyor line at the testing tanks of the Aermotor plant in Chicago. From Nautec Corporation, *Annual Report Fiscal Year ended June 30, 1961* (New York, 1961), [11].

do the work," as one former Braden-Aermotor employee recalled.[85] The merger of the two companies, with their distinct ways of doing things, proved disruptive at first. Many Braden employees viewed Aermotor with contempt and as an interloper. The Braden factory ran into problems when attempting to integrate the old machinery from Aermotor's Chicago plant with its own, mainly converting them from belt-driven to plugged-in electric power.[86] Robert W. Schuetz, Braden president, encapsulated the integration difficulties in a May 1965 employee newsletter: "Aermotor products was [sic] just a name to most of us a year ago. We all have experienced many frustrations, many set backs and many growing pains to find out what and how to manufacture this line. . . . We are not out of the woods yet, but all indications seem to say we are going to succeed in good products in sufficient quantities to meet customer demands."[87]

More change was yet to come in 1965. On August 31, the stockholders of Nautec approved a plan presented by the corporate board earlier in the month to spin off the Braden Winch and the Aermotor Water Systems divisions into a separate company to be known as Braden-Aermotor Corporation. On September 17, Nautec distributed stock of the new corporation to its shareholders on the basis of one share of Braden-Aermotor for each Nautec share. This left only two divisions under Nautec, namely Bertram Yacht and Ivy Hill Lithograph. Dean was named vice president and general manager of the Aermotor Water Systems division of the newly formed Braden-Aermotor Corporation.[88]

Until 1965, Aermotor's branch houses had experienced little impact from the shift in corporate ownership to Nautec. Just the names of the letterheads, invoices, and paychecks had changed. In December 1965, the company moved its Illinois sales force from Elk Grove Village to a new shared Braden-Aermotor office at Downers Grove.[89] In addition to this branch, Aermotor by the mid-1960s had branches in Amarillo and Dallas, Texas; Cordele, Georgia; Harrisburg, Pennsylvania; Kansas City, Missouri; Minneapolis, Minnesota; and Omaha, Nebraska. The company also had distributors in seventeen cities to serve areas not covered by the branch houses.[90] Although the new Braden-Aermotor Corporation claimed to meet delivery schedules to its branches in 1964 and 1965, "in spite of some material shortages and supplier price increases influenced by the impact of the Viet Nam conflict upon many manufacturing industries," the closure of the Chicago factory, relocation of equipment, and training a new workforce at the plant at Broken Arrow took its toll on quality and production.[91] Marvin Isvik, the Omaha branch manager, recalled:

> The largest problem was with the windmills. They never were able to make the mills at the same level. For example, they had a machine in Chicago, made there in the factory, a six-spindle machine to drill holes in the [wheel] hub and tap the threads in them. It did them so that the wheel arms would all screw in just right. At Broken Arrow they did not know how to operate this machine. They set it in the yard and allowed it to rust while they drilled and tapped the holes by hand. How could they line up right?[92]

With none on the premises, Braden-Aermotor used other foundries in the region to cast windmill gearboxes and other parts, but limited-quantity runs made it difficult to obtain certain parts in a timely manner.[93]

Although the company continued to manufacture and sell its full line of six-, eight-, ten-, twelve-, fourteen-, sixteen-, and twenty-foot windmills, the tail stencil was simplified to read just "AERMOTOR" in an aqua-blue color. The new corporate name, "Aermotor Water Systems, a Division of Braden-Aermotor Corporation," was deployed on brochures and letterheads, along with a simple logo of a capital A centered within an oval that was split by a white background, symbolizing air, and blue water.[94] Interestingly, orders taken at Broken Arrow for repair parts still used Noyes-era telegraphic names, such as "Abactor" for an assembled eight-foot motor for a three-post tower and "Dacapo" for the twelve-foot assembled motor for the four-post tower.[95] Parts for earlier Aermotors, like the 602, were no longer available through the company, and customers were advised to buy new 702 motors.[96]

Even by the mid-1960s, windmills remained important machines for pumping water in many parts of the developing world where electricity was still scarce. Aermotor continued to export windmills abroad, but in certain markets import tariffs or governments' refusal to issue import permits made it nearly impossible for them to succeed. For the past half century, some importers-turned-competitors attempted to reproduce the oil-bath-style Aermotors both in mechanics and appearance. Yet copying the Aermotor windmill was often easier said than done,

Staff and salesmen gather for a dealer appreciation day at the Cordele, Georgia, branch in the late 1960s. From left to right are Robert Culbertson, Buddy Clements, Bill Wenger, Jeannie Adkins, Lenora Cape, Harry Helms, and Harold Willingham. Author's collection.

since the machine involved many lightweight and complex castings, a frustration for foundries. For example, it required a seven-part mold to produce one Aermotor hub.[97] There was also Aermotor's use of nonstandard screw thread sizes, such as 13/32-inch rather than 3/8-inch, which were made from its own proprietary taps and dies and further complicated attempts to replicate the windmill abroad.[98]

One of those to build a near duplicate of the old Aermotor 602 oil-bath windmill gearbox was Metters Ltd. of Australia, but the mill retained a bell-shaped hub.[99] Another Australian company, John Danks & Company in Melbourne and Sydney, was an early importer of Aermotor windmills, but out of frustration with the rising cost to import and knowledge of Aermotor's introduction of the 702 in 1933, the company about two years later started manufacturing a replica 602 windmill under the name Coo-ee.[100] W. D. Moore & Company of O'Connor in western Australia, which bought US-made Aermotors from Danks prior to 1935, became a primary importer of Aermotor mills from 1935 to 1945. However, the cost for the imported six-foot Aermotor 702 motor (without tower) in 1937 was A$10 more than the comparable Danks 602, and the eight-foot 702 motor was A$47, compared with A$31 for the equivalent Danks mill. After World War II, Frederick Moore and Douglas Joslin decided that W. D. Moore should manufacture a version of the 602 Aermotor because, through experience in selling parts, they believed the 602 held up better than the Aermotor 702.[101] The company made six-, eight-, ten-, and twelve-foot Model 602 windmills, using the name "Aermotor" on its tail vanes and sales literature. However, it's unclear from the historic record whether W. D. Moore & Company had a license from Aermotor in Chicago to use the commercial name. The company manufactured these windmills from 1946 to 2000.[102]

The experience was similar in South Africa, another large windmill market, where local manufacturers lobbied their national government during the 1930s to curtail imports of foreign-made windmills. One of the country's largest importers of Aermotors during the first half of the twentieth century was a Cape Town agriculture implement dealer, L. Paul Andrag, and his five sons, Martin, Alfred, Paul, Walter, and Hellmut. The company, P. Andrag & Sons, imported the Aermotor motors but manufactured its own towers. When it became apparent in 1963 that Aermotor was closing its Chicago plant and moving production to Broken Arrow, Andrag saw its chance to manufacture Aermotors in South Africa by purchasing a swath of used factory equipment from Aermotor. Martin and Hellmut Andrag and engineer Gerd Rose traveled to Chicago in May 1963 to meet with Aermotor's secretary and treasurer, Melvin Jensen, to negotiate the machinery purchase.[103] Frustration and misunderstanding between the two companies arose quickly. In a letter dated April 4, 1964, Hellmut Andrag complained to his brother Martin about Aermotor's failure to keep its side of the bargain:

> [T]he negotiations with Jensen are difficult, since when he fails to keep promises about machines previously agreed upon, he would simply say "it is beyond my control" or "unavoidable delays." On the other hand, we must make the best out of the situation and we rely on his assistance for the removal and packaging of the machines. The other side is that we now have to purchase a number of new machine tools, which will be more suitable than the old ones from Aermotor. Perhaps that is also a "blessing in disguise." Our problem is of course to finance the new factory installation, which problem now becomes bigger.[104]

Andrag was initially promised sixty-three machines from the Chicago plant at a value of $1,604. But that number was reduced to thirty-eight machines after a disagreement with Braden. Andrag ended up paying $1,000 for the machinery, which included four eccentric presses, a horizontal "bulldozer" press for making blades, a line punch, two large lathes, a surface grinder and planer, and foundry patterns for the eight-foot Aermotor gearbox and sundry parts. While book value for the machines was inexpensive, Andrag paid nearly $22,000 in 1964 to crate and ship them back to South Africa.[105] Andrag manufactured only the eight-foot windmill, called the "ANDRAG AERMOTOR" on the tail, at its Stikland factory for sale in South Africa. It made about 1,200 windmills from 1965 to 1973, when production ended. (Only one original Aermotor machine—the line punch—is still used by Andrag, now known as Agrico, a large South African agricultural implement manufacturer, at its Lichtenburg center pivot factory.)[106]

With the move of production to Broken Arrow completed at the end of 1965, Aermotor projected manufacturing about 5,000 windmills for the market.[107] Yet Braden-Aermotor Corporation saw windmills as a small piece of its overall objective to carve out an expanded position in the pump market. It noted that the push of population into rural areas resulted in an $80 million market for pumps in 1966, and the company saw new markets, such as water purification, pollution control, and desalination systems, as the way forward.[108] One of its popular submersible pumps was the SB model. This four-inch pump had threaded cast iron bowls and Lexan impellers. This was shortly followed by the SA revision of the model S sub-

P. Andrag & Sons of Belleville, South Africa, manufactured a Model 702 variant, as depicted in this brochure image (ca. 1964), for the southern Africa market from 1964 to 1973. Courtesy Aermotor Windmills, Aberdeen, South Africa.

mersible pump, which had a locked impeller stack. Then the company introduced its AJ convertible jet pump line, which allowed it to reduce production costs and inventory by consolidating several different pump models into one new series.[109] Braden-Aermotor also continued to press its line of Bilby and observation towers to government agencies, commercial towers for televisions and radio stations, and home towers for citizen band and shortwave radios. Its three-post antenna towers could be purchased in heights of 20 to 250 feet. Braden-Aermotor in 1966 introduced a 257-foot flare tower, used to burn off gas in oil fields. However, not many were made and sold by the company mostly due to the cost of materials and third-party galvanizing, which made the flare towers overpriced for the market.[110]

Securing reliable and cost-effective domestic sources for castings and galvanizing windmill parts became increasingly difficult for Braden-Aermotor in Broken Arrow. By the mid-1960s, the company had licensed its 702 windmill production to manufacturers in Australia, South Africa, and Argentina to cover those markets more efficiently than attempting direct exports from the United States. Agar, Cross & Co., a large Argentine importer of Aermotor windmills from 1892 to 1952, came under pressure to find local foundries and machine shops to manufacture Aermotors for the country's market. Agar arranged for Fabrica de Implementos Agricoles S.A. (FIASA) of Buenos Aires to take over its manufacturing activities in 1963. By 1965 FIASA had on its Bragado factory grounds a gray iron foundry and machine shop. The company also added an electric hot-dip galvanizing plant, enlarged machine and assembly space, sawmill and box making, and equipment to manufacture steel structures, including windmill and transmission towers. In 1966 Braden-Aermotor made the decision to grant FIASA the licenses to manufacture its 702 mill. However, unlike previous licenses, FIASA purchased from Braden-Aermotor most of the patterns, tooling, and equipment to fabricate the windmills. The Aermotor licensing agreement gave FIASA full rights to manufacture and sell the Model 702 windmill under the Aermotor name and other names that FIASA might select, taking Aermotor out of US windmill manufacturing altogether and requiring it to import all its branded windmills from Argentina by 1968. Even a majority of Aermotor's overseas windmill sales for the next fifteen years were managed by FIASA.[111]

One of the first FIASA-manufactured eight-foot Aermotor windmills arrived in a wooden crate and was sent to the Omaha branch office where manager Marvin Isvik, salesman LaVerne Roberts, and Richard Kuhl assembled the mill on a stub tower on October 7, 1966. Isvik recorded their observations and experiences in a letter to Braden-Aermotor in Broken Arrow on November 17. They first noted that putting together the wind wheel pieces was "a time consuming task, but the parts fit together well" and were about the same weight as the former Chicago-made wheel sections. They noticed the teeth of the large gears were "considerably heavier and thicker" and the bearing bar was not properly filed to allow it be pulled tight against the mill frame. The men found that the oil collector was "not bent as we bend it on assembly, but the clearance between collector and inside surface seems to be close." They also highlighted that a half-dozen other parts were either slightly longer or shorter.[112] Isvik concluded:

> The over-all appearance of the mill is good. The wheel does not have quite as much shine and the bolts used to assemble the wheel detract from the appearance. The parts fit well. The gears are well meshed and there is no excess clearance. The paint color is more nearly the color we used a few years ago.[113]

In addition to manufacturing the eight-foot Aermotor, FIASA produced the six-, ten-, twelve-, fourteen-, and sixteen-foot mills. Before the move of production to FIASA, Braden-Aermotor had stopped manufacturing the twenty-foot windmills. The company had about five crates of twenty-foot windmills in Broken Arrow that came from the former Chicago plant, but they were never machined. Most of the wooden molds for the twenty-foot windmill part castings were by then in a state of deterioration. Braden-Aermotor's remaining twenty-foot windmill parts were sold to a large well-drilling firm in Roswell, New Mexico, in 1967, where they were quickly snapped up by West Texas and New Mexico ranchers who still used these large mills.[114]

Aermotor had advertised extensively in trade publications for nearly eighty years, but it was not a big proponent of customer giveaways, which were viewed mostly as frivolous cost. The company occasionally produced pens, pencils, and pocket knives with advertising printed on them for branches to hand out to customers. However, one of its standout gifts came in 1967 with the initial order of 500 small metal windmill kits from Aero Manufacturing Company in Lincoln, Nebraska. These models included 137 pieces of thin galvanized metal strips for the tower, four angled for the corner posts; a cut and pressed galvanized wind wheel with a lubricant-impregnated bearing in the center; and a cast head to which a galvanized tail vane was attached. On the vane in red was the Aermotor logo, plus "AERMOTOR Water Systems[,] Since 1888[,] Broken Arrow, Oklahoma." Assembled, the windmill stood sixteen-and-three-quarter inches tall. The branches were

Ron Garrett, Ron Seibert, and Hershel Hicks (in the front row) head a regular sales training seminar at Braden-Aermotor Corporation in Broken Arrow, Oklahoma, ca. 1967. Courtesy PACCAR Winch Division, Broken Arrow, Oklahoma.

encouraged to sell them to Aermotor dealers for $4.75 apiece, with volume discounts ranging from $4.45 each for orders of eleven to twenty-five, up to $3.85 for fifty-one or more. Branches were also encouraged to display them around their offices. Aermotor called them "[a]n attractive conversation piece in stores, offices, homes and schools." This type of advertising also helped to detract from the fact that Aermotor windmills were no longer made in the United States. From late 1967 to December 12, 1974, Aermotor branches ordered 21,516 of these windmill kits with the "Broken Arrow" tail, according to Aero Manufacturing.[115]

While Aermotor appeared to be on the upswing as it settled into its Broken Arrow location, Braden-Aermotor Corporation in 1968 was about to undergo a significant shakeup within its senior management. Seymour Heller, who had overseen the integration of both Braden Winch and Aermotor over the past three years as chairman, resigned on December 19, 1967. The board named Paul Hershey to succeed him. Robert Schuetz, for now, remained president, and Wendell Dean continued as assistant secretary in charge of the Aermotor division through 1968. In October 1968, Hershey and the board installed James E. Beebe as the new president, and the company changed its name to Braden Industries, dropping the Aermotor name from its corporate identity. For the sake of its windmill, electric pump, and tower sales, it kept the Aermotor name and A-style logo. That same year Braden Industries claimed to receive a steady flow of windmill heads from Argentina at its Broken Arrow plant, where it continued to manufacture the relevant towers. The company also planned to realize improved profits from windmill sales within its Aermotor division, since the production of the mills was now effectively outsourced.[116] "Greater emphasis is being placed on the manufacture and distribution of steel towers during the coming year and marketing efforts will center on the petroleum and petrochemical industries as well as state and federal government agencies in the areas where sales and profits can be increased," the company said in its 1969 annual report.[117]

However, it was understood that Beebe was hired to shake up the ranks inside Braden Industries and focus on improving the company's beleaguered financial performance. Shortly after the release of the company's 1969 annual financial report, Beebe terminated Wendell Dean as head of the Aermotor division. Dean, who had joined Aermotor in Chicago in 1952 as sales manager, was often the public face for Aermotor in Broken Arrow and was not afraid to speak his views and opinions to senior Braden management. He also promoted collegiality among the Aermotor managers, salesmen, and factory workers, including inviting them to his home for social events.[118] After Dean's departure, the Aermotor division at Broken Arrow consisted of James E. Fetters, named manager of marketing, and Stanley A. Anderson, who was made branch and product coordinator. The division continued to support eight branches across the country.[119]

Even the tower business, long a hallmark of Aermotor's engineering prowess, was being phased out at Braden Industries in 1969, with Beebe citing "the limited profit potential and the need to concentrate manufacturing efforts into more productive areas."[120] At the end of the decade, the manufacture of Aermotor towers was awarded to a small company in Enid, Oklahoma, about two and a half hours northwest of Broken Arrow. The company, Log-Master, was started in the early 1960s by former Phillips University economics professor Robert R. Nigh, who started out by manufacturing downhole water well electronic monitoring equipment for water exploration in the US Southwest and developing countries. Along with Log-Master, Nigh operated Industrial Iron Works, also in Enid, to make oil field tanks and fire escapes in addition to handling sheet-metal work. From Braden, Industrial Iron acquired the tooling and machinery to produce an array of Aermotor towers, including those for windmills and the observation and Bilby towers used by the federal government.[121]

Overall, Beebe showed little interest in boosting Aermotor's windmill and tower business. He believed the company's success heading into the 1970s would be based on industrial winch and electric pump technologies, even though Aermotor was still ranked the world's top producer of windmills. In a talk before the Tulsa Society of Investment Analysts on November 19, 1969, Beebe warned of the water-pumping windmill's demise, stating, "[T]o be realistic about it . . . we recognize that the future for windmills is, in fact, about like it was for blacksmiths and buggy whips . . . not very bright. A shrinking rural population, and spreading electrification, will ultimately spell the end of the windmill market."[122]

CHAPTER SEVEN

Saving an Icon

The world's most powerful post–World War II capitalist nation, the United States, began to show cracks in its economy at the start of the 1970s. The Vietnam War proved costly, and the country's manufacturing and export dominance came under threat from new players in Asia, namely Japan, Hong Kong, South Korea, Singapore, and Taiwan, with their abundance of cheaper but highly skilled labor. Inflation and unemployment in the United States rose in tandem, along with interest rates, leading to a phenomenon that economists at the time called "stagflation." Political turmoil, both domestically and overseas, also took its toll on the US economy and the American people. Bloody fighting was on the rise across much of the developing world. In 1973 Americans felt the pain of having their petroleum supply severely curtailed when the Arab nations imposed an oil embargo on the United States, Europe, and other nations that backed Israel. President Richard Nixon also resigned in August 1974 after the Watergate scandal. Americans' optimism, which punctuated much of the 1950s and 1960s, plummeted.

By 1970 Aermotor's business domestically was largely focused on electric domestic water well pumps for homes, farms, and small businesses. The children of World War II veterans continued to purchase homes in suburbs that were not tied to public water mains, but instead sourced water from drilled wells on their properties. These required varying types of pumps depending on the depth of the wells. With the exception of some remote farmers and ranchers across the Great Plains and arid western states, in addition to the pockets of austere Amish and Mennonite communities in the Northeast, the days of the water-pumping windmills appeared to be numbered in an age of electrical dependence. In 1972 it was estimated that electric pumps outsold windmills by two hundred to one, and the choice of mills was limited to Aermotor, based in Broken Arrow, Oklahoma; Dempster Mill Manufacturing Company of Beatrice, Nebraska; and Heller-Aller Company in Napoleon, Ohio.[1] Ken McAdams wrote in a 1974 article, "Electric pumps now serve water to most farm and ranch homes. And the Aermotor Company, whose name is almost synonymous with windmill, has moved to Argentina, where, I suppose labor costs are less and there's a better market."[2] Aermotor's windmills came to the United States in wooden crates from the FIASA plant at Bragado, near Buenos Aires.

Internally at Aermotor, now defined as a division of Braden Industries, Inc., in Broken Arrow, there was plenty of strife between management and the labor union. From March 7 to May 29, 1970, hourly workers on the assembly line, who were dissatisfied with the failure to negotiate a new contract at the start of the year, struck. Executives, supervisors, and office personnel were called out from behind their desks to maintain partial production schedules and product shipments on the factory floor, while contract negotiations continued from April to May, when a new three-year labor contract was signed. James E. Beebe, president of Braden Industries, attempted to assure shareholders that, although the company's sales were down 6 percent, net income during the first six months was up 126 percent to $120,000, compared with the same period in 1969. Demand for water systems products during the first half of 1970 remained "brisk." "Management is pursuing many internal programs for cost reduction, greater liquidity and increased productivity, as well as instituting more comprehensive marketing methods, the beneficial results of which should become dramatically apparent during the last half of this year," Beebe said.[3] However, recovery that year remained sluggish with the lingering effects from the twelve-week strike and a slowing national economy being felt. The company commented that its sales by the third quarter of 1970 were down $1.25 million, and "the thing we need most to do now is to tighten our belts, bow our backs and do our best to recover where we can."[4]

The Aermotor division introduced several new lines of jet and submersible pumps during the early 1970s, including the DCJ convertible jet pump in 1970; the DSJ shallow-well jet pump, which had many interchangeable parts with the DCJ, in 1971; the DMJ two-stage jet pump of 1972; the revised model S submersible pump, known as the SD, in 1973; and the SE 30 and 50 plastic high-capacity submersible pumps.[5] At the same time, Braden Industries had introduced a number of new winches. Factory space at Broken Arrow was tightening, and there was little room to expand physically, since the land next to the plant had been donated earlier to the local high school for athletic fields. Aermotor also continued to struggle to fit in at Braden Industries. Although it had been nearly a decade since Aermotor relocated its operations from Chicago to Broken Arrow, the company was still viewed as "the second child," as one senior manager recalled.[6]

In 1974 Braden Industries was enticed to relocate its Aermotor division to Conway, Arkansas, where it gained access via tax breaks and other state incentives to set up pump production in a brand-new 49,000-square-foot building.[7] Conway, located about thirty miles north of Little Rock off Interstate 40, was founded in 1875 by Asa P. Robinson, who had overseen the construction of a railroad line through the area four years earlier. The central Arkansas town grew and by the mid-1960s hosted several large employers, including bus manufacturer IC Corporation, Baldwin Piano Company, and Nabholz Construction.[8] Trucks transported the equipment for manufacturing electric submersible and jet water well pumps from Broken Arrow to Conway over a period of months, along with about ten

The Aermotor Division of Braden Industries relocated to Conway, Arkansas, in 1974, enticed to the location by tax incentives and the ability to set up pump production in a new 49,000-square-foot building. Photograph by Dick Wegehoft, courtesy Susan Scott, Little Rock, Arkansas.

people, including Aermotor's heads of engineering and accounting. This left room to hire and train many new factory employees from the local community, without connection to the machinists' union established at the Broken Arrow plant.[9] Aermotor officially occupied its new plant in Conway on December 2, 1974.[10]

The move to Conway also allowed Aermotor to implement a number of new manufacturing processes for its burgeoning pump business. For instance, it introduced an electrostatic epoxy powder coating process to replace its earlier method of painting exterior product surfaces. To make the epoxy powder adhere, castings were given a positive electric charge before being introduced to a chamber of negatively charged epoxy particles suspended in the air. The attraction of the opposite-charged materials resulted in an even coating of epoxy that was superior to traditional paint spraying. The coated materials then entered an oven of 400 to 450 degrees Fahrenheit to melt and cure the epoxy, creating a barrier to oxidation. At the new plant Aermotor also used a device that placed castings under intense pressure to detect any cracks or leaks that otherwise might go unnoticed during regular visual inspections. The company even deployed a special machine to ensure the straightness of pump shafts down to five one-thousandths of an inch. Lastly, Aermotor tested each unit under pumping conditions for capacity, head pressure, head shutoff pressure, and amperage pulled by the motor before final packaging and shipment to the branches.[11]

For the factory worker in Conway, Aermotor implemented a split workweek consisting of two groups of employees—one group worked Monday through Thursday and the other Tuesday through Friday, each with ten-hour shifts and receiving a three-day weekend. The company wrote about the new work schedules in 1976, stating:

Other than the obvious 3-day weekend, the plan offers multipurpose benefits for both Aermotor and the Aermotor Team. Each member of the Aermotor Team saves one day per week in driving to work (some members

At the Conway factory, each submersible pump was built from the motor up by one individual at his or her workstation. Photograph by Dick Wegehoft, courtesy Susan Scott, Little Rock, Arkansas.

Meeting of branch managers and field salesmen at Conway in 1976. Aermotor used these annual meetings to review the latest trends and developments in windmills and electric pumps. Courtesy Stanley A. Anderson, Skiatook, Oklahoma.

commute as much as 40 miles); if ever there is need for overtime, it can be accomplished on Monday and/or Friday, with a 2-day weekend maintained; and 10 additional hours output per week is achieved from machines which have heavy time requirements.[12]

Arkansas's right-to-work status promised Aermotor an end to financially devastating labor unrest that had periodically plagued its operations in Broken Arrow by giving employees the choice of whether or not to join a union rather than it being a mandatory part of their employment on the factory floor. Aermotor, for its part, kept the unions at bay by offering good wages and benefits as well as a generous bonus program. One part-time worker from the 1970s recalled the benefit of the Conway building having air conditioning, "which at the time was a rarity and with summer temperatures in the 90s made the plant very attractive to potential workers."[13]

Aermotor also continued its longtime policy of holding annual sales meetings, during which salesmen from the branches and field offices came to Conway for two days of discussions about market conditions, new products, and better ways to manage one's sales territory. Spouses were also invited to the gatherings and were offered tours of local attractions. Generally more than fifty people attended these annual events.[14] The meetings concluded with a dinner during which awards were given to the branch managers and field agents with the most sales. The reigning salesmen for Aermotor in the mid-1970s included Marvin Isvik, manager of the Omaha, Nebraska, branch, and LaVerne Roberts of the same location. LaVerne Roberts's sons, Jack, manager of the Fort Wayne, Indiana, branch, and Ray, salesman for the Rockford, Illinois, subbranch, both of whom joined the company in the mid-1970s, rapidly showed their ability to sell electric pumps and windmills.[15]

Since 1968, Aermotor's windmills had been transported from FIASA's factory in Buenos Aires by steamship to the port of New Orleans. There they would be transferred to barges for shipping up the Mississippi River to an inland canal that connected the river with Tulsa, Oklahoma. In Tulsa, Aermotor maintained a warehouse for its windmills. From this facility, complete windmills and parts were shipped to the company branches and distributors throughout the United States. The process, however, was extremely inefficient and frustrated Aermotor salesmen. In 1974 the company undertook a major overhaul of how it received its imported windmills. It closed the Tulsa facility and hired Beck Atkinson, a longtime Texas pump dealer, to manage a new windmill warehouse in Dallas. There he straightened out the inventory and made subsequent orders from Argentina. Windmill shipments that year started to arrive at the port of Houston and then were transferred to Dallas via truck or rail.[16]

By 1976 Aermotor had solidified its senior management team at Conway, which consisted of Jim Fetters, vice president of marketing; Stan Anderson, coordinator; Russ Jones, plant manager; Joe Allen, materials manager; Euell Johnson, manager of quality control; Dick Wegehoft, chief engineer; Bill Duwe, manager of research and development; and Andy Dunlap, national sales manager. Sales and product distribution were conducted from branch offices in Omaha; Fort Wayne; Dallas; Fresno, California; Norcross, Georgia; Minneapolis, Minnesota; and Harrisburg, Pennsylvania. A Conway sales branch was also set up a couple blocks from the manufacturing plant. A number of the company salesmen conducted business from small field or home offices. Aermotor had about twenty-five salesmen canvassing the country in the mid-1970s. The sales team prided itself on delivering products to its customers—mostly well drillers—the same or next day up to a maximum of two to three days, depending on the distance from the branch

James E. Fetters, who started with Aermotor in the mid-1950s and led the company's transition to licensing windmills to FIASA in Argentina by the late 1960s. From *VI Pipeline* 7, no. 3 (March 1977): 1.

warehouse to customers' locations.[17] In the middle of the decade, Aermotor had grown into one of the largest suppliers of submersible and jet pumps, especially to support residential construction in suburbs where there was a lack of connection to public water sources. Other suppliers of electric pump equipment took notice. One of those firms was Valley Industries of St. Louis, Missouri, a company with a forty-year history in manufacturing steel products, such as tubing, for the water and oil well drilling industries.

In 1972 Valley experienced a 38 percent increase in its water well casing sales due to rural home construction, which required increasingly deeper wells to reach fresh water. Before 1968, the average water well depth was 150 feet, but by the early 1970s, wells were going deeper. It was also estimated that by 1970 Americans would pump as much as 370 billion gallons of water per day from the ground. To go along with its robust well casing products and to offer a more complete package to well drillers, Valley in 1972 began testing its own four-inch-diameter submersible water well pumps. About one hundred were built and deployed by the end of that year.[18] Valley said "initial reaction from the field has been highly encouraging," and it expected "strong demand" for the pumps once full production took off, which was expected by mid-1973.[19] As many as 5,000 Valley submersible pumps had been sold to well drillers by the end of 1973, and the company subsequently set up the Valley Pump Division in St. Louis under general manager David Suey, shortly after named division president, to oversee this rapidly evolving business.[20] Valley's Waterwell Products Group, which included its well casings business, had earned a profit of $9.9 million, or 20 percent of the company's overall earnings, by the end of 1974. Within a year and a half of first introducing its submersible pump, Valley sold 17,000 units, which prompted it to introduce additional water well–related products, including a new high-powered turbine pump for farm irrigation, community water supply, and transporting coal slurry through pipelines.[21] With the US water well industry valued at a total of $250 million during this period, the company believed its prospects for increasing market share were great and announced in July 1975 that it would spend about $750,000 over the following twelve months to expand its pump manufacturing.[22] Valley also added to its portfolio by acquiring other firms. In November 1975, Valley acquired Hydraulic Products Company, which manufactured components for its turbine pumps, and Retco Manufacturing, Inc., which added line-shaft turbine pumps to its business. Both firms were based in Lubbock, Texas.[23] The following year Valley set its sights on acquiring the submersible pump line of the Aermotor division from Braden Industries, which it did in February 1977 for $2.1 million, and forecast that by the end of that year its $8 million in pump sales would increase by more than twofold. During the first nine months as part of Valley, Aermotor contributed more than $1 million in sales, or more than half of the company's total pump business.[24]

Aermotor's new Conway plant contained some of the most advanced pump testing and manufacturing equipment for the period. For pump research, the company had both production and research and development test pits, each measuring

An electric pump testing station at Aermotor's Conway plant during the mid-1970s. From *VI Pipeline* 7, no. 3 (March 1977): 2.

ten feet by ten feet by fifteen feet, and a mock test well with a six-inch-diameter casing with fixed water level. Aermotor had two groups within its engineering department, one responsible for redesigning existing pump products and another for developing new products. Between late 1975 and 1976, the company, still a division of Braden Industries, had contracted with three consulting firms to make recommendations on new pump markets, such as eight- and ten-inch submersible turbine pumps, end suction pumps, and submersible sewage pumps. When Valley took over Aermotor, it set out to boost investment in the research and development equipment and staff, warning, "If the present staff size is maintained the new products will be limited. . . . This will result in a very noticeable lull in a new product introduction in 1979 and 1980."[25]

Valley continued to focus on developing a fuller line of pumps, including larger industrial-size units, or turbine pumps. It relocated its engineering and sales activity of the Valley Pump Division to Conway in 1978. The company also opened a new factory in Madison, South Dakota, to build turbine pumps and better serve the agricultural market in the north-central and Plains states.[26] However, Valley quickly found out that sales of turbine pumps depended heavily on sometimes unpredictable factors such as grain prices and rainfall. The year before, its sales of these pumps fell due to falling grain prices and heavy rain in the central United States. Heavy pumps sales, on the other hand, realized continued demand from builders of municipal water, sewage plants, and industrial water systems.[27] In 1978 Valley took the bold step to manufacture components in house that it once purchased from third-party suppliers. For example, it installed plastic-molding equipment to make components for its submersible pumps. "Valley is now able to supply waterwell drillers with not only the steel or plastic casing but also the submersible pump to bring the water out of the ground, power cable, tanks and specialized drilling tools—everything but the drilling rig and the hole in the ground," the company said in 1978.[28] Before the end of the decade, Valley was the largest supplier in the country for steel water well casing and supplied an estimated 6 percent of the US market for submersible water well pumps. These pumps ranged in size from the 3- to 5-horsepower residential submersibles made in Conway to 500-horsepower fabricated submersible pumps at Valley's Madison plant for irrigation and municipalities and the Lubbock plant's large 500-horsepower line shaft turbine pumps for agriculture and industrial applications. In addition, the company's heavy-duty water-pumping turbines attracted interest from government buyers for large-scale irrigation and water supply projects in Libya, Saudi Arabia, and Nigeria during the late 1970s.[29]

The Aermotor acquisition also gave Valley a stake in windmills with wind wheel sizes ranging from six to sixteen feet in diameter, which continued to be imported from FIASA's factory in Buenos Aires. It was estimated that about 150,000 multibladed water-pumping windmills were pumping water throughout the United States in the mid-1970s.[30] Aermotor's traditional windmill sales reached about $1.3 million in 1976, with Dempster and Heller-Aller as its primary US competitors in this market. Valley claimed to sell three out of every four windmills used to pump water by the late 1970s.[31] The bulk of its windmills sales continued in the Midwest, Texas, and Southwestern states. In 1978 Valley rewarded its most prosperous branch manager for pump and windmill sales, thirty-seven-year Aermotor veteran Marvin Isvik in Omaha, Nebraska, with the construction of a new 12,000-square-foot facility, which included a four-door truck dock for the warehousing area. The warehouse offered a thirty-inch-diameter test pit with heavy electrical and hydraulic connections to test large submersible pumps. By the late 1970s, the Omaha branch served distributors and dealers in Nebraska, Iowa, eastern Idaho, South Dakota, Colorado, Wyoming, Montana, northern Missouri, and northern Kansas.[32]

However, the public's increased interest in the potential for windmills to generate electricity and reduce reliance on increasingly costly fossil fuels caught Valley's attention. Behind the scenes, Valley's senior management envisioned Aermotor developing an array of "wind energy" products for individual and commercial electric applications.[33] But by the mid-1970s the wind energy field in the United States and Europe was already crowded with upstart firms, as well as the reemergence of former wind turbine makers, such as Jacobs Wind Electric, Winpower, and Winco (formerly Wincharger Corporation), and rebuilders of earlier machines.[34] Aermotor was no stranger to electricity-generating windmills, having sold limited numbers of both 32-volt and 110-volt units for charging batteries from about 1917 to 1933, when they were discontinued.[35]

In June 1977, Aermotor drafted a proposal to build a one-kilowatt wind turbine at its Conway plant for testing at Rockwell International's Rocky Flats site in Golden, Colorado. The company's goal was to take a twelve-foot Aermotor windmill, less the pumping mechanism, and adapt it to turn a generator. The wind generator would use a traditional forty-foot-tall, four-post, galvanized windmill tower. The wind wheel hub shaft would drive a series of four gears, allowing the generator shaft to turn fifteen times per full rotation of the wind wheel. In minimum wind velocities of seven to ten miles per hour, Aermotor said the windmill should be capable of generating 120 volts for storage in a DC battery system. Aermotor estimated the cost to build the first prototype, including personnel, design, and machining, at $98,328, with a second one costing significantly less at $20,378. The projected retail cost per kilowatt for the machine, according to Aermotor, would be about $1,830. However, to make the windmill commercially viable, the company said it would require a minimum annual production of 1,000 units. Aermotor believed it could control the cost of the wind turbine by using imported FIASA parts and off-the-shelf components, such as the generator and related electronics. Jim Fetters, by now Valley's Aermotor division manager, assigned James Herr, a former McDonnell Aircraft Company test engineer who joined Aermotor at Conway in 1975, as manager of the test program. Also involved were Larry Dyer, plant superintendent, and Euell Johnson, quality control manager. Under the request for proposal to Rockwell International, Aermotor said the design effort, Phase I, would take six months, with the fabrication and testing, Phase II, lasting an additional sixteen months. The company exuded confidence in its ability to deliver an electricity-generating windmill, stating in its proposal, "Aermotor has an established marketing, distribution and service organization with experience in marketing Aermotor windmills; Aermotor can readily market this WTG (wind turbine generator) once development has been completed."[36] There is no record that the electricity-generating windmill passed the proposal stage.

In 1978 Valley was selected by McDonnell Aircraft Company of St. Louis to be part of a federal government-sponsored effort to design, build, and test the viability of medium-size electricity-generating windmills. McDonnell Aircraft's wind turbine design, called the Giromill (short for cyclogiro windmill), consisted of three vertical blades rotating around a central tower. The angles of the blades to the wind were individually modulated to receive the wind most efficiently and to produce an optimum level of electrical output. The Giromill's merry-go-round-style wind wheel allowed it to rotate whichever direction the wind blew. Several years earlier, McDonnell Aircraft performed analysis of twenty-one different Giromill configurations, including systems with theoretical outputs of 120, 500 and 1,500 kilowatts. It was estimated that a 500-kilowatt Giromill system set up at a site within a wind mean of twelve miles per hour could produce electric power at 4.05 cents per kilowatt hour, which was at the time 18 to 39 percent less than that of conventional windmills, according to McDonnell Aircraft. The company also subjected Giromill models to wind tunnel tests.[37] Rockwell International, project manager for the US Energy Research and Development Administration (a predecessor of the Department of Energy), was satisfied that the Giromill concept had potential and awarded McDonnell Aircraft a $1.5 million contract to field-test the 40-kilowatt prototype in August 1977.[38] William Duwe,

Valley's research and development director for the Aermotor Division, was relocated from Conway to McDonnell Aircraft's St. Louis headquarters to spend the next nine months designing the fixed tower and other structural components as part of the Giromill's Phase I development.[39] The result of Duwe's design work was a four-post tower—similar to those of Aermotor's windmill and fire towers—measuring sixty feet tall to support the vertical wind blade system, which had a diameter of fifty-eight feet and vertical blade height of forty-two feet. Each blade weighed about 500 pounds. The blade system, developed by McDonnell Aircraft engineers, rotated on a twenty-four-inch-diameter steel tube that was centered vertically inside the steel tower. The generator and electrical connections were located at the base of the tower. During the construction, called Phase II, the Giromill's steel components were fabricated at Valley's plant at Tallulah, Louisiana, with electrical and other components, including the blades, shipped to the site.[40] The prototype was assembled at Tallulah in late February 1980 and weighed a total of 32,983 pounds. It was set into motion for its first brief test on March 3, reaching the rotational design speed of 33.5 rotations per minute. Duwe noted in a Valley employee newsletter at the time, "We ran the Giromill 35 times between March 3 and March 14, and it performed as expected."[41] In April 1980, Rockwell officials inspected and approved the unit. It was then disassembled and shipped to Rocky Flats for re-erecting and long-term testing, starting July 3.[42]

Valley touted the Giromill with an artist rendering of the machine on the cover of its 1978 annual report and enthusiastically told shareholders that the 40-kilowatt windmill had the ability to supply electricity to sixteen homes or to replace the natural gas required to power large crop irrigation systems in the Midwest. It was also envisioned that a 40-kilowatt windmill could provide power to smaller isolated communities in mountainous areas, on islands, or along coastal regions with adequate winds. David Suey, president of the Valley Pump Division, said that the company planned to "grow right along with the development of new and more efficient windpower systems and applications." Valley secured a licensing agreement from McDonnell Aircraft that would have allowed it eventually to build Giromill-style windmills with generating capacities of up to 500 kilowatts, or enough electrical power for 200 homes.[43]

At the test site, however, problems with the Giromill quickly became apparent. Robert V. Brulle, McDonnell Aircraft's engineer in charge of the Giromill program, said the winds at Rocky Flats were inconsistent. It was not uncommon for there to be days of no wind, followed by sudden sustained breezes. Brulle noted that the Giromill survived a storm with sustained winds in excess of one hundred miles per hour. On the technical front, the windmill blade actuator suffered routine breakdowns, affecting test quality. On December 24, the Giromill's lightning rod broke off and damaged a blade and support arm, and then the day after Christmas the rotor on the stub tower blew over. The damages were repaired and the unit returned to operation on April 24, 1981. The routine setbacks led to the test being discontinued in the summer of 1982. Reflecting back on the Giromill more than twenty-five years later, Brulle wrote, "Although we showed an effi-

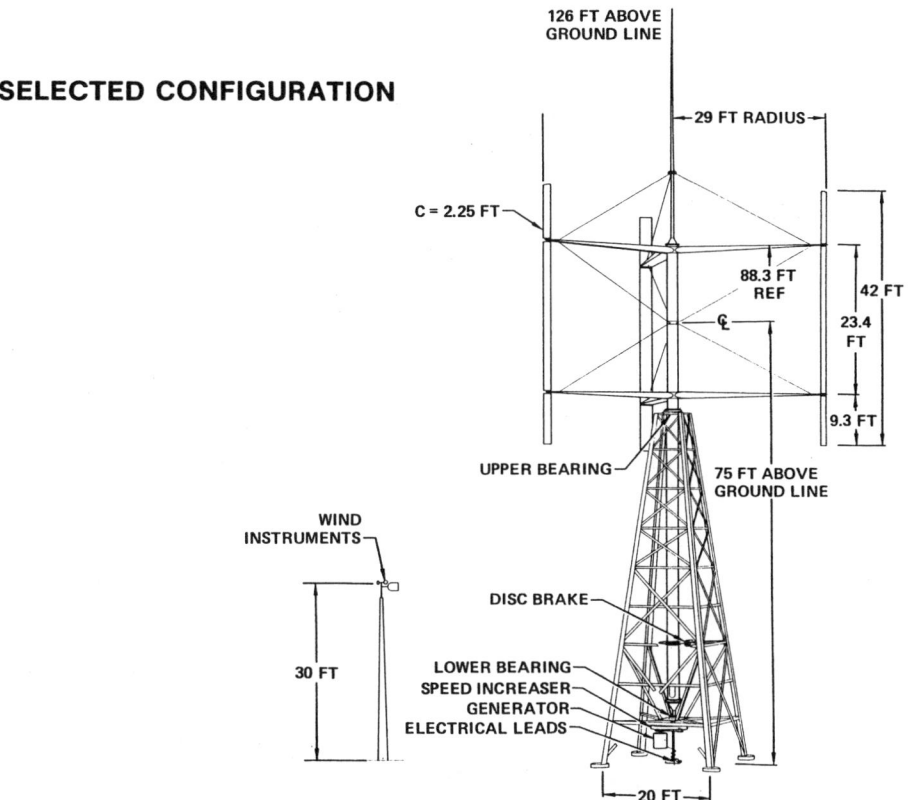

SELECTED CONFIGURATION

126 FT ABOVE GROUND LINE

29 FT RADIUS

C = 2.25 FT

88.3 FT REF

42 FT

23.4 FT

9.3 FT

UPPER BEARING

75 FT ABOVE GROUND LINE

WIND INSTRUMENTS

DISC BRAKE

30 FT

LOWER BEARING
SPEED INCREASER
GENERATOR
ELECTRICAL LEADS

20 FT

Design for the electricity-generating Giromill, a joint project of Valley Industries' Aermotor Division and McDonnell Aircraft Company in the late 1970s. The forty-kilowatt windmill was not put into commercial production. Courtesy David P. Suey Sr., Beatrice, Nebraska.

ciency rating higher than conventional horizontal-axis windmills, it did not balance out the complexity between the Giromill and a conventional windmill."[44] For Valley and its Aermotor Division, the Giromill's termination was a blow to their hopes for the windmill, and this was the last effort made by Valley to pursue an electricity-generating windmill.

Aermotor had not made any significant changes to its Model 702 water-pumping windmill line since it ended production of its twenty-foot-diameter windmill and sold off the inventory in 1967.[45] The rule of thumb for windmills was the larger the wind wheel, the deeper the pumps could operate or, for shallower wells, larger volumes of water could be more quickly brought to the surface. Many of the sixteen-foot and larger windmills were found on ranches and Indian reservations across West Texas, New Mexico, and Arizona to reach water sources deep underground, sometimes more than 1,000 feet below the surface. The Navajo tribe, which was a large buyer of fourteen- and sixteen-foot windmills, asked Aermotor in 1976 if there was any chance the company might return to manufacturing twenty-foot windmills, perhaps to meet an order of about seventy-five machines for delivery over a three-year period. The Navajos have the largest reservation in the United States, covering 27,673 square miles across a swath of northeastern Arizona, northwestern New Mexico, and portions of Utah and Colorado. Aermotor remarked at the time of the tribe's request that any reentry to

In late 1976, FIASA delivered to Aermotor in Conway what it claimed to be its one hundred thousandth Model 702 windmill produced for the world market. The windmill, with an eight-foot wind wheel, was erected at the plant for display. Photograph by Dick Wegehoft, courtesy Susan Scott, Little Rock, Arkansas.

manufacturing twenty-foot windmills would require a "legal arrangement, iron clad" to avoid any unexpected cancellation of the order, a move Aermotor was unwilling to make.[46] Yet the company felt compelled in 1978 to consider development of an eighteen-foot-diameter wind wheel. Aermotor conducted performance tests for an eighteen-foot windmill against a Challenge Company mill of similar size. (Challenge of Batavia, Illinois, had gone out of business shortly after World War II.) However, the windmill never got off the drawing board, as Aermotor determined there was not a sufficient market to justify the added expense of production at FIASA and importation of these units into the United States.[47]

In late 1976, Aermotor received its 100,000th windmill from FIASA, an eight-foot Model 702. The mill arrived at the port of Houston in a crate marked "MOLINO 100,000 FABRICADO POR FIASA." The mill was shipped to the Dallas warehouse, where it was forwarded to Conway and erected on a sixty-foot tower. Stenciled on the tail of the windmill was "100,000." More than one hundred guests attended the installation ceremony at Aermotor's Conway headquarters. The company wrote at the time: "In addition to reminding us of our close relationship with our Argentine friends, it will become the focal point for various employee activities including a picnic area."[48]

Workers making vane sheets at the FIASA factory in Bragado, Argentina, ca. 1975. Courtesy FIASA, Bragado, Argentina.

Internally and among some of its longtime customers, however, was a growing dissatisfaction with the quality of Aermotor's windmills, such as premature wearing of bearings, gears, and other cast parts.[49] Jim Fetters and his managers worked vigorously with customers and FIASA to fix the problems. However, by early 1979, Aermotor began watching longtime managers and employees leave the company. LaVerne Roberts and his sons Jack and Ray in the branch offices departed for other companies, but soon thereafter formed Roberts Pump and Supply Company in Grand Island, Nebraska, and thirty-year company veteran Stanley Anderson resigned to take a job as vice president of marketing at upstart windmill manufacturer Wind Baron Corporation of Scottsdale, Arizona.[50] This sudden loss of institutional knowledge in the field left Aermotor filling positions with newly minted and largely inexperienced windmill salesmen. Before his departure, Anderson oversaw the compilation of detailed instructions for assembling the company's windmills and towers.[51] Fetters told his windmill salesmen in June 1979 that if "you keep a few key points in mind, you'll be able to answer the vast majority of [customer] inquiries," namely that Aermotor's windmills were designed solely to pump water and attempts to use them for other applications voided the warranty; that salesmen should focus customers on the windmills' pumping capacities rather than on their rate of horsepower and shaft torque, which were "irrelevant in selecting a pumping windmill"; and that the basic design of the Model 702 Aermotor had not changed since the early 1930s—"few companies can offer repairs for products made more than 45 years ago"—and the Model 602 (manufactured from 1916 to 1933) could still be updated.[52]

In July 1979, Valley's senior management at St. Louis initiated a plan to bring back windmill production to the United States, claiming quality and delivery problems with FIASA.[53] However, certain Valley Pump and Aermotor division managers immediately criticized the plan as ill conceived. David Suey, president of the Valley Pump Division, resigned from Valley Industries in August that year and became executive vice president (and president several years later) of

David P. Suey Sr., who resigned in August 1979 as president of the Valley Pump Division, was critical of the company's efforts to break windmill manufacturing ties with FIASA. Courtesy David P. Suey Sr., Beatrice, Nebraska.

Dempster Industries in Beatrice, Nebraska. Before leaving, Suey wrote a letter to Valley Industries president Patrick Gilligan, warning him of the disastrous consequences related to canceling the company's licensing agreement with FIASA and that it would not stop the Argentinian company from manufacturing windmills.[54] Despite the protests, Valley Industries moved forward with ending its relationship with FIASA via a simple telex and, to handle its new windmill manufacturing, began a $2.5 million expansion of the Conway plant.[55] An immediate problem faced by Valley was how to handle a large inbound order of windmills and components from FIASA, which Valley had to honor via irrevocable letters of credit. These shipments continued to arrive at Conway in 1980 and were sold as the company retooled its operations to manufacture its own windmills.[56] To consolidate further and support both the electric pump and windmill lines, the Brentwood, Missouri, operation of Valley Pump Group was moved to Conway the year before and the company appointed Daniel F. Benson as its Aermotor division president.[57]

Valley brought to Conway a group of its pump engineers to redesign the Model 702 windmill. The main goal of the engineers was to find ways to reduce costs to restart manufacturing windmills in the United States.[58] For example, the traditional labor-intensive sheet-metal bonnet covering the gearbox was replaced with a molded fiberglass version held in place to the pitman guide by a plastic nut. The company noted that the fiberglass bonnets withstood rough handling better than their metal counterparts and resisted gaping punctures from gunshots.[59] Inside the gearbox and wind wheel hub, the engineers replaced Aermotor's longtime use of babbitt bearings with self-lubricating, Telfon-impregnated sleeve bearings, called Garlock bearings after manufacturer Garlock Bearings, Inc. Valley engineers determined that Garlock bearings could operate without oil for more than 5,000 hours, while babbitt bearings without lubrication would fail within two hours.[60] Valley's windmill, marketed as the Model 802, also abandoned the traditional Aermotor red paint on the gearbox and replaced it with a powder-coated bright green color. The front of each gearbox had a brass identification tag, with a serial number, and "Aermotor" was stenciled in green on the tail.[61] Like FIASA, the windmills were built in six-, eight-, ten-, twelve-, fourteen-, and sixteen-foot wheel sizes.[62]

The first new US-made windmills rolled off the Conway assembly line on March 25, 1981.[63] However, it wasn't until June 1981 that Valley officially dedicated its new windmill operation at Conway, with Arkansas governor Frank White, Valley Industries president Patrick Gilligan, and other dignitaries in attendance.[64] Valley said the Conway plant had the capability to produce 4,000 windmills a year. That first year of windmill production Valley claimed to manufacture 80 percent of all Aermotor windmills in the United States. However, the company still had a large number of FIASA-built Aermotors in stock and often mixed the two lines when they shipped them to branches during the first year of selling the 802. In May 1982, Benson told the *St. Louis Post-Dispatch* that Aermotor was now "producing a much higher-quality mill" than what it had been paying FIASA to produce and noted that Aermotor had reduced its production costs and improved

One of Valley Industries' U.S.-made Model 802 windmills on display at the American Wind Power Center in Lubbock, Texas. The company's Conway plant made the Model 802 in the early 1980s. The windmill, which had a green-painted mainframe and lettering on its tail, experienced numerous technical problems. Photograph by author.

delivery times to customers.[65] The new Valley Aermotors ranged in price from $831 for the six-foot windmill to $8,213 for the sixteen-foot unit, plus the cost for a tower.[66] By 1982, the company estimated that windmill sales accounted for upward of 15 percent of its $31 million annual pump sales.[67]

FIASA could not ignore the potential of losing its sales in the United States and quickly made changes to retain its position in the US windmill market. Before the end of the 1970s, the Argentine company pared down its workforce, renovated its foundry, updated windmill part patterns and tooling, and enhanced its quality control department at the Bragado factory. FIASA even hired personnel from firms such as Massey-Ferguson and Somisa to help boost its engineering and manufacturing quality.[68] In early 1980, it made arrangements with two Texas pump equipment distributors, Amarillo Service and Supply and Alamo Machinery, to purchase and import windmills and parts directly from FIASA. Before the end of the year, however, the principals of FIASA contacted Jim Fetters, who had been laid off by Valley in late 1979 and was living in Fort Lauderdale, Florida, and pitched him the idea of forming a company to import FIASA windmills into the United States. Fetters contacted Beck Atkinson, once head of the Aermotor Dallas warehouse during the early 1970s and one of the owners of Amarillo Service and Supply, to join him. Together they formed Essex Associates, Inc., in Dallas, which operated a 12,000-square-foot warehouse to serve as the primary importer of FIASA 702 windmills for distribution in the United States, Mexico, and Canada. To distinguish themselves from the new US-made Aermotors, FIASA windmills were painted blue with "FIASA" cast into the back of the gearbox and stenciled in black on the vane sheet.[69] Essex management claimed to maintain an inventory of "several hundred windmills, as well as towers, accessories and parts" at its Dallas location, and had seventeen active distributors, mostly in the US Southwest, with a goal of adding more agencies in the next several years.[70]

To the dismay of the Valley Pump Group, the rapidly deployed Model 802 Aermotor immediately showed weaknesses in the field. Valley's engineering department, then headed by Robert H. Kain, received numerous complaints from windmill distributors and dealers. Other weaknesses with the machines were highlighted through employee observations and findings during manufacturing and in-house evaluation programs.[71] As one observer described it, "They had to reinvent the mill from scratch and used 'new technology' and ignored what had worked for years. They shut out all of the old Aermotor windmill hands and used all new people. They didn't solicit any input from any of the windmill people, the manager of the windmill warehouse, the salesmen, the warehousemen, the old Aermotor customers, no one."[72] Perhaps the biggest setback in the making for the new US-built Aermotors was that its repair parts did not fit on the older mills. "This was a critical error because the largest percentage of windmill sales was due to repair parts for old windmills," David Suey said.[73]

With the exception of the Garlock bearings, the company tweaked nearly every mechanical aspect of the 802 windmill within the first two years of introducing it to the market. By August 1982, the hub and shaft assembly of the mill had been revised to provide a shoulder in order for one to positively locate the hub axially on the shaft. Thus by accurately locking the hub to the shoulder on the shaft, the company also ensured the hub was flush on the end of the mainframe and offered a tight oil seal. For the wind wheel assembly, Valley introduced lead washers to ensure that arms could be threaded tightly into the hub. The company also said both the inner and outer wheel bands had been revised to ensure easier assembly. The pinion gears of the 802, which handled loads three times heavier than the large gears, were changed from cast iron to high-strength carbon steel to prevent premature wear. Valley strengthened its 802 pump-rod assembly, which initially proved inadequate in high wind conditions. This was done by replacing nylon bushings with steel bushings and by adding spring washers. Valley explained that the spring washers "ensure the assembly is tight and takes up the high load when the pump rod goes from the down stroke to the up stroke." Further, the assembly was held in place by spot-welding the nut to the rod. Valley replaced the plastic bushing on the brake band with a larger steel bushing for improved contact. The company fixed a problem with its initial tailbone casting, which often caused the furling rod of the windmill to become dislodged during operation. Many complaints were received from installers about the difficulty in hooking the vane spring to a hole in the end of the rod when tension was being applied to the spring. Valley fixed this by replacing the hole with a hook. The company also introduced a skirt to the bottom of the fiberglass hood to improve the seal between the hood and gearbox, where observations of oil leaks were initially recognized on the first 802s. Other significant improvements were made to the 802 mast pipe. In its initial mast pipe redesign, Valley engineers eliminated the separate mainframe turntable ring. The company explained that this change was due to previous complaints about the Model 702 turntable rings working loose. However, Valley experienced returns of 802 mast pipe assemblies because the upper pipes broke where they were welded in the lower pipe. The company remedied the problem by making the assembly of the upper and lower pipes a pressed fit to remove the potential for movement between the parts and by using stronger welds.[74]

The fixes to the problem-plagued 802 model did not stop there. Over the next year and a half Valley engineers solicited additional feedback on how to improve the windmill's performance and durability from its distributors and dealers. It accepted some recommendations, such as press-fitting the windmill hub to the shaft, tightening tolerances between the tailbone casting and brake lever furl arm retainers, leak testing mainframe castings at the factory, and obtaining heavier fiberglass hoods and replacing the rivets to the hood handle with screw-and-nut combinations. To correct the balance between the new, lighter galvanneal-coated sheet-metal 802 wind wheel blades when repairing older Aermotor wheels with heavier dipped galvanized blades, the company suggested adding weight to them. Other recommendations were brushed aside as unnecessary. Some dealers noted that the 802 windmill wheel arms were longer than those on comparable FIASA mills. Valley engineers responded that "we are making parts to the old Chicago standards not to FIASA standards." Valley did not budge on introducing metal hoods to its 802, stating that "the fiberglass helmet has been developed to the point where it is now a better helmet than metal."[75]

Behind the scenes, the Valley Pump Group's parent—Valley Industries—was under pressure from its shareholders to improve its financial performance. The company, like many of its US counterparts in the early 1980s, was weathering a severe economic recession. In addition to water pumps, the Valley Industries portfolio of businesses included oil and gas pipes, industrial products, and coal mining, all of which suffered during the recession lasting from July 1981 to November 1982. Its business lines showed some promise of financial recovery in the second half of 1983. Valley's heavy pump sales, however, remained slow due to the lagging industrial recovery, and agricultural pump markets were dampened by the poor farm economy, widespread wet weather across the country, and the federal government's crop set-aside program.[76] The economic hit to Valley Industries, which recorded a net loss of $26.5 million in 1984, caused it to sell off all of its businesses other than steel pipe. Valley Pump Group was sold to Mueller Company of Decatur, Illinois, for $9.8 million.[77]

When Mueller purchased the assets of Valley's Aermotor Division, it immediately shut down windmill manufacturing in Conway due to significant financial losses, product liability, and lawsuits associated with the product. Mueller contacted Essex and asked the firm to handle warranty claims for replacement windmills and parts of FIASA manufacture to its customers at a negotiated price, which Essex did for a number of months. The mills were shipped as "warranty replacement" from Essex's Dallas warehouse and painted the blue FIASA color with a plain vane sheet that simply read "Argentina" stenciled in black. FIASA even considered purchasing the Aermotor windmill line from Mueller, but decided against it due to disagreement over the proposed sale price and technical differences between FIASA's Model 702 and the new 802 Aermotor. Mueller continued to shop the Aermotor windmill operation around and in March 1986 sold it to a group of investors in San Angelo, Texas, headed by James F. Lane, who at one time was a small distributor of Aermotor windmills and pumps under the name Lane Dublin Company. The investors included area ranchers and bankers. They purchased all the assets of the former Aermotor windmill manufacturing operation in Conway and the right to use the name "Aermotor Windmill" for sales in the United States. More than thirty truckloads of machinery, raw castings, finished parts, and windmills, along with original drawings, were transported to San Angelo. The Aermotor line of electric pumps, however, remained with Mueller.[78]

The newly formed Aermotor Windmill Corporation, led by Lane, who was also president of Lane Company, a manufacturer of sprinkler, pipe, and irrigation products, set up windmill assembly operations in a 20,000-square-foot space of the former Dresser Manufacturing building on North Bryant Boulevard in San Angelo. The company hired half a dozen employees at startup and appointed Bennett W. Reed as plant manager. The plant initially operated a single shift from 8 a.m. to 5 p.m., Monday through Friday. Lane added upward of a dozen employees within two years. The company contracted with San Angelo Die Casting and Manufacturing Company to make some of its windmill parts. Aermotor also began making arrangements with foundries for castings. The first Aermotor windmill from San Angelo was assembled on May 15, 1986. Lane predicted that by 1987 the company would sell upward of 1,500 windmills, and "double that within three years."[79] (Actual numbers of windmills sold by Aermotor in the late 1980s could not be verified by the author.) There were enough inventoried parts in the first two years that Aermotor could assemble most windmill orders. By the second year, the company began machining some locally cast gearboxes and painted all its parts primer red, with "San Angelo, TX Aermotor U.S.A." stenciled on the windmill vane sheets in the same color.[80] The windmill sizes ranged from six feet to sixteen feet, costing $1,100 for the smallest model and up to $6,500 for the largest. However, its most popular size was the eight-foot wind wheel.[81] The company found almost immediate interest in its windmills from countries in Africa, the

Middle East, and South America, leading Lane to conclude that developing countries "are the future of the windmill."[82]

By the early 1990s, the city of San Angelo wanted the property at North Bryant Boulevard for other purposes and pressured Aermotor to relocate. One of Aermotor's shareholders had a 20,000-square-foot building available at Dan Hanks Lane on twenty-two acres, and the operation was moved there in 1992. At the new facility Aermotor boosted its workforce to seventeen.[83] "We had little turnover in the lower jobs in the shop, such as those working the drill presses. Lathe operators were pretty steady," recalled Calvin Schovajsa, a machinist at Aermotor. "A good hand could make about $7 an hour. A punch operator didn't draw as much." Most of the employees at the San Angelo factory had grown up on nearby farms and climbed windmill towers and changed the oil when they were young, Schovajsa said.[84]

Aermotor's shareholders, however, became dissatisfied with Lane's return on investment and forced him to resign in the early 1990s. They filled his position with Frank McQuerry, but he died a couple of years later due to complications from surgery.[85] Prices for Aermotor windmills had also started to climb, with the six-foot windmill costing $1,695 and the sixteen-foot mill $8,580 in 1992.[86] The company was then bought in 1995 by a San Angelo machine shop owner, Mike Hardegree, who immediately added 20,000 square feet of warehouse space, expanding the overall operation on Dan Hanks Lane to 40,000 square feet. Hardegree told a San Angelo newspaper in 1996 that Aermotor shipped windmills to 226 dealers in the United States, Mexico, Puerto Rico, Colombia, Brazil, Africa, and Japan, with more overseas markets coming on board.[87]

During the mid-1990s the windmill market continued to undergo changes. Dempster Industries of Beatrice, Nebraska, was the last of the US-based windmill manufacturing competitors to Aermotor, as the Heller-Aller Company in Napoleon, Ohio, was shuttered after its president, Max Kelley, died suddenly in 1995. (Heller-Aller sold some FIASA windmills in markets east of the Mississippi River for about three years in the early 1980s.)[88] FIASA's master US importer, Essex, closed and liquidated operations at the end of 1992 with cofounder Fetters citing a desire to retire and Essex's warehouse lease nearing expiration. But FIASA remained active in the US market by selecting three master distributors: Austin Pump and Supply of Austin, Texas; Amarillo Pump and Supply, Inc., of Lubbock and Amarillo, Texas; and Roberts Pump and Supply of Grand Island, Nebraska. H. Beck Atkinson, a cofounder of Essex, continued to serve as the point person between FIASA in Argentina and the new master distributors in the United States. Each of the three purchasing firms chose its own territories for the sale of FIASA windmills. Amarillo Pump claimed the "central part" of the country from coast to coast (Atlantic to Pacific), while Austin Pump took the southern portion of the country from coast to coast, and Roberts Pump distributed in the northern plains, upper Midwest, and Pacific Northwest. Within five years, this arrangement proved unworkable for both FIASA and the three distributors.[89] In 1998 FIASA settled on a new master US importer for its windmills—American West Windmill

Company of Abernathy, Texas, today a subsidiary of Gicon Pumps & Equipment. For its US-bound windmills, FIASA switched from blue to bright red paint, with "FIASA" cast on the rear of the gearbox and "American West Windmill Company" stenciled in red on the vane sheet.[90] There was also windmill manufacturing competition emerging from Mexico during the 1980s. Molinos de Vento, S.A., of Chihuahua, Mexico, made windmills similar in appearance to the Aermotor Model 702 with wind wheel sizes ranging from eight to sixteen feet in diameter, but most of these windmills were sold in Mexico and other Latin American countries.[91] Another Mexican windmill maker, Felizardo Elizondo Guajardo, S.A. de C.V., founded in 1953 and based in Apodaca, produced a 702 variant that it also called the "Aermotor"; it manufactured them with wind wheel sizes ranging from six to twenty feet in diameter. However, to avoid running afoul of US trademark laws, the windmill has been sold in the United States as the Wind Engine 702.[92] In addition, another group of competitors—domestic rebuilders—were making their mark in the windmill business and a foe to all the manufacturers. These firms, often operators of small rural machine shops, scooped up large volumes of worn-out windmills from dealers and ranchers and refurbished them for sale back into the market. From these operators, thousands of Aermotor Model 602 and 702 mills were revived and placed back on top of towers.[93]

Financial problems, diminished employee morale, and cheaper windmill imports and rebuilds were taking a heavy toll on Aermotor in San Angelo, and by 1998 the company had reached the brink of collapse. A number of windmill builders approached the absentee owners about purchasing the operation, but the asking price and associated risks turned most of them away from making a deal.[94] Kees Verheul, a retired engineer and rancher, was made aware of Aermotor's plight by a man whom he hired in 1998 to take down nine inactive windmills on one of his West Texas ranches. He admittedly knew nothing about windmills, other than their long history of pumping water in the parched Southwest and across the Great Plains. Verheul made his fortune running an oil well drilling tool manufacturing company in Houston, Texas, which his father had founded in 1937. After taking over the company in 1961, he expanded Oilfield Machine and Supply Company's annual sales from $1 million to $70 million within twenty years. He sold the business to a Canadian company in 1975 and stayed on under contract to the new owner until 1983, when he retired at the age of forty-seven. Verheul next took up ranching in West Texas but grew bored with the business and by the late 1990s was looking for a new outlet, something that could rekindle his passion to work with his hands and cut metal. He decided to drive from his 19,000-acre ranch in Spur, Texas, to San Angelo to visit the Aermotor plant. Four days later, he offered the owners $1 million and the windmill company was his. "I didn't buy Aermotor because of windmills. . . . I liked the looks of the company," he said, adding that he delighted in the new "challenge."[95]

Verheul knew the company was in trouble when he took it over. The morale among the eighteen employees, as well as windmill production, was low. "I realized the company was in terrible condition when I purchased it," he recalled. "I never

Kees Verheul (right), the Texas businessman, engineer, and rancher who is credited with saving the Aermotor Windmill Company from financial disaster when he acquired the San Angelo, Texas, firm in 1998. Courtesy Kees Verheul, Port O'Connor, Texas.

thought I made a mistake on buying it. I could see a potential for improvement."[96] Verheul's hands-on management style and understanding of metal-working machinery gained him an appreciation among the shop employees. He scoured the country for used machinery to improve the manufacturing process, including acquiring a large press, small mill, engine lathe, turret lathe, horizontal boring mill, another drill press, and a computer-controlled milling machine. The company also embraced metallurgical improvements in both castings and steel parts from suppliers.[97] Within the first five years of owning the company, Verheul spent upward of $2 million on retooling the plant.[98] Verhuel claimed that he was less interested in the number of windmills produced and sold, which included six-, eight-, ten-, twelve-, fourteen-, and sixteen-foot wind wheel sizes, and focused attention on maintaining the mechanical integrity that had made the Aermotor one of the most recognized brands in the industry's more than 140 year history.[99] However, the company made important changes to the way it conducted windmill sales, specifically ending a policy of direct sales that emerged prior to Verheul's acquisition and selling only to authorized dealers. He called direct sales in the windmill business "certain death." Verheul also reduced the company's windmill prices about 20 percent to bring them within 5 percent of competition from Argentina and Mexico.[100]

By 2006 the production and sales changes instilled at Aermotor by Verhuel were starting to pay off. That year the company sold more than 1,000 windmills, a 69 percent increase over 2005 sales and the best year in more than a decade. From San Angelo, windmills were shipped to dealers across the United States. The Navajo Nation in Arizona was viewed as one of the company's biggest customers.[101] Sales overseas also rose, with an increased emphasis on the African market where remote villages without access to electricity could use the free wind to pump their water.[102] Verheul summed up his view of the Aermotor windmill in an interview with a San Angelo newspaper in 2007: "It's so simple, it's brilliant. . . . It will last 100 years with proper maintenance."[103]

CHAPTER EIGHT

Experimenters and Restorers

The Aermotor name in the minds of many farmers and ranchers has become synonymous for water-pumping windmill, just as Band-Aid is to wound care, FedEx to overnight shipping, and Xerox to photocopying documents. This name identification is testament to the company's founders, La Verne W. Noyes and Thomas O. Perry, who cleverly picked the name for the windmill and never deviated from it. Even those who carried on the manufacture of these windmills long after the deaths of the founders realized the value of the name and kept it prominent on the galvanized tail vanes. It is common today for the casual observer to mistake another make of windmill and call it an Aermotor. In 1989, for example, the Nebraska Department of Motor Vehicles introduced a new license plate with black lettering on a white, blue, and orange screened background that had a black silhouette of a windmill in the upper-left corner resembling an Aermotor. This infuriated longtime windmill manufacturer Dempster Industries, based in Beatrice, Nebraska. The Dempster windmill, as many pointed out, has a distinctive two-part vane sheet contrasted with the solid-sheet tail of the alleged Aermotor on the state license plate. The *Omaha World-Herald* newspaper quoted the state director of the Department of Motor Vehicles as stating, "It's just a generic windmill," reinforcing the notion that the Aermotor design, for many people, typified what a US windmill on the Great Plains should look like.[1] This aura created around Aermotor as the quintessential windmill has been cemented by countless photographic images taken of farms and ranches, cowboys, and cattle to illustrate books, articles, and postcards during the 1950s, 60s and 70s, often with an effort to romanticize the rugged lifestyle of the Old West.

The popularity of the Aermotor windmill, specifically the Model 702 oil-bath machine, is further compounded by the fact that many thousands of them still dot the rural landscape, especially in the Southwest and Great Plains—more than any other windmill. With annual oil changes and some basic maintenance, these windmills can easily pump water for fifty to sixty years before succumbing to mechanical failure or being taken out of service. Some of the oldest and longest-operating Aermotor windmills for livestock watering are found on Amish farms in the Ohio Valley and Pennsylvania. As one former Aermotor windmill installer in Benton, Kansas, explained it, many farmers and ranchers chose to have a windmill repaired or rebuilt before they would ever consider buying a new one, while they would ordinarily replace other farm equipment. "And there's been no other piece of equipment on the farm that's been more abused than the windmill," the installer said.[2] Since so many Aermotor windmills were manufactured, it is still relatively easy to scavenge parts from older units to repair others. This encouraged a number of machinists during the past three decades to become rebuilders of Aermotor windmills and return them to service in the field, often at prices below those for newly manufactured mills.

New Mexico State University at Las Cruces for the past forty years has offered an annual workshop to those individuals interested in learning how to properly install and maintain windmills for practical water-pumping purposes. The 1973 Arab oil embargo dramatically raised energy prices and turned many people's attention back to wind power. Ranchers and farmers were not immune and eyed their old water-pumping windmills—many left inoperable on their towers for decades—as a means to reduce their electricity costs. A group of ranchers in the Southwest approached the university about introducing a course on how to erect, maintain, and repair windmills. The university's College of Agriculture took notice and asked Danish native and instructor Mogens I. Rasmussen to conduct a survey on the history and future of windmills before making the commitment to the course. Rasmussen estimated in 1974 that there were about 175,000 windmills still standing in the United States, mostly in the West, of which half were in either operable or reparable condition. While there were people experienced in maintaining windmills, Rasmussen found that the professional ranks had greatly thinned since the 1950s as electric water pumps assumed a dominant role in farming and livestock management. These findings were enough for the university to introduce its week-long Windmill Technology Certification Workshop in 1975. News about the course spread throughout the country, and there has been no shortage of attendees—about twenty-five per year—from ranchers and farmers to water management employees of the various American Indian reservations, overseas missionaries, landscapers, and hobbyists.[3] When Rasmussen died in the mid-1980s, the workshop was taken over by James Dean, then professor and agriculture mechanization instructor of the NMSU Ag and Extension Education Department. After Dean's retirement, the workshop was continued by Carlos Rosencrans with assistance from one of Rasmussen's early students, Craig Runyan, now emeritus. Through Dean, the workshop established a close relationship with Aermotor Windmill Corporation in San Angelo, Texas. Since 2007 the company has sponsored the now four-day annual workshops by providing windmill materials; it even encourages its new dealers and their employees to attend.[4] The workshop is also supported by a cadre of windmill professionals throughout the West and South that volunteer instruction. Other patrons of the workshop, including former attendees, contribute benevolent donations to ensure its continuation.[5]

New Mexico State University at Las Cruces, under the direction of Professor Mogens I. Rasmussen, began a weeklong training program for windmillers in 1975. Pictured is the class of 1976, with Professor Rasmussen (far right). Courtesy New Mexico State University's College of Agricultural, Consumer and Environmental Sciences, Las Cruces.

As competitors have faded away, the US-made Aermotor and related-style imports, such as those manufactured by FIASA of Argentina, Felizardo Elizondo Guajardo, S.A. de C.V., of Mexico, and Iron Man Windmill Company in China, have become the de facto working windmills for those seeking them to pump water on today's farms and ranches. The choice and size of the mill is generally based on well depth and output as well as unit price. Felizardo Elizondo Guajardo and Iron Man Windmill are the only firms that manufacture a Model 702 variant with a twenty-foot-diameter wind wheel.[6] Aermotor and the other firms have a maximum wheel diameter of sixteen feet. Yet for most of these manufacturers, the eight-foot-diameter wind wheel remains the most popular unit. "I strongly believe that the Chicago Aermotor-originated 702 is the finest windmill ever produced and that there is little that can be done to improve it," said Tom Conlon, founder and chief executive officer of Iron Man Windmill Co., in a 2012 e-mail.[7]

Interestingly, the water-pumping performance of the Aermotor windmill has changed little since the 1930s. In the late 1980s the US Department of Agriculture's Conservation and Production Research Laboratory in Bushland, Texas, conducted a year-long test on the performance of water-pumping windmills. The lab obtained two newly built Aermotor and Dempster windmills, both with eight-foot wind wheels, and placed them on thirty-two-foot towers. The windmills had nearly identical gear ratios and were set to stroke lengths of seven inches. The only

In the late 1980s, the US Department of Agriculture's Conservation and Production Research Laboratory at Bushland, Texas, conducted tests aimed at improving the pumping performance of traditional mechanical windmills. Courtesy R. Nolan Clark, Amarillo, Texas.

difference between the two windmills was that the Aermotor had eighteen blades while the Dempster had fifteen. Identical pumps were used, then exchanged halfway through the experiment. Pumping performance was compared at two pumping lifts, 65 and 100 feet, during the test. The USDA lab determined the correct regulating spring adjustment to ensure that the pump speeds of the two windmills were similar. The researchers found that for the same regulating spring setting (same hole from windmill gearbox), the Dempster mill pumped more water and operated faster than the Aermotor with nearly the same efficiency. Yet when the spring setting was adjusted so the pump speeds (strokes per minute) were comparable, the two units then pumped nearly the same amounts of water. The USDA lab reported in its findings that the peak overall efficiency of the Aermotor measured 18 percent, while that of the Dempster was 13 percent. The peak efficiency was reached at a wind speed of ten miles per hour "when the windmills were slowly turning and the torque was high." The lab noted that the Aermotor's larger number of wind wheel blades may have contributed to a higher torque at low-wind speeds. Overall, the USDA found that "performance of multibladed windmills is highly dependent on proper adjustment of the furling spring tension and selection of the optimum pump speed."[8]

During the same period as the performance tests, the USDA studied the potential of a variable-stroke pumping mechanism for windmills as opposed to the standard single-acting pumps, which have a fixed stroke length and require continuous operating torque from the windmill rotor. USDA researchers at Bushland noted that single-acting pumps fail to take advantage of the extra power available at higher wind speeds, maxing out at about thirty to thirty-five stokes per minute at about twenty-two miles per hour. The agency proposed designing a spring-loaded mechanism that would change the stroke length as the wind speed increased. Although this idea was tested using two eight-foot Dempsters, USDA officials said it could have equally applied to the Aermotor or other mechanical windmills. The USDA's variable-stroke test pump was composed of two hydraulic cylinders, a pressure accumulator, an oscillating lever arm, and other hardware. Trials showed that the test pump increased the water flow rate at a wind speed of twenty-two miles per hour from almost seven gallons per minute to about twelve gallons per minute when compared with the single-acting pump operating at the same wind speed. While the USDA demonstrated its hydraulic variable-stroke pump design had the ability to "greatly improve the performance of many mechanical water pumping windmills," chief researcher R. Nolan Clark later commented that the complexity and additional cost associated with such a system could not be supported by traditional windmill manufacturers. Thus, single-acting pumps remain the industry's mainstay.[9]

The US Bureau of Land Management in Vale, Oregon, used an Aermotor to test a spring counterbalance mechanism to operate water-pumping windmills at low-wind speeds. The bureau used two forty-eight-inch-long by two-inch-wide New Holland bail wagon springs connected between the top of the tower and the pump rod. As many as four springs, counterbalancing up to 100 pounds apiece, were used to test the system for the deep-well pumping tests. The bureau used a scale made from a hydraulic cylinder and pressure gauge to measure the upstroke and downstroke loadings. To allow the test windmill to turn more freely in the wind, bureau researchers also replaced the Aermotor's mainframe support friction bearing with a roller bearing. Test observers expressed concerns to the bureau about the spring counterbalance mechanism, stating that conventional windmills were "not designed to take the loading on the downstroke and, if the windmill is loaded on the downstroke, accelerated and unusual wear will result." The bureau countered this argument by pointing out that these windmills are already loaded on the downstroke due to the weight of the pump rods and that "no accelerated or unusual wear" occurred during the tests.[10] However, none of the windmill builders at the time, including Aermotor, were convinced to incorporate a spring counterbalance into their designs.

Installing a water-pumping windmill is an expensive undertaking for most farmers and ranchers. In 2013 an eight-foot Aermotor with a thirty-three-foot tower—the company's most popular windmill—and the related pump hardware cost about $8,000. The installation cost by a licensed contractor would be about $2,000. The cost of a four-inch PVC-cased and gravel-packed well to a depth of 200 feet comes to about $16 per foot, or $3,200. Thus the cost of a fully installed windmill can easily reach upward of $13,200.[11] The USDA's Natural Resources Conservation Service through its Environmental Quality Incentives Program offers some financial assistance to those landowners installing windmills or solar-powered electric-submersible pumps if the technology is incorporated into an approved plan to improve environmental quality. For example, several windmills spread across a pasture help keep livestock from congregating around a single water source, which may result in destruction of erosion-controlling vegetation. During 2013, the Environmental Quality Incentives Program in Texas paid $417.96

This Aermotor windmill with a twelve-foot-diameter wheel on a thirty-three-foot tower was erected in October 2007 on the 2J Trusts ranch in Moore County, Texas, with financial assistance from the US Department of Agriculture's Environmental Quality Incentives Program, which is administered by the Natural Resources Conservation Service. Courtesy Natural Resources Conservation Service, Lubbock, Texas.

per foot of a windmill's wheel diameter to participating landowners, and it did not matter if the windmill was installed over a new well or an existing one without a functioning pump or mill. In 2011 and 2012, the Natural Resources Conservation Service in the Texas Panhandle and South Plains assisted about 170 ranchers in fifty-one counties with installing windmills or solar-powered electric-submersible pumping plants. The agency did not break out the number of windmills versus other pumps that were installed through the program.[12]

The biggest frustration for most windmill users occurs below ground with the pump rod and cylinder. Leather cups or seals inside the cylinders wear out— generally within a two- to three-year period—and must be replaced. To get to the leather cups inside the pump, the rod and plunger must be pulled to the surface. Three replacement leather cups cost only about $17, but if the farmer or rancher relies on a serviceman to perform this work, it generally costs between $250 and $350 per windmill. Without proper maintenance, the pumping mechanism inside the cylinder will fail to lift water and may even seize, causing potential damage to the rest of the windmill. Newer materials, such as cups made of polyethylene or polyurethane, promise to reduce the frequency of pump maintenance.[13] The windmill should also have its oil changed once a year, which can be done either by the farmer or rancher himself or by a serviceman. An eight-foot Aermotor, for example, takes two quarts of lightweight nondetergent oil at a total cost of about $14.[14]

While Aermotor and its competitors have stuck to manufacturing windmills for water-pumping purposes, some installers have used them as a foundation to develop other wind-energy-based applications. Ken and Sharen O'Brock, who started O'Brock Windmill Distributors of North Benton, Ohio, in 1968, have distributed and installed a variety of Aermotors converted to pump air instead of water. The air is compressed on the up-and-down stroke of the pump rod and can be deployed for pond aeration and workshop air compressors. Some of the biggest customers for these wind-based air pumps are the Amish, whose religion forbids them from using mechanical devices powered by electricity or petroleum-based energy sources. O'Brock's compressed air pumps have also been used for industrial applications, such as on Union Carbide's capped landfills at Harriett's Bluff, Georgia, to bubble groundwater for the purpose of breaking down hazardous chemicals and in the reverse for vapor extraction at a DuPont Chemical Company landfill in Frankfort, New York. Both projects were completed in the mid-1990s. During the same period the O'Brocks converted five eight-foot Aermotor windmills on twenty-seven-foot towers from reciprocal to rotary motion to operate as oil skimmers for the US oil company Amoco (now BP plc). These windmills remain in operation today. In 2012 the O'Brocks set out to design an Aermotor-based, two-stage pneumatic cylinder compressor to supply more than one hundred pounds of pressure to large storage tanks for powering tools in Amish workshops.[15]

However, a long sought after alternative use for traditional multibladed windmills is to convert them from a mechanical device to an electric generator. Some windmill manufacturers, including Aermotor, introduced electricity-generating

An Aermotor windmill with compressed air pumps designed by Ken O'Brock. Three were erected in 1990 by O'Brock at the Union Carbide landfills at Harriett's Bluff, Georgia, to bubble ground water for breaking down hazardous chemicals. Courtesy Ken O'Brock, North Benton, Ohio.

variants of their mills in the early 1900s. However, widespread commercial success of these types of windmills was never achieved for two main reasons. First, the companies attempted to convert their mechanical platform, built for low-wind-speed, heavy-torque applications, such as pumping water, instead of designing a new platform around the electric generator itself. Second, multibladed windmills tended to absorb much of the wind's usefulness in the blades themselves, known as "wind congestion." A number of upstart wind power developers at the time took an interest in the sleeker, lightweight propeller design used on early airplanes. These propellers allowed wind to pass through, helping to give an airplane lift. Used in the context of wind-based electricity generation, these types of propellers could operate efficiently in winds ranging from eight to twenty-five miles per hour, and connecting these propellers directly to generator shafts allowed for a higher and more consistent flow of electricity to the storage batteries. In areas with fickle or low winds, even the nimble two- and three-bladed wind generators could not compete with the compact gas-powered farm electric light plants. Many wind and farm electric plants, however, were swept aside with the federal government rural electrification push in the 1930s to early 1950s.[16] It was not until the early 1970s energy crisis and environmental movement that a resurgence of public interest in small wind-electric generators emerged. Aermotor and traditional windmill manufacturers still in business at the time received numerous inquiries about building electricity-generating variants of their windmills, but they never took the step commercially. Past lessons and a flood of upstarts with a specific focus on wind-electric machines made the effort unattractive. There were examples of tinkers who made efforts to convert their water-pumping windmills to charge batteries, but it was not enough for companies like Aermotor, Dempster, and Heller-Aller to set up production for wind-electric generators. In the late 1990s and early 2000s, another wave of household interest in wind power swept through the country, this time based on government and personal interest in proliferating so-called "green" energies. Once again, inquiries were made, and the traditional windmill manufacturers held fast.

Unaware of Aermotor's early past with wind-electric machines, Carlos Fernandez-Bueno, a surgeon, farmer, and self-taught machinist in Dickerson, Maryland, looked at his second-hand Aermotor windmill on top of a forty-foot tower one day in 2004 and wondered how he could convert its mechanical energy into useful electric power for use on his farm. For the next two years, he tried various sizes of generators and methods for attaching them behind the Aermotor's wind wheel. Most of the generators, like those for vehicles, required thousands of rotations per minute to generate electricity, which the wind wheel of a water-pumping windmill could never produce. Fernandez eventually located a Servo generator from a CT scanner that was going to scrap. He then figured a way to attach the 110-pound generator to the Aermotor's gearbox and used different wheel ratios to drive a belt between the wind wheel and generator shafts. With this setup, he successfully generated sufficient and consistent DC power to supply a battery bank, which he then converted to useful alternating current for lighting his horse barns.[17] It was not long thereafter that Fernandez received a call from Kees Verheul asking to see his windmill. Fernandez recalled:

> He had flown in from San Angelo. He had rented a car and showed up [at the farm] . . . He had a voltage meter and the wind was blowing. It was a nice breeze. He was doing measurements and voltages and stuff like that. He looked at me and said "I'm really impressed. This is the first time I have ever seen something like this work." I said, "Really?" And he said "Do you have a patent on this?" I told him no, not really. He goes, "You need to get a patent on it." At this time, I didn't know who he was. I thought he was someone who just came through curiosity, that type of thing. As we got into the afternoon and went to lunch, he said, "You know, I own Aermotor." I go, "Really?" He goes, "What is it that you can do with this?" I go, "Well, obviously, this has a long ways to go, but this thing has a lot

of torque and if we get the right ratio you know we can probably make it work." And he being an engineer, he was really intrigued with the whole thing.[18]

Fernandez continued communicating with Verheul, and Aermotor agreed to send him a sixteen-foot mill, the company's largest machine, to test and develop his wind-electric generator further. Over the next two years, he replaced the belt drive with a simple chain sprocket system and deployed an inverter system that eliminated the need for batteries and offered a direct tie to the power grid.[19] His patent-pending and Underwriters' Laboratories–certified American Classic Wind Power windmill has a rated capacity of 6 kilowatts at a wind speed of 26.7 miles per hour and is constructed with individual UL-certified electrical components, including generator and grid-tied inverter. From a distance, the windmill looks like a traditional sixteen-foot Aermotor pumping water. The gearbox, generator, platform, and tail weigh about 2,400 pounds and can be erected on heavy-duty traditional windmill towers ranging from forty-seven to eighty feet tall.[20] The units cost about $55,000 apiece installed in 2011, but with available federal and state tax credits at the time, that price could be reduced by almost half. Fernandez estimated the return on investment of the American Classic Wind Power unit comes in about ten years.[21] In 2011 Fernandez installed his first two units for homeowners in Adamstown and Finksburg, Maryland. While he was unable to reach an agreement with Aermotor to produce and sell these electricity-generating windmills, the company granted Fernandez the right to use its windmill platform and make the necessary changes, as long as the wind wheel and tail, reading "Aermotor, San Angelo, TX USA" are not changed and all components used on the American Classic are manufactured in the United States.[22]

Many people, however, simply appreciate US windmills for their nostalgic qualities and like to restore and install them on their properties just to spin in the breeze and not pump water. This interest is no different from that of people who like to collect and restore old cars and tractors, for example. When Frank Whelan of Hartford, Wisconsin, asked his wife Vivian why she had to collect windmills—why not something smaller—she told him, "I wasn't a 'teacup and saucer' girl; . . . I was a farm girl who did farm work and loved the outdoors and windmills!" Today the Whelans have two restored Aermotor oil-bath windmills on their property, and they are far from alone.[23] During the past three decades, sizeable collections of restored windmills have sprung up across the country. Some of these collectors have amassed large gatherings both of open-gear and oil-bath windmills by numerous makers, such as the Windmills at Riverside Farm in Poland, Indiana, which is operated by Neal Yerian; Roger Bailey's collection in McCool Junction, Nebraska; the Emick Ranch collection in Lamar, Colorado; Terry Rodman's collection in Jasper, Minnesota; and the Windmill Farm Bed and Breakfast, founded by Chuck and Ruby Rickgauer at Tolar, Texas, just to name a few. Most collections are carefully displayed for preservation purposes and often available for viewing upon permission of their owners. Some of the windmills in these collections are

Carlos Fernandez-Bueno of Potomac Wind Energy in Dickerson, Maryland, designed this electricity-generating windmill with a rated capacity of six kilowatts using the sixteen-foot Aermotor mainframe, wind wheel, tail, and standard tower. It was erected in nearby Adamstown, Maryland, in 2011. Courtesy Walter J. Leskuski Jr., Frederick, Maryland.

extremely rare due to the fact that so few were made or survived the elements. There are also windmill-centric museums, some of which are municipally supported, scattered throughout North America, including the American Wind Power Center in Lubbock, Texas; Batavia Riverwalk in Batavia, Illinois; Etzikom Museum and Historic Windmill Centre in Etzikom, Alberta, Canada; Kregel Windmill Factory Museum of Nebraska City, Nebraska; Mid-America Windmill Museum in Kendallville, Indiana; Shattuck Windmill Museum and Park at Shattuck, Oklahoma; Spearman Windmill Park of Spearman, Texas; and Windmill State Wayside in Gibbon, Nebraska.[24] Antique windmill restorers and collectors gather annually

The Windmill Museum at Lubbock, Texas, was established in 1993 by Billie Wolfe and Coy Harris. A 30,000-square-foot gallery building displays over 110 restored windmills. Another 70 wind machines are exhibited outside and cover the history of American wind power from 1621 to the present. Photograph by Coy Harris, Lubbock; courtesy American Wind Power Center, Lubbock.

in the United States for the International Windmillers' Trade Fair and have access to historic information through the *Windmillers' Gazette*, a quarterly publication, and Internet-based forums such as the *Vintage Windmills Discussion* group for restoration tips and advice. Aermotor windmills are part of many collections large and small, with the most valued being the pre-1900 open-gear pumping and power mills, the first oil-bath Model 502, and the hard-to-find Model 612. For the beginning windmill collector, an oil-bath Model 602 or 702 is a good start due to the high availability of parts for restoration.

Outside North America, windmill collections that contain Aermotor mills may also be found. In western Australia is the Morawa District Historical Society Museum, which contains one of the country's largest collections of both domestic and imported windmills. The museum began to amass its windmills in the early 1990s through the efforts of volunteer curator and historian Malcolm Walter. This effort was followed in 2002 with the launch of a quarterly newsletter, *The Windmill Journal: History of Australian & New Zealand Windmills*, which has since been carried on by coeditors Bruce Hewitson and Walter's wife, Helen, after his death from an automobile accident in 2010. More than forty windmills are now available on display at the museum. South Africa, another significant location for water-pumping windmills, has the Fred Turner Folk and Culture Museum at Loeriesfontein in the North Cape region. This museum contains more than twenty different types of windmills.[25] The International Molinological Society, a group of more than 500 members with an interest in preserving the history of milling technologies powered by wind, water, and muscle, publishes twice a year a scholarly journal, titled *International Molinology*, and meets every four years, mostly in Europe, to present research.[26] Texas-based historian T. Lindsay Baker has attended a number of these gatherings and presented papers over the past two decades that have introduced society members to North America's rich contribution to the advancement of windmill technology.[27]

Interestingly, vintage Aermotors and related variants of this windmill may still be found pumping water for livestock and garden irrigation in various parts of the world. On small islands scattered throughout the Mediterranean Sea, fresh water is rarely found flowing on the surface and thus US-style windmills, like the Aermotor, were imported years ago to pump fresh water to the surface. Some of these countries have made efforts to restore these wind machines to preserve their heritage and as an alternative energy source for agriculture. In 2009 the island government of Gozo in the Malta archipelago, subsidized 75 percent of the labor and equipment costs, or up to €1,000, to bring back into service eighteen Aermotor windmills.[28] Likewise, numerous Aermotor windmills, among others, may be found in operational condition around Protaras, Cyprus, supplying water to gardens. However, with the proliferation of direct-tied and solar electric water pumps, these windmills—like their US counterparts—are under constant threat of neglect and decay. This has been the case for those used for many decades in salt production at Brazil's Cabo Frio and Lagoa de Araruama in the coastal state of Rio de Janeiro. Many Aermotor windmills, including open-gear models, can still be found pumping briny water from one pond to another to assist in the evaporation process for harvesting salt crystals. This way of life, however, is quickly disappearing under Rio de Janeiro's rapid urbanization.[29]

CHAPTER NINE

Still Turning

By 2005, the US economy appeared prosperous despite signs of rapidly rising oil and home prices and wars being fought simultaneously in Iraq and Afghanistan. US manufacturers fared well at this time from the declining value of the US dollar, making their products more competitive both in domestic and overseas markets. During this period, Aermotor Company in San Angelo, Texas, experienced positive windmill sales to its dealers, who in turn erected them on farms, ranches, and residential properties throughout the country and even the world. Some people purchased windmills simply to display them on their properties, not to use for pumping water. For Kees Verheul, owner of Aermotor, 2006 marked the high point in his time bringing the once-ailing company back to health. It was also the year he decided to sell the business. "The reason I sold Aermotor was that I have achieved what I had set out to do; to turnaround the company and make a quality product; [and] to get my estate in order since I was in my seventies," he said.[1]

Verheul had quietly maintained a succession plan for the company after he hired Bob Bracher in 2002. Bracher was born in 1948 and earned his bachelor's degree in agriculture journalism from Texas A&M University. He worked for an agricultural publication for a few years until he was called back to work on the family farm in Uvalde, Texas, shortly after the death of his father. He farmed full time for the next fifteen years. In 1986 he started selling agricultural, ranch, and rodeo equipment, which brought him into contact with Verheul. Recognizing Bracher's business acumen, Verheul asked him to help improve Aermotor's marketing of windmills. Bracher started out working part time for the company before joining full time as an outside sales manager. "The more I got into it, the more I liked it," Bracher said.[2]

After Aermotor moved to San Angelo in 1986, there was growing frustration among installers about how the company sold its windmills. The company's new owners began selling windmills directly to ranchers from the factory. While the purpose of selling direct and taking out the perceived middleman was an attempt to bring the cost of the US-made Aermotor windmill more in line with those from foreign competitors, it quickly proved disastrous. Many installers—for a long time

Bob Bracher, who joined Aermotor in 2002 to improve marketing and build export sales, served as president from 2005 to 2012. Courtesy Bob Bracher, San Angelo, Texas.

the front end for sales of windmills—turned their backs on Aermotor for Model 702 variants made by FIASA of Argentina and Felizardo Elizondo Guajardo, S.A. de C.V., in Mexico. Bracher reversed this practice of selling windmills directly to farmers and ranchers and instead returned to channeling Aermotor's marketing through professional water-well drillers.[3]

Bracher was also tasked with pursuing overseas sales. "I knew there was a need for windmills in third world countries. When I came to Aermotor, we were only selling a few windmills a year internationally and they were to Mexico," he explained. Under his sales leadership, Aermotors were soon being shipped to Nicaragua, Colombia, and Venezuela in Central and South America; Nigeria, South Africa, and Tanzania in Africa; Germany, Belgium, Sweden, and Malta in Europe; Afghanistan and Saudi Arabia in the Middle East; and to Far East countries such as Vietnam, Taiwan, the Philippines, and New Zealand. Bracher estimated that under his watch, 5 to 10 percent of Aermotor's annual windmill sales were attributed to exports.[4] Ideally, however, he said Aermotor needed to aim for 50 percent of its windmill sales coming from exports.[5]

In August 2005, Bracher was promoted to president and James Dockal, a nineteen-year employee, was appointed executive vice president in charge of sales. Verheul credited Bracher with "dramatically" growing the company since 2002 and expanding its dealer network to several hundred. "I look forward to continued growth with Bob at the helm," Verheul said at the time of Bracher's promotion.[6] Six years later, Verheul explained in an interview that he had been "grooming" Bracher to take over the company, adding, "Looking back, I don't know of anything I would have done differently."[7] Verheul sold the company to Bracher and three partners from nearby El Dorado, Texas—Jim Kosub, Clint Griffin, and T. Cy Griffin—in 2006 for an undisclosed amount of money.[8] He retired to Port O'Connor on the Texas Gulf Coast.[9] Aermotor would go on to report one of its best years in 2006–2007, with sales rising to $4 million, up from about $700,000 in 2002.[10]

Today the 40,000-square-foot manufacturing facility in San Angelo offers a large dock door on one side of the building to receive tractor-trailer and flatbed loads of materials, such as sheet metal, unfinished castings, and galvanized angle iron. Aermotor has not done any of its own foundry work since the company left Chicago in the early 1960s. Today it contracts this work to third-party plants throughout Texas, Oklahoma, and California. Aermotor maintains a strict rule that all parts going into its windmills be sourced and manufactured in the United States. For example, Aermotor buys sheet metal in four-by-ten-foot sheets from California Steel Industries in Fontana, California; rough castings of hubs and gears from Fraser & Fraser in Coolidge, Texas; and mainframes from Central Machine & Tool Co. in Enid, Oklahoma. A decade ago, Aermotor began requiring ductile iron to be used in its mainframe casts because it is less porous and holds up better than traditional gray iron castings. About seventy-five feet past the entry door and entrance to the front office are racks, baskets, and pallets to organize and hold this inbound inventory. Immediately nearby are pallets of sheet metal. On a row of shelves near the same area are dies, many originating from the 1930s, which are still used to make blade pieces, turntables, and tower sections.[11]

Straight ahead to the far end of Aermotor's building is a mixture of both early and modern metal-working machines for grinding, turning, drilling, bending, and punching holes. For example, sheet metal is punched into required shapes, such as wind wheel blades and tails, by a machine referred to by the employees as the "dinosaur." This 500-ton, air-powered press, which was made by Dreis & Krump Manufacturing Company, came from Aermotor's Chicago plant. The plant also contains a 1949 hob-back machine that cuts the teeth into gear castings. While operating the older equipment requires highly skilled machinists, the company in recent years has added three computer-controlled milling machines to handle more complex work. These were substantial investments made by the company starting in 2010. The automated horizontal lathe cost Aermotor about $110,000 alone, but it increased the company's production by 30 percent. The company makes every effort to minimize metal waste. For example, after the blade and helmet pieces are cut from the sheet metal, the excess or "drops" are used to make other parts, such as sail ribs and upper furl ring assemblies. Metal shavings from drill presses, lathes, and grinders are crated and sold for scrap.[12]

Finished parts are returned from the machining area to the center of the plant on handcarts for assembly. Cut sheet-metal blades, for example, are run through a roller to give them the proper and consistent curvature long characteristic of Aermotor wind wheels. A single worker assembles each wind wheel. Since the sheet metal is already coated in protective zinc, wheel sections no longer need to be galvanized, unless specifically requested by the customer. Aermotor has two workers whose sole job is to make the windmill helmets. The process hasn't changed much since the company began making its first oil-bath windmills in 1915. From start to finish, each helmet takes about forty-five minutes to an hour to bend, shape, and solder. Another worker joins the mainframes with their components, including hubs and shafts, bearings, small and large gears, pitman arms, guide wheels, oil rings, and pitman guides. Many of these parts are held in place by cotter and shaft pins. A heavy hammer is still used to strike the top of each steel pitman guide to make sure the geared mechanism is snug in the mainframe. Each windmill

A worker at Aermotor's San Angelo plant assembles a windmill mainframe and hub in 2012. Photograph by author.

A worker shapes and solders a sheet-metal windmill hood at the San Angelo plant in 2012. The process is much the same as it was when Aermotor's oil-bath windmills entered the market in 1915. Photograph by author.

is assembled on the stub tower and tested by manually turning the wind wheel before being partially disassembled and packed onto pallets for shipping.[13]

Aermotor installs Garlock bearings in all of its new windmills but offers traditional babbitt bearings as repair parts to rebuilders of earlier oil-bath models. Garlock is made from a Teflon-based composite, while babbitt consists of a combination of lead and tin melted together. Aermotor claims that Garlock is more rugged than the softer babbitt material, and without oil the "windmill gearing could run for a month," while an unoiled babbitt bearing would seize the windmill in just a couple of days. While Aermotor babbitt bearings are manufactured in the United States, the cost of the material to make them is impacted by world market prices for lead and tin. When severe flooding struck Thailand in the fall of 2011, for example, smelters were damaged, causing prices on both lead and tin to jump and making the cost of babbitt more expensive.[14]

Outside the main Aermotor factory is a smaller building that is used to stencil the oxidized red paint lettering "Aermotor, San Angelo, TX USA" onto the vane sheets. This building is also used to cut wood boards and assemble them into various-size crates for shipping wind wheel sections and tails. The crates are

Stencils are used to paint the Aermotor name on the various sizes of windmill tail vanes. Photograph by author.

stenciled "Aermotor" on each of their sides. Behind the main plant building is a laydown yard to store steel angle and other windmill tower components. Aermotor punches the bolt holes of the steel tower pieces on site before trucking them in batches to Fort Worth, Texas, for dip-galvanizing at US Galvanizing. Once this process is finished, the pieces are returned by truck to the San Angelo plant inventory.[15]

Windmill mainframe and stub tower orders are packaged and strapped onto pallets in the same area of the plant where materials for assembly are received. These outbound orders are loaded into truck trailers for delivery to dealers and installers throughout North America. In 2011 and 2012, Aermotor claimed to work with as many as 500 dealers, who collectively ordered about 2,000 new and replacement windmills. These dealers are either licensed contractors in the water-well business or distributors that sell to the industry. About 4 percent of total domestic sales in 2012 went to installers on American Indian reservations scattered throughout the US Southwest.[16] The company states that an order received from a customer in the morning will be shipped out the same day because there are sufficient windmill parts in inventory or production.[17]

In 2011, Aermotor opened its first branch in nearly twenty-five years—at Kearney, Nebraska. The operation, which includes a 4,000-square-foot warehouse, covers the windmill-rich territory of Nebraska, Kansas, Colorado, Wyoming, and

South Dakota. From this facility, the company also started rebuilding some used Model 702 and 802 windmills for sale, but this activity remains small for Aermotor, as larger rebuilders, such as Dakota Windmill in Hurley, South Dakota, and Muller Industries in Yankton, South Dakota, dominate this market.[18]

Since 2008, new windmill sales have softened across North America for myriad reasons. First, the global economic recession that started in the United States that year triggered collapsing real estate prices, rising fuel costs, and a slowdown in consumer spending. Numerous manufacturers, including Aermotor, struggled financially against diminished customer orders between 2009 and 2010. Aermotor furloughed several of its production line employees during the depth of the recession, cutting its overall headcount from twenty-five to twenty-two. Although it was a small number of affected employees, Bob Bracher still called it one of the most difficult actions he had to do as president of the company.[19] Aermotor benefited from a small uptake in windmill orders in 2011 from West Texas and Southwest ranchers facing one of the worst droughts in the region since the 1950s. However, the drought caused ponds, streams, and wells throughout the region to dry up, forcing many cattlemen to sell their herds to counterparts in the upper Midwest for finishing. Some ranchers attempted to hold on to their herds by resorting to burning prickly pear cacti to hydrate them, a practice dating back to the mid-1700s to counter drought conditions.[20]

However, the biggest threat to the windmill industry in recent years has been the proliferation of other remote water-pumping technologies. One of those technologies, and the quickest to emerge on the market, is solar-powered water pumps. These water pumps trace their roots to the early 1970s and closely follow the development of photovoltaic solar panels in the decades that followed. Initial research focused on photovoltaic-powered DC pumps. By the mid-1980s the technology was being introduced commercially by manufacturers to farmers and ranchers. While these early pumps were less expensive than traditional windmills, they pumped at shallower well depths—less than 120 feet—and frequently broke down.[21] Since the 1990s, the durability and capabilities of solar pumps have increased substantially through improved electronic circuitry, digital controls, and the addition of various pump types, such as helical rotor, diaphragm, and centrifugal. There are also aboveground solar pumps that draw water from a pond, for example, and discharge it into a pipeline that then can transport the water over miles to where it's needed. The cost of these more sophisticated pumps has subsequently increased and in most cases is now equivalent in price to new mechanical water-pumping windmills. One of the most significant solar pump advances came in the mid-1990s with the introduction of the Grundfos SQ Flex pumps. Grundfos of Denmark designed these pumps with the flexibility for either DC or AC use. Specially designed digital circuitry gives the pump controller the ability to analyze and adjust the amount of incoming voltage from the solar panels to ensure that the pump operates at an optimal level. When the sun is not shining, the farmer or rancher can still fill water storage tanks by switching on a portable AC fuel-powered generator, for example, and the SQ Flex pump can accom-

A solar pump installed during the summer of 2012 in southwest Wyoming, which has the capacity to pump ten gallons of water per minute. Courtesy Pronghorn Pump & Repair, Glenrock, Wyoming.

modate that change in the power source. When the pump controller electronics senses that the AC power has been interrupted—for example, when the portable generator uses up its fuel and shuts off—it then switches back to the photovoltaic panels to deliver DC power to the pump. The SQ Flex technology caught the attention of the Museum of Modern Art in New York, which included the pump in its "SAFE, Design Takes on Risk" exhibit in 2006, and it is now in the museum's permanent collection. Other manufacturers, such as Bernt Lorentz GmbH of Henstedt-Ulzburg, Germany; Franklin Electric, based in Fort Wayne, Indiana; SunPumps in Safford, Arizona; and Bison Solar Pump of Balko, Oklahoma, have begun introducing similar flex technologies and other advancements into their solar pumps. Another increasingly attractive quality of solar-based water pumps among farmers and ranchers is that their installation and maintenance can be done from the ground. There is no need to climb a windmill tower and conduct this work high in the air. Solar pump installation is also lighter work in that plastic pipe is typically used on shallow-well pumps and can be placed into or lifted out of the well hole by hand. It is not uncommon for solar pumps to be installed over former windmill wells.[22]

Despite the cost variable, windmills have some operational advantages over solar pumps based on geography, well depth, and weather conditions. First, unlike solar pumps, windmills are not dependent on sunlight to operate. The wind may blow day or night. Second, precautions must be taken with solar pumps in cold climates to keep water lines from freezing and burning up the pump, whereas a windmill's wooden pump rod serves as a shear pin and breaks if there is ice in the lines rather than harming the pump mechanism. In addition, windmills are structurally more rugged than solar pumps, particularly in their ability to stand up to vandalism caused by bullets. Solar pumps with their glass photovoltaic panels are vulnerable to bullet damage, especially on public grazing land that is shared with hunters. When a bullet strikes the solar panel, the protective glass plate breaks, causing an immediate reduction in voltage output from the solar cells and diminished water flow from the pump. The risk of dehydrating livestock is great in that it may take days before the damaged solar panels are detected and repaired. And the cost to replace a solar panel is in the hundreds of dollars. A windmill, on the other hand, may suffer a bullet hole to its sheet-metal tail but will still continue to operate. Attempts have been made in recent years to install solar pumps on the expansive Navajo reservation in the US Southwest. There, too, the pumps suffered a high rate of vandalism from bullets. Consequently windmills remain the predominant pumping source for meeting both domestic and livestock water requirements on the reservation, with the Navajo Nation operating more than a thousand of them.[23]

Another competitive technology to mechanical water-pumping windmills, though better suited to higher-volume water applications such as village water supply and irrigation, is a system called wind-electric water pumping. This technology was developed by the USDA in the late 1970s as an alternative to natural gas– and diesel-fueled irrigation pumps in the US Midwest, which were becoming increasingly expensive to operate. In a wind-electric pumping system, the three-phase motor on a submersible pump is driven at variable speed directly by the wind turbine's three-phase AC alternator. Most of these early wind turbines, however, quickly proved unreliable, and interest in this form of rural water-pumping had diminished by the early 1980s. In the late 1980s, the USDA Agricultural Research Service in Bushland, Texas, began working with Bergey Windpower Company of Norman, Oklahoma, in support of a US Agency for International Development–funded village water supply project in Morocco. The Bergey wind turbines were simpler and more reliable than the previous turbines used, and this allowed the researchers to focus on the pumping application. Bergey and the USDA refined the technology, developed control electronics, and field-tested several different pumping systems with simulated pumping depths of 100 to 400 feet. In 1988, Bergey installed the first commercial wind-electric water-pumping system in Ain Tolba in northeast Morocco to replace an inoperable diesel-powered water pump for the village water supply system. A few years later, the USDA began working with the Bergey 1.5-kilowatt, three-bladed wind generator, which was then connected to a 1.1-kilowatt electric motor and powered a 740-watt submersible pump motor. This ten-foot-diameter wind turbine was of similar size to mechanical windmills, offering the USDA a basis of comparison between the two types of machines. More than 700 hours of tests were conducted, with the system operating at seven different pumping heads ranging from 55 feet to 190 feet. At the West Texas test site, the USDA found the wind-electric system's average daily water volume exceeded the mechanical windmill by about 1,000 gallons, or 45 percent more water, almost year-round. The agency concluded that wind-electric pumps operated better than mechanical windmills when the average wind speed exceeded eleven miles per hour but were comparable when wind speeds ranged from eight to eleven miles per hour.[24] Other operational benefits to emerge from improved wind-electric pumps in the 1990s were the ability to deliver higher water vol-

A Bergey ten-kilowatt wind-electric water-pumping system installed at Ain Tolba in northeast Morocco in 1988, eliminating the community's reliance on diesel engines to pump water. Courtesy Bergey Windpower Company, Norman, Oklahoma.

umes and reduced maintenance costs (no leathers to change). Since wind-electric pumping systems today use modern high-reliability small wind turbines, they require far less maintenance and have better storm protection. Another benefit to wind-electric systems has been the ability to erect the turbine towers away from wells since power can be transmitted through electric cables to the well pumps at distances of up to a half mile. This technical attribute is particularly important in hilly terrain where turbines need to be placed at the highest elevations to optimally catch the wind, while wells may be located in the valleys. By the late 1990s, several hundred wind-electric pump systems, from 1 to 10 kilowatts in size, had been installed in more than twenty, mostly developing countries.[25]

The overall result of shrinking sales of windmills in the United States—Aermotor estimates that there are now only about 4,000 units sold a year, which includes those of competitors—has encouraged the company increasingly to look overseas for sales opportunities, and thus it has recently expanded its export activities.[26] Orders from abroad come to Aermotor at San Angelo in a variety of methods and from an array of customers. The company sales staff generally receives export queries by e-mail and telephone from foreign aid agencies, nongovernmental organizations, charities, and agricultural equipment distributors. Once the orders are approved, which can take months to finalize due to the coordination of complex financial arrangements, the shipments are generally trucked from the San Angelo plant to Houston, where a designated freight forwarder has them loaded into ocean containers that are then stowed on vessels for overseas delivery. By 2012 the company had contracts for routine exports to dealers in Nigeria, Peru, South Africa, Russia, and Romania, with more in the works.[27]

The United Nations reported in 2010 that globally eight out of ten people living in rural areas still lack access to clean drinking water. This problem is particularly acute in sub-Saharan Africa and Oceania, where women and children routinely walk long distances each day to collect life-sustaining water, enduring

An Aermotor windmill with a twelve-foot wind wheel being installed on a ranch south of Moscow, Russia, in 2007 for watering cattle. Courtesy Aermotor Company, San Angelo, Texas.

both physical strain and lost time that could otherwise be spent obtaining an education or improving one's social and economic well-being. Many sources of water also experience seasonal shortages and contamination. It is estimated that every twenty-one seconds a child dies from illnesses related to consumption of water from unclean sources, such as open wells, ponds, rivers, and streams.[28] In some regions the only chance to obtain clean water is by tapping it from underground aquifers. This requires drilling wells and installing hand pumps since electricity in most rural areas of developing countries is unavailable. Aermotor is often approached by humanitarian organizations to supply water-pumping windmills to villages in underdeveloped countries.[29] These windmills are appreciated by aid groups and the local populace for their ruggedness and low maintenance. They have proved their ability to stand up to abuse from violent storms and vandalism. Other pumping technologies have been less successful in developing countries. Generators, for example, require expensive petroleum-based fuels, while solar pumps are easily stripped of their photovoltaic panels by thieves. As Chief Mobolaji Saint Matthew-Daniel of the Nigerian village Sankara explained in a 2013 news report, "The windmill is high up and well installed into the ground so people cannot vandalize it." The village operates a new Aermotor windmill, which supplies enough water to sustain about 200 people and 500 cattle.[30]

In recent years, mechanical water-pumping windmills have also been installed in some of the most dangerous parts of the world, such as Afghanistan and Iraq. In 2011, Aermotor received formal recognition from the State of Texas and the US military for its supply of a windmill to a village in the Afghanistan province of Logar, south of Kabul. A year earlier, while on assignment to assist the agricultural development staff in Afghanistan, Craig Runyan, president of Williamsburg, New Mexico–based SolandAer and a longtime instructor for New Mexico State University's Windmill Technology Certification Workshop, proposed the use of windmills for pumping water to the Agriculture Development Team attached to a National Guard unit from South Carolina. They were immediately interested, and the team leader, Sergeant Major Michael Hall, contacted the nonprofit TNY in the United States to request support for the windmill project in Logar. TNY offered $5,000 in financial support. Hall then contacted Runyan, who had returned to New Mexico, by telephone and asked what could be done with that amount of

Michael Guy Morrow became head of Aermotor Windmill Company in June 2012. Photograph by author.

money. Runyan immediately called Aermotor, knowing that $5,000 would only cover about half the cost of a complete water-pumping windmill system, to see what the company could do. Bracher replied, "Anything in the house, anything for our troops." After identifying the well and determining the best wheel and tower sizes, Runyan ordered a ten-foot Aermotor and a twenty-seven-foot tower with pump rods, cylinder, and stuffing box to be sent to the Agricultural Development Team in Logar. Aermotor even covered the $10,000 cost to ship the windmill from San Angelo to Bagram Airfield in Kabul. The windmill was installed in the summer of 2011 and immediately began supplying water for household, agricultural, and government-building use.[31]

Citing health reasons, Bracher decided in June 2012 to sell his interest in Aermotor back to his three investor partners and retire to ranching. He appointed Michael Guy Morrow, who had been hired by the company a year and half earlier to focus on sales and marketing, to take over following his departure.[32] For the previous ten years Morrow had worked for American West Windmill Company of Abernathy, Texas, the primary US importer of FIASA-manufactured windmills. While working as a territory salesman for American West in Broken Bow, Nebraska, Morrow was approached in 2010 by Preferred Pump and Equipment to serve as its branch manager in Grand Island, Nebraska, selling Aermotor windmills, and that was when he met Bracher.[33]

During his time at American West, Morrow learned about the mechanical and commercial problems of both the US-built Aermotor and FIASA-made windmills. Customers often complained about Aermotor's brake bands prematurely breaking. Morrow fixed this problem by directing manufacturing staff to shorten the brake band button, beefing up the bar, and taking the threads off, thereby preventing it from hanging up and snapping. Aermotor also fixed quality issues associated with small gear loss on the hub shafts and designed a new timeable spoke for use on all sizes of its windmills. Morrow also oversaw the installation and programming of computer-controlled milling machines at the plant.[34]

By the start of 2012, Aermotor had become the last of the country's once-vibrant industry of mechanical water-pumping windmill manufacturers. Dempster Industries of Beatrice, Nebraska, a windmill company founded in 1878, ten years before Aermotor, closed its operations at the end of 2011. The company had mostly ended manufacturing of new windmills by the start of the 2000s and focused on its fertilizer spreader production, for which it was well known in the agricultural community. Dempster, however, continued to sell windmill parts until its last employee, Rosemary Heble, was discharged by the owners on December 28, 2011. She had reportedly worked for the financially troubled company unpaid since the end of 2010. In early 2013 Canadian businessman Ryan Mitchell acquired the assets of the newly formed Demspters LLC for $1 million in an effort to restore the company's manufacturing of windmills, submersible pumps, fertilizer spreaders, and recycling trailers, and rehired a handful of former longtime Dempster Industries employees, including Heble, to work for him. There were also industry rumors that Dempster Industries planned to manufacture wind-electric generators. When Mitchell applied for a $300,000 development loan from the City of Beatrice in late 2013, he met with resistance for myriad reasons, including his previous unsuccessful business ventures in Canada, a corporate registration of Dempsters LLC in Wyoming instead of Nebraska, and lack of clarity into the extent of his ownership of the company's property and assets. His loan request was denied. The once-proud Dempster factory buildings have continued to crumble.[35] With dimming prospects Mitchell auctioned off 90 percent of the factory equipment on April 4, 2014, hoping to focus future manufacturing on the company's Alleycat recycling trailers and spreaders at another location.[36]

For Aermotor, the demise of former competitor Dempster has not been a point of celebration but rather a stark reminder that the company has to work even harder to avoid a similar fate. Competition from overseas-manufactured windmills and the proliferation of solar-powered water pumps have kept Aermotor's windmill sales flat. The company has about twenty-five employees overall, and while it does not publicly share windmill production figures, since 2012 it has continued to manufacture and sell an estimated 2,000 windmills a year. Shortly after taking over management of Aermotor in 2012, Morrow and his fellow business partners—Jim Kosub, Clint Griffin, and T. Cy Griffin—formed San Angelo Supply Company. While a separate entity from Aermotor, the business unit is located in the same building as the windmill manufacturing. "We thought why not offer local contractors all their water-well products as well. Most contractors do much more water systems work than they do windmill work," Morrow said. He also explained that the business is under pressure to diversify. "The windmill business is not growing. It is declining with the solar pump explosion over the last ten years," he said. San Angelo Supply sells Franklin Electric's SolarPAK line in addition to pressure tank lines, pump cables, PVC and galvanized pipe, valves,

Aermotor Windmill Company plant at San Angelo, Texas. Courtesy Don Treadwell, San Angelo, Texas.

fittings, and other related parts to install the systems. Commenting on the impact of this product addition on the company's bottom line, Morrow said, "I never knew just how much the solar pump had cut into windmill sales until we started seeing how fast they go out the door once we got the Bison Solar Pump line. We are just tipping the iceberg." When the company began selling these pumps in 2012, it focused on West Texas and Nebraska markets. With the products in high demand, Bison Solar Pump experienced difficulty at first keeping up with sales, but since 2013 it has met San Angelo Supply's orders. In another recent move, San Angelo Supply uses the computer-based milling machines of Aermotor to produce some oil-field products. "We are taking it slow and being cautious because we don't want to overload our employees and hurt our windmill production," Morrow said.[37]

Despite competitive threats of new remote water-pumping technologies, mechanical windmills continue to maintain their presence on ranchland across the Southwest and Great Plains. Casual observers, as well as historians, artists, photographers, and journalists, remain captivated by the stoic gracefulness of these machines as their blades turn in the wind. Windmills also stand as a physical representation of human mechanical prowess of years ago to raise life-giving water otherwise hidden beneath the earth's surface.[38] Although a shadow of what it once was during its Chicago heydays, Aermotor remains the most widely recognized windmill on the market, which is testament to the company's founders, La Verne W. Noyes and Thomas O. Perry, who through sound engineering and clever branding popularized the Aermotor and ensured its place as an icon of the American West.

APPENDIX

How an Aermotor
Oil-Bath Windmill Works

An Aermotor, like most traditional US water-pumping windmills, consists of three main parts: the windmill, including wind wheel, hub and shaft, mainframe, gearing, and vane assembly; a tower; and drop pipe, sucker rod, and cylinder. Each piece of the windmill, like a three-legged stool, is useless without the other. Each part is engineered with a specific purpose to ensure efficient and long-lasting operation in the field.

Wheels on Aermotor windmills, no matter the age or size, consist of eighteen contoured, trapezoidal, sheet-metal blades held in place by special metal clips onto two concentric steel rings and secured by steel spokes that are attached to the hub. The wind wheel starts turning when a sufficient wind, of eight to ten miles per hour, strikes the curved blades slightly off center. Keeping the slightly off-center wheel facing the wind is a V-shaped sheet-metal tail vane attached to the back of the mainframe of the mill by a hinged steel tailbone and vane spring. The mainframe, which supports the wheel sections and tail, swivels on a mast pipe. When the wind becomes too great—for example, reaching above twenty-five miles per hour—the force of the wind starts to furl the mill; turning it on the mast pipe so both the tail and wheel are parallel to the direction of the wind. The tension of the vane spring increases as the mill is furled and returns the mill to face the wind as the velocity decreases. Tension on the vane spring can be adjusted to furl faster or slower in the wind depending on the workload of an individual mill. If the vane spring is connected to the outermost hole in the tailbone, or the one closest to the vane, the tension is then greatest on the spring and therefore holds the wheel into the wind at higher velocities. The windmill can also be taken out of service by a manual furling device or mechanical regulator.

To access the wind, windmills are erected on tops of wood or metal towers. All-metal Aermotor towers are known for their distinctive attributes, namely their loop steps, which are connected to a single corner post and towers with the dovetailed tops that hold the mast pipe in place. Today the company's towers range in height from twenty-one feet to forty-seven feet, although taller towers are available from other manufacturers. For the best access to the wind, the tower needs to be at least fifteen feet taller than any nearby wind obstruction, such as buildings and trees, within a radius of 400 feet. A tower can be assembled on the ground

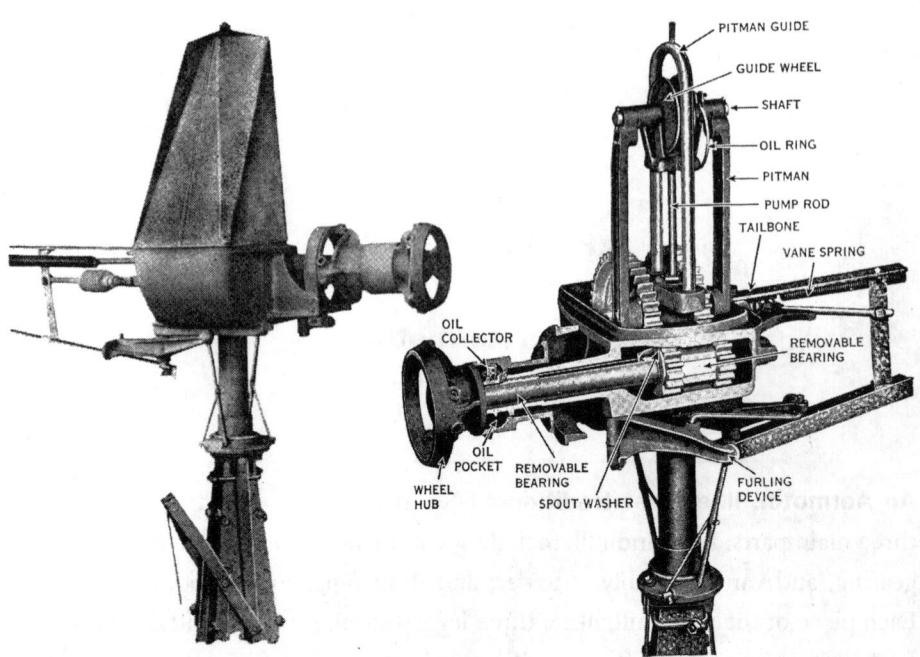

An assembled Aermotor Model 702 motor (left) and cutaway showing the internal parts (right). From Aermotor Company, *How to Choose Your Water Pumping System* (Chicago, ca. 1954), 30.

and hoisted into place by a crane, winch, or boom truck or can be built up section by section from the ground. Either way, it is important for the tower to be centered over the well, anchored securely to the ground, and plumb. Aermotor recommends that anchor posts at the bottom of its towers for fourteen-foot mills or less be set into holes about two feet in diameter and four feet nine inches deep; the sixteen-foot mills should have anchor post holes six feet six inches deep. The anchor post in the holes should be covered with at least two feet of concrete or tightly packed rocks before being topped off with dirt. From the ground to the stub tower, the angled steel corner posts come in sections bolted together and are held in place by a series of horizontal girts. The ground-level section of a four-post tower has crisscrossed angle iron braces on each side. The upper sections are pulled together similarly by crisscrossed round steel braces. On Aermotor "widespread" tower bases, the cross braces are reconfigured on one side to allow easier access to the well pump and for removing or placing pipe in the well. Near the top of the tower and just below the wind wheel is either a wooden or metal platform on which a person can stand during maintenance and oil changes. To reach the head of the windmill, looped steps are attached to one leg of the Aermotor tower for climbing.

The wind wheel, when in motion, drives the hub and shaft, which are connected and housed within the cast iron mainframe. Inside the gearbox is the shaft with two small pinion gears attached and evenly spaced by a babbitt or Garlock bearing. These small gears mesh with two larger gears, which drive two pitman arms that are held together at their opposite ends by a short shaft with roller guide supporting the yoke. As the gears drive the pitman arms, they transfer their rota-

Section of the Model 702 wind wheel. From Aermotor Company, *How to Choose Your Water Pumping System* (Chicago, ca. 1954), 31.

tional motion to the reciprocating action needed to operate the pump rod. The up-and-down motion of the pitman arms is kept in place by a looped steel pitman guide, to which is also bolted the protective sheet-metal helmet of the windmill. A metal ring supported by the pump rod yoke picks up oil from one of the large gears on the downstroke of the pump rod and carries it to the cross shaft, keeping all the upper parts lubricated. (The large gears stay continuously lubricated by an oil bath contained in the bottom of the mainframe. A six-foot Aermotor oil-bath mill requires a quart of lightweight, nondetergent oil, while the eight- and ten-foot mills take two quarts; the twelve-foot, one gallon; the fourteen- and sixteen-foot, two gallons; and the twenty-foot, five gallons.)

Attached to the yoke, the pump rod is suspended through the center of the mast pipe. Inside the tower it attaches to a swivel, where it connects to a series of narrow rods—wood is recommended—which in turn are bolted end to end and attach to the sucker rod inside the steel drop pipe of the well. At the bottom of the well the sucker rod screws onto the plunger on a pump cylinder. These pumps consist of a steel rod, one check valve, and one plunger valve. The check valve screws onto the bottom of the cylinder, and above is a plunger valve that is attached to the sucker rod. Water flows into the cylinder through the check valve during the upstroke of the plunger valve, and then on the downstroke the check valve closes, pushing the water through the plunger valve and into the drop pipe. The plunger valve then closes on the upstroke, pushing the water higher while refilling the cylinder below through the check valve. With each stroke, water fills the drop pipe until it reaches the surface and discharges into a tank or pond via a raised splash pipe. If the water is destined for an overhead tank, for example, then a stuffing box is used to force the water higher until it discharges into the holding tank. As a general rule, when calculating the length of a cylinder to use, the length

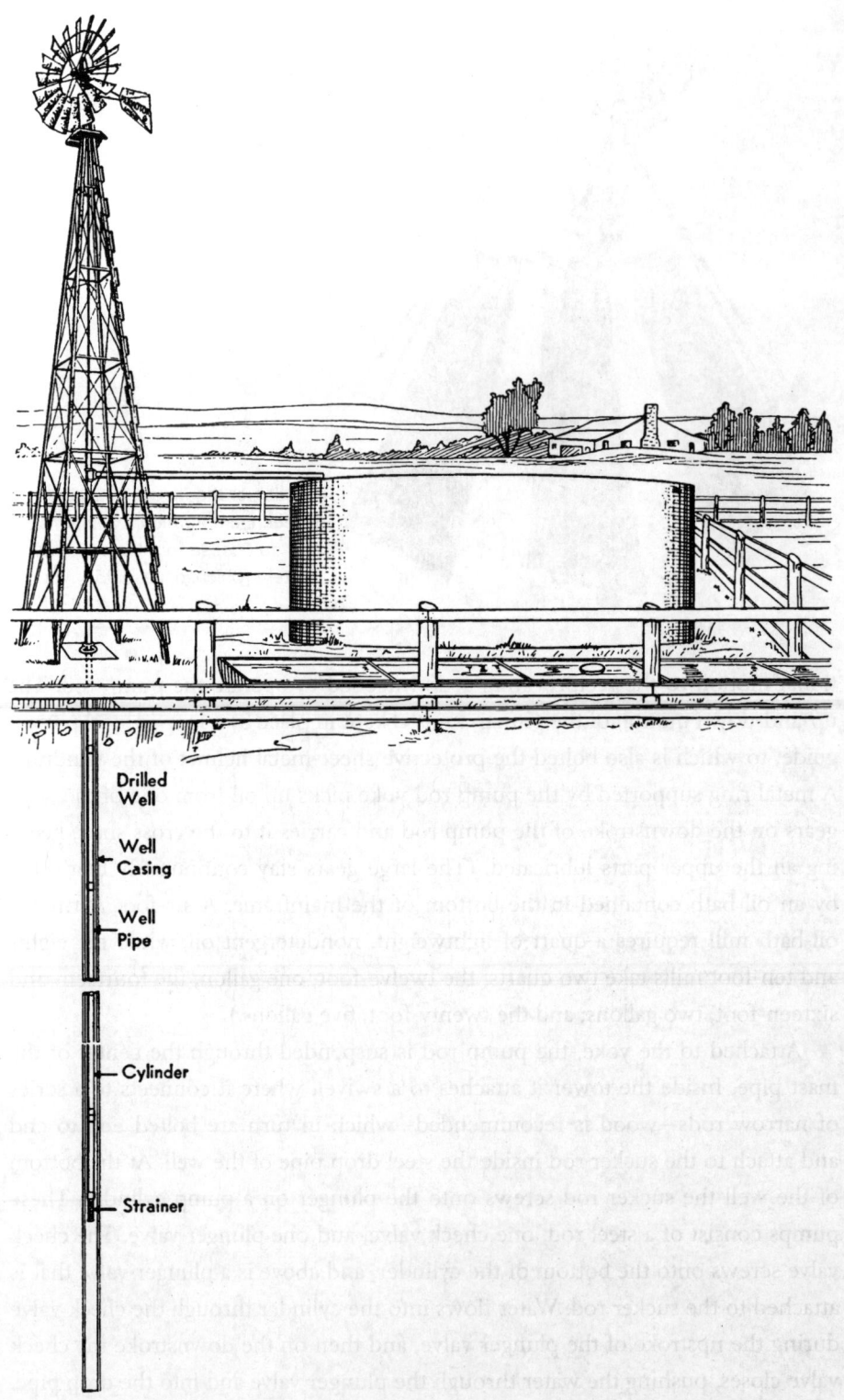

Drilled
Well

Well
Casing

Well
Pipe

Cylinder

Strainer

Typical setup of an Aermotor windmill for a ranch or farm. From Aermotor Company, *How to Choose Your Water Pumping System* (Chicago, ca. 1954), 40.

of the pump stroke should be equal to or greater than (in inches) the diameter of the Aermotor's wind wheel (in feet); for example, an eight-foot mill's stroke equates to about eight inches.

In light or variable winds, the actual amount of water pumped by a windmill is considerably less than its theoretical capacity. With the Aermotor vane spring set at maximum tension, the six- and eight-foot mills when running at full speed make about thirty-two strokes of the pump per minute; the ten-foot, twenty-six strokes; the twelve-foot, twenty-one strokes; the fourteen-foot, eighteen strokes; and the sixteen-foot, sixteen strokes. Aermotor pitman arms can be adjusted from the standard long stroke to a shorter one, but the pump capacity will be reduced by 25 percent in exchange for a gain of 33.3 percent in well depth. When considering the use of a windmill, the Beaufort Scale of Wind Force, maintained by the National Weather Service, should be consulted for an area's average wind speed; well depths and potential output should be known; and water requirements should be understood. For example, a milk cow consumes about thirty-five gallons of water a day, whereas a steer requires fifteen gallons. A horse takes in about twelve gallons of water a day, while smaller livestock, such as hogs and sheep, drink between two and four gallons daily.

NOTES

CHAPTER 1

1. Wulff, *Traditional Crafts of Persia*, 284–85; Harverson, *Persian Windmills*; Baker, "Overview of Horizontal Windmills," 2.

2. Pernoud, *Die Kreuzzuge*, 363.

3. Wulff, *Traditional Crafts of Persia*, 285; Berman, "Preservation of Records for Thirteenth Century Windmills," 227.

4. Bauters, "Oldest References to Windmills in Europe," 113–15.

5. Brouwers, e-mail message to author.

6. Ibid.

7. Denewet, "Niet Nederland maar Vlaanderen was de bakermat!"

8. Brouwers, e-mail message to author.

9. Ibid.

10. Devyt, *Westvlaamse windmolens*, 30–32, 71.

11. Van Duijnhoven, "Irrigating Time."

12. Hills, *Power from Wind*, 136.

13. Traill, ed., *Social England*, 120.

14. Ibid.

15. Hans Korsman, "De Noord-Hollandse achtkante binnenkruier."

16. Brouwers, e-mail message to author.

17. Ibid.

18. Needham, *Science and Civilization in China*, 4:556–60; Hong-Sen Yang, *Reconstruction Designs*, 85; Baichun Zhang, "Ancient Chinese Windmills."

19. Gregory, *Industrial Windmill in Britain*, 113

20. Hills, *Power from Wind*, 92–93; de Little, *Windmills of England*, 57–61; Reynolds, *Windmills & Watermills*, 98–100.

21. Smeaton, "An Experimental Enquiry," 138–74.

22. Freese, *Windmills and Millwrighting*, 8; Nijs and Brouwers, "Wieksystemen."

23. Wailes, *The English Windmill*, 94; De Decker, "Wind Powered Factories."

24. Nijs and Brouwers, "Wieksystemen."

25. Bauters and Pouw, *Van Zadelsteen tot Zetelkruier*, 225.

26. Gregory, *Industrial Windmill in Britain*, 136.

27. Goslin, *Relative Advantages of Wind*, 7.

28. Ogden and Burke, "Windmill at Flowerdew Hundred," 4–5; Cline, "Landmark Windmill Finds New Home"; Harris, e-mail message to author. (In the mid-1970s, English millwright Derek Ogden was commissioned to build an eighteenth-century-style English post mill on the site of the original Flowerdew windmill. The replica windmill was opened to the public in April 1978. In November 2009, the windmill was donated to the American Wind Power Center in Lubbock, Texas, where it was reconstructed and remains on display today.)

29. Norberg, "Fort Ross Shows Off New Russian-Built Windmill"; Birkland, "Replica Russian Windmill Gifted"; Fort Ross Windmill, "Fort Ross Windmill."

30. Quinn, *Saltworks of Historic Cape Cod*, 15–17; Church, "Padanaram Salt Works," 490; Gillis, "Sea Breezes to Salt," 9.

31. Marks and Coleman, *History of Wind-Power on Martha's Vineyard*, 11.

32. Ibid., 15.

33. Watkins, "A Common Crystal," 175.

34. Gillis, "Sea Breezes to Salt"; Hayward Area Historical Society, "Short History of Hayward."

35. Wilson, *History of Crisfield*, 41–42.

36. John Mullens & Sons, *Divining-Rod*, 15–16.

37. Dick, "Water," 222.

38. Ramsower, *Equipment for the Farm*, 150–52.

39. Fuller, *Underground Waters*, 7, 39–41.

40. Bowman, *Well-Drilling Methods*, 88–89; Baker, *Field Guide to American Windmills*, 25–27, 89.

41. *Well-Drilling Methods*, 90–92; *A Field Guide to American Windmills*, 90.

42. Ramsower, *Equipment for the Farm*, 155; Baker, *Field Guide to American Windmills*, 90–91.

43. Baker, *Field Guide to American Windmills*, 18–19.

44. Dickerman, *How to Make the Farm Pay*, 681.

45. Allen, *American Cattle*, 11.

46. Lusk, "Water Sources in Early West Texas," 58.

47. Dick, "Water," 242.

48. Cook, "Water Stop," 14; Baker, "Windmills and Railroad Water Systems," 2–5.

49. "Un Moteur à Vent," 17; Walter and Girard, *Éolienne Bollée*, 22; B[rouwers], "Windmotoren," 19.

50. Walter and Girard, *Éolienne Bollée*, 22–23.

51. Rogier, e-mail message to author, February 15, 2013.

52. Baker, "Patents as a Key"; Patent No. X7739, James Kerr, US Department of Commerce, Patent Office.

53. Ward to Miss Elizabeth T. Ward, September 1, 1850.

54. "Improved Windmill," 236; Johnson, *Wind as a Motive Power*, 1–37.

55. "Improved Windmill."

56. "Inventions Wanted in Texas," 132.

57. [Munro-Fraser], *History of Contra Costa County*, 482.

58. Tustin, "Recollections of Early Days in California," 8; Patent Nos. 28,423 and 97,136, W. I. Tustin, US Department of Commerce, Patent Office; "Tustin's Improved Adjustable Windmill," 241; Bancroft, *History of California*, 7:96.

59. Manning, "Windmill in California," 33.

60. Ardrey, *American Agricultural Implements*, 142–43.

61. Patent No. 11,629, Daniel Halladay, US Department of Commerce, Patent Office.

62. Ibid.

63. Hurt, "Windcatchers and Eyecatchers," 27.

64. Jordan, "Evolution of the American Windmill," 9.

65. Sanford, "Windmills Saved Thirsty West," 13.

66. U.S. Wind Engine and Pump Company, *Is the Halladay Mill Durable?*

67. Baker, "Large-Diameter Halladay Standard Windmills," 3.

68. Robinson and Schielke, *John Gustafson's Historic Batavia*, 51–53; Popeck, interview.

69. *Halladay Standard Wind Mill.*

70. *H. C. Chandler & Co.'s Railway Business Directory and Shippers' Guide, for the*

State of Illinois, 44, 46, and "Halladay's Improved Wind Engine for Pumping Water and Grinding Grain" unpaged [advertisement].

71. *Illinois State Gazetteer for the Years 1864–65*, 146.

72. *Illinois State Gazetteer, 1878*, 708; Webb, *Great Plains*, 337; Swenson, "Wind Engines in Western Illinois," 72.

73. US Department of the Treasury, Bureau of Statistics, *Statistical Abstract, 1902*, 526.

74. Webb, *Great Plains*, 338; Wade to Walter Prescott Webb, July 5, 1927; Baker, "Turbine-Type Windmills," 39.

75. Baker, "Large-Diameter Halladay Standard Windmills."

76. Baker, "Windmills with Variable-Pitch Blades," 2–5; Baker, "Windmills and Towers of the Temple Pump Company," 6–7.

77. Wolff, *Windmill as a Prime Mover*, 75.

78. Ardrey, *American Agricultural Implements*, 143.

79. U.S. Wind Engine and Pump Company, *Descriptive Catalogue of U.S. Wind Engine & Pump Co.*, 41–43; Baker, "Windmills and the Union Pacific Railroad," 3–5.

80. Webb, *Great Plains*, 339.

81. Ibid.

82. Ibid., 340; Barbour to Walter Prescott Webb, July 15, 1927.

83. Hurt, "Irrigation of the West," 36; Hendrix, "Windmill Monkeys," 51–52; Kirk, "The Windmill Man," 16, 98–99.

84. Hill, "The Wind-Mill of To-Day," 161.

85. U.S. Wind Engine and Pump Company, *Descriptive Catalogue and Price List*, 5.

86. Ibid., 17.

87. Baker, *Field Guide to American Windmills*, 26–28.

88. "W. I. Tustin, Patentee and Sole Proprietor," 33.

89. [Pennsylvania State Agricultural Society], *Report on Wind Engine Test*, 2.

90. Ibid., "Office of U.S. Wind Engine & Pump Company, Batavia, Illinois. To Our Agents (March 6, 1885)," 2.

91. Ibid., 4–5.

92. U.S. Wind Engine and Pump Company, *Halladay Standard Pumping & Geared Wind Mills*.

93. U.S. Wind Engine and Pump Company, *Descriptive Catalogue and Price List*, 2.

94. "Windmill Letter," 30.

95. Barbour, *Wells and Windmills in Nebraska*, 31–35; Barbour, *Homemade Windmills of Nebraska*, 5–12; Danker, "Nebraska's Homemade Windmills," 14–15.

96. Wilcox, *Irrigation Farming*, 364–67; Davidson and Chase, *Farm Machinery*, 299–302; Baker, "Windmills That Get Down and Wallow on the Ground," 2–4; Baker, "Battle Axe Homemade Windmills," 2–3.

97. Wilcox, *Irrigation Farming*, 367–68; Barbour, *Wells and Windmills in Nebraska*, 56–66; Barbour, *Homemade Windmills of Nebraska*, 58–76.

98. Davidson and Chase, *Farm Machinery*, 302–3.

99. Holly, *Modern Dwellings*, 40.

100. Wolff, *Windmill as a Prime Mover*, 3.

CHAPTER 2

1. Ardrey, *American Agricultural Implements*, 215.

2. Baker, *Field Guide to American Windmills*, 34.

3. Patent No. 132,602, James A. Risdon, and Patent Nos. 207,189 and 220,083, Samuel W. Martin, US Department of Commerce, Patent Office.

4. Mast, Foos and Company, Springfield, OH, *Illustrated Catalogue of Mast, Foos & Co., Springfield, Ohio, Manufacturers of the Iron Turbine Wind Engine*, 6.

5. Mast, Foos and Company, *Improved Iron Turbine Wind Engine*, 1; Mast, Foos and Company, *Mast, Foos & Co., Springfield, Ohio, U.S.A., Manufacturers of Iron Turbine Wind Engines*, 3.

6. Collar, "Iron Turbine Wind Engine."

7. Mast, Foos and Company, *Mast, Foos & Co., Springfield, Ohio, U.S.A., Manufacturers of Iron Turbine Wind Engines*, 7–12, 14.

8. Ibid., 2.

9. "Windmills and Wind-Engines," 14; E. C. Leffel, *Croft's Improved Iron Wind Engine*; Patent No. 224,817, Henry Croft Sr. and Henry Croft Jr., US Department of Commerce, Patent Office.

10. Baker, *Field Guide to American Windmills*, 35–36; Baker, "Pioneer Metal Windmills," 2–6; Eide, "Free as the Wind," 14–17; Kirkwood Manufacturing Company, *Descriptive Catalogue*.

11. "Windmills and Wind-Engines."

12. "Windmills and Wind-Engines"; Patent No. 223,379, William H. B. Page, US Department of Commerce, Patent Office.

13. "Steel Wind Mills," 12.

14. Ibid.

15. Ibid.

16. Ibid.

17. Thomas O. Perry, "Thomas Perry in Household of Gideon Perry."

18. Thomas Osborne Perry to the University of Michigan, Ann Arbor. (Hereafter cited as Perry survey.)

19. "Editorial Notes," 364.

20. Perry survey.

21. Ibid.

22. Patent No. 146,548, Thomas O. Perry, US Department of Commerce, Patent Office.

23. Ibid.

24. Chaney, *University of Michigan*, 30.

25. Perry, *Experiments with Windmills*, 12.

26. Ardrey, *American Agricultural Implements*, 155.

27. Perry, *Experiments with Windmills*, 19.

28. Ibid., 22–23.

29. Ibid., 27.

30. Ibid., 27–67.

31. Ibid., 90.

32. Ibid., 93–94; Dole, "History and Development of the Windmill," 28.

33. Wood, "Windmills," 791; Bodmer, *Hydraulic Motors*, 267–68.

34. Newell, *Letter of Transmittal*, 9; Perry, *Experiments with Windmills*, 21.

35. Smith, "Memories."

36. Chaney, *University of Michigan*, 31.

37. Thomas O. Perry, letter to the editor.

38. "La Verne Noyes Family Record."

39. [Noyes], *Descendants of Reverend William Noyes*, 29.

40. "Bibliographical Note," La Verne and Ida Noyes Collection.

41. Patent No. 161,817, La Verne Noyes, US Department of Commerce, Patent Office.

42. Nichols, *Iowa State College of Agriculture and Mechanic Arts*, 125.

43. Patent No. 175,736, La Verne W. Noyes, US Department of Commerce, Patent Office.

44. Baker, "The 'Aermotor Man' and His Haying Tools," 3–5.

45. Patent No. 199,378, La Verne Noyes, US Department of Commerce, Patent Office.

46. Baker, "The 'Aermotor Man' and His Haying Tools," 5.

47. Ibid.

48. Nichols, *Iowa State College of Agriculture and Mechanic Arts*, 125.

49. Patent No. 243,955, La Verne W. Noyes, US Department of Commerce, Patent Office.

50. Patent Nos. 335,085 and 349,660, ibid.

51. Noyes, *A Score of Ways*.

52. Sears, Roebuck and Company, *Sears, Roebuck and Co.[,] Incorporated[,] Cheapest Supply House on Earth*; Redhead, Norton, Lathrop & Co., Wholesale Stationers & Jobbers, advertisement, *Bushnell's Des Moines City Directory*, 544.

53. Noyes, *A Score of Ways*.

54. Nichols, *Iowa State College of Agriculture and Mechanic Arts*, 125.

55. Thomas O. Perry, letter to the editor.

56. Dole, "History and Development of the Windmill," 28.

57. Aermotor Company, *The Aermotor*.

CHAPTER 3

1. Perry survey; Baker, *Field Guide to American Windmills*, 36–37.

2. Baker, "Product History," 7; Aermotor Company, *8 Ft. $25 40 Ft. Steel Galvanized Fixed Tower*.

3. "L. W. Noyes–The Aermotor Co.," 43; Baker, "Product History," 7.

4. Aermotor Company, *To Aermotor Agents. Bulletin No. 1*, 1.

5. Baker, "Product History," 7; Aermotor Company, *8 Ft. $25 40 Ft. Steel Galvanized Fixed Tower*; Aermotor Company, *To Aermotor Agents. Bulletin No. 1*, 3.

6. Aermotor Company, *To Aermotor Agents. Bulletin No. 1*, 1.

7. Ibid.; Aermotor Company, *To Aermotor Agents. Bulletin No. 4*; "Building Permits"; Baker, *Field Guide to American Windmills*, 38, 64–65.

8. Ardrey, *American Agricultural Implements*, 156–58.

9. Aermotor Company, *7th Annual Catalogue*, 21.

10. Ibid.

11. Patent No. 375,378, Thomas O. Perry, US Department of Commerce, Patent Office.

12. Patent Nos. 431,991; 11,181 (reissued); and 485,883, Thomas O. Perry, US Department of Commerce, Patent Office.

13. "The Aermotor Wind Mill," 488.

14. Smith and Winchester, Boston, *Smith & Winchester, Illustrated Catalogue*, 19; Baker, "Tilting Windmill Towers," 2.

15. Aermotor Company (Buffalo, NY), *Aermotor Company[,] Buffalo*, 76; Smith and Winchester, Boston, *Price List and Testimonial Circular*, n.p.

16. "The Aermotor Wind Mill"; Baker, "Tilting Windmill Towers."

17. Aermotor Company, *To Aermotor Agents. Bulletin No. 4*, 3.

18. Aermotor Company, *To Aermotor Agents, Bulletin No. 6*, 1.

19. Aermotor Company, *7th Annual Catalogue*, 135–42; Baker, "Tilting Windmill Towers," 2–4.

20. Baker, "Tilting Windmill Towers," 4.

21. Aermotor Company, *To Aermotor Agents. Bulletin No. 1*, 2; Patent No. 457,819, La Verne W. Noyes, US Department of Commerce, Patent Office.

22. Patent No. 457,820, La Verne W. Noyes, US Department of Commerce, Patent Office.

23. Aermotor Company, *8 Ft. $25 40 Ft. Steel Galvanized Fixed Tower*.

24. Aermotor Company, *To Aermotor Agents. Bulletin No. 2*, 1.

25. Baker, "Aermotor Windmill Towers," 7; Patent No. 591,418, La Verne W. Noyes, US Department of Commerce, Patent Office.

26. Aermotor Company, *Eleventh Annual Descriptive Catalogue*, 28; Baker, "Aermotor Loop-Step Tower Ladders," 6–7.

27. Aermotor Company, *Thirteenth Annual Descriptive Catalogue*, 26.

28. Ibid., 26–27.

29. Baker, "Aermotor Windmill Towers," 7; Baker, "Aermotor 'Trussed Tripod' Steel Towers," 8–9.

30. Aermotor Company, *Aermotor Towers Are Towers of Strength*; Aermotor Company, *Aermotor Trussed Tripod Towers*; Patent Nos. 889,395 and 12,842 (reissued), La Verne W. Noyes, US Department of Commerce, Patent Office.

31. Aermotor Company, *To Aermotor Agents. Bulletin No. 1*, 2.

32. Baker, "'Suburban Outfit' Windmill Towers," 2–3; Patent No. 523,864, La Verne W. Noyes, US Department of Commerce, Patent Office.

33. Aermotor Company, *To Aermotor Agents. Bulletin No. 5*, 3; Aermotor Company, *7th Annual Catalogue*, 106.

34. Baker, "Galvanized Steel Tanks and Troughs," 3–5; Patent No. 575,369, La Verne W. Noyes, US Department of Commerce, Patent Office; Aermotor Company, *Aermotor Stock Watering Troughs*; Baker, "Freeze-Proofing Windmills," 5–6; Baker, "Tank Heaters for Stock Watering," 5–7.

35. Aermotor Company, *Aermotor 4-Wheel Steel Truck*.

36. Patent No. 495,510, Thomas O. Perry, US Department of Commerce, Patent Office.

37. Brigolin, e-mail message to author, March 5, 2014.

38. Aermotor Company, *Twelfth Annual Descriptive Catalogue*, 4–5; Brigolin, e-mail messages to author, September 26, 2013, and October 28, 2013.

39. Baker, "Power Windmills," 3–7; Baker, "Power Aermotor Windmills," 3.

40. Aermotor Company, *Twelfth Annual Descriptive Catalogue*, 16–17.

41. Patent No. 616,134, La Verne W. Noyes, US Department of Commerce, Patent Office.

42. Aermotor Company, *Twelfth Annual Descriptive Catalogue*, 3.

43. Aermotor Company, *7th Annual Catalogue*, 43, and *Twelfth Annual Descriptive Catalogue*, 9.

44. Aermotor Company, *Power Aermotors Keep the Boys on the Farm*, 4, 6.

45. Aermotor Company, *Stock Ledger* (May 7, 1890–June 27, 1893); "New Incorporations."

46. Aermotor Company, *Stock Ledger* (May 7, 1890–September 8, 1919), 100–105.

47. Patent Nos. 451,225; 499,394; and 502,528, Thomas O. Perry, US Department of Commerce, Patent Office.

48. Aermotor Company, *To Aermotor Agents. Bulletin No. 3*, 1.

49. Ibid.

50. Aermotor Company, *The Following Is a Contract*.

51. Nichols, *Iowa State College of Agriculture and Mechanic Arts*, 126; Baker, "Marketing of Wind Engines," 35.

52. "L. W. Noyes—The Aermotor Co."

53. Aermotor Company, *The Aermotor*.

54. Aermotor Company, *To Aermotor Agents. Bulletin No. 2*, 3.

55. Ibid.

56. Baker, "How Windmill Companies Used Branch Houses," 5; Aermotor Company, *No. 28*[,] *Buffalo Branch, Agency Contract*.

57. Aermotor Company, *To Aermotor Agents. Bulletin No. 4*, 4.

58. Aermotor Company, "Twenty Aermotor Branch Houses," 86.

59. Aermotor Company (Kansas City, MO) to J. E. Bonebrake, October 23, 1895.

60. Aermotor Company, *$25.00 Present*.

61. Aermotor Company, *To Aermotor Agents. Bulletin No. 1*, 4.

62. Aermotor Company, *7th Annual Catalogue*, 86.

63. "Showed Their Colors."

64. Moore, "Early Aermotor Windmill Company History in California."

65. Sageser, "Windmill Irrigation," 32.

66. Eastman, "Windmill Irrigation in Kansas," 183–87; Baker, "Irrigating with Windmills," 217.

67. Murphy, *Windmills for Irrigation*, 34.

68. Aermotor Company, *Aermotor Irrigation*, 1.

69. Patent No. 765,036, La Verne Noyes, US Department of Commerce, Patent Office.

70. Baker, "Survey of Windmill Regulators"; Patent Nos. 648,988 and 648,989, La Verne Noyes, US Department of Commerce, Patent Office.

71. "Twenty Aermotor Branch Houses," 86.

72. "A High Windmill," 292; Aermotor Company, *8ft. Aermotor $25[,] 12ft. Aermotor $50[,] 16ft. Aermotor $100*.

73. "A Magnificent Show," 19.

74. "List of Exhibitors in Implement Annex," 18; Johnson, ed., *History of the World's Columbian Exposition*, 44.

75. "Here and There at the Fair," 24; Aermotor Company, *The Accompanying Cut*; Baker, "Industrial Sabotage in 1893," 8–10.

76. "Where the Wheels Go Round," 32.

77. Ibid.

78. Aermotor Company, *The Accompanying Cut*.

79. Aermotor Company, *Galvanized after Completion*; Grille and LeLarge, *L'Agriculture et les Machines Agricoles*, 132–33.

80. "Wind Mill Manufacturers Meet at Chicago," 22; Achilles, *Made in Illinois*, 250.

81. "Meeting of Wind Mill Manufacturers," 20.

82. "1899–Chronological," 51–52.

83. "Manufacturers Meet Jan. 24."

84. "Want Laws to Check Bribery."

85. "Conspired to Defraud"; Moore, "Early Aermotor Windmill Company History in California."

86. Baker, "Export of Wind Engines," 105.

87. US Department of Commerce and Labor, Bureau of Statistics, *Windmills in Foreign Countries*, 75–77.

88. Aermotor Company, *Eleventh Annual Descriptive Catalogue*, 5; Aermotor Company, *Water Supply Bulletin*, 15.

89. "R. B.," review of A. Chatterton, "Utilité des moulins à vent aux Indes," 1043.

90. "Paris Exposition," 18.

91. Baker, *Field Guide to American Windmills*, 101.

92. Baker, "Export of Wind Engines," 106; "The Aermotor Windmill[,] the Oiling Problem Solved," 963.

93. "The Tilting 'Aermotor,'" 16,810.

94. Van Sante-Baetens, *Aermotor*; Rogier, e-mail message to author, June 26, 2012; Patent Nos. 241,014 and 260,192, R. Van Sante-Baetens, Royaume de Belgique Brevet D'Invention, Le Ministre de L'Industrie et du Travail.

95. Gagey, "Les moulins à vent," 91–102; Ghazi, e-mail message to author.

96. Malcolm Walter, "Metters Ltd[.]," 3–5; Moore and Walter, "W. D. Moore & Co.," 2–3; Walter, "Imported Windmills," 6–8; Walter, "Windmills Imported into Australia," 9; Helen Walter, e-mail messages to author, July 31, 2012; August 17, 2012; September 4, 2012; September 20, 2012; and October 15, 2012.

97. "The Tilting 'Aermotor'"; James Martin & Company, "James Martin's Aermotor," 31.

98. Patent No. 11,555/08, La Verne W. Noyes, Australia Department of Patents.

99. Dutrieu, *L'Eau à la Campagne*; Patent No. 228,607, Jules Dutrieu, Royaume de Belgique Brevet D'Invention, Le Ministre de L'Industrie et du Travail.

100. Rogier, e-mail message to author, June 26, 2012.

101. Aermotor Company, *7th Annual Catalogue*, 5. Aermotor reprinted this testimonial in its product catalogues for the next seventeen years.

102. Aermotor Company, *Eleventh Annual Descriptive Catalogue*, 5.

103. Patents Nos. 472,809; 523,842; and 523,843, La Verne W. Noyes, US Department of Commerce, Patent Office.

104. Aermotor Company, *Eleventh Annual Price List*, 66, 89; Aermotor Company, *Twentieth Annual Aermotor Repair List*, 2; Patent Nos. 608,657; 609,457; 614,233; and 725,763, La Verne W. Noyes, US Department of Commerce, Patent Office.

105. Aermotor Company, *Eleventh Annual Descriptive Catalogue*, 9.

106. Ibid., 10–13.

107. Aermotor Company, *Galvanized Steel Aermotor with Removable Shaft Arms*; Aermotor Company, *Twentieth Annual Aermotor Repair List*, 4; Patent No. 897,390, La Verne W. Noyes, US Department of Commerce, Patent Office.

108. Aermotor Company, *In the Present Removable Arms*.

109. Aermotor Company, *Aermotor Selling Points*, inside front cover.

110. Aermotor Company, *Twentieth Annual Aermotor Repair List*, 4.

111. Aermotor Company, *Eleventh Annual Descriptive Catalogue*, 15.

112. "Windmill Firm to Move Plant"; "Aermotor Company Go to Chicago Heights," 20.

113. Aermotor Company, "Special Directors Meeting, April 20, 1905," 15–18.

114. Aermotor Company, *Aermotor Selling Points*, 1–2.

115. Aermotor Company, *Eleventh Annual Descriptive Catalogue*, 4.

116. Aermotor Company, *Thirteenth Annual Descriptive Catalogue*, 13.

117. US Department of Commerce, Patent Office, "Aermotor Company, of Chicago, Illinois. Trade-Mark for Windmills."

118. Aermotor Company, *Eleventh Annual Descriptive Catalogue*, 40–41.

119. Patent Nos. 485,881 and 489,989, Thomas O. Perry, and Patent No. 563,794, La Verne W. Noyes, US Department of Commerce, Patent Office.

120. Aermotor Company, *Eleventh Annual Descriptive Catalogue*, 17–18.

121. Aermotor Company, *Aermotor System Air Pressure Water Supply*, 1, folder.

122. Cook, "Water Stop," 13–14.

123. Ibid., 15.

124. Putnam, *Gasoline Engine*, 36; Baker, "'As Old As the Hills,'" 2.

125. Baker, "'As Old As the Hills,'" 5.

126. Ibid.

127. Aermotor Company, "Aermotor Pumping Devices," 9; Aermotor Company, *Aermotor Gasoline Pump Air Cooled*.

128. Aermotor Company, *Aermotor Heavy Back-Geared Gasoline Pumping Engine*.

129. Aermotor Company, *Price List[,] Aermotor Gasoline Engines*, 2.

130. Aermotor Company, *Engines for Pumping and Power*.

131. Aermotor Company, *Aermotor General Purpose Gasoline Engines*; Wendel, *American Gasoline Engines*, 13.

132. Aermotor Company, *110 Miles of Steel Towers*.

133. Patent Nos. 821,126; 836,465; 836,836; 870,053; 872,758; and 933,493, Daniel R. Scholes; and Patent No. 877,587, La Verne W. Noyes, US Department of Commerce, Patent Office.

134. Dubois, *Systematic Fire Protection*, 50–51; Baker, "Forest Service Towers," 7–9; Aermotor Company, *Aermotor Observation Towers*; Thornton to Baker, February 5, 1986.

135. Dubois, *Systematic Fire Protection*; Baker, "Forest Service Towers," 8.

136. Vana, e-mail message to author; Forest Fire Lookout Association, "Lookout Resources."

137. Aermotor Company, *Aermotor Galvanized Steel Bell Towers*.

138. "Machinists Out in Many Cities."

139. Aermotor Company, "Resolution Respecting Workman's Compensation Law," 40; Illinois Workers' Compensation Commission, *Chronology of Workers' Compensation Legislation*, 1–5.

140. Noyes, *Occasional Verses*.

141. Ida Noyes to La Verne Noyes, December 19, 1886.

142. *In Memoriam[,] Ida E. Smith Noyes*.

143. "Recent Realty Sales and Leases."

144. "Address, L. W. Noyes, '72," 12–13.

145. Noyes, "The Manufacturer and the Farmer," 8.

146. "Biography. Noyes, Mrs. La Verne W. (Ida Elizabeth Smith)" file, Chicago Historical Society.

147. "Where Chicago Millionaires Play"; *Extracts from a letter by La Verne Noyes*.

148. Goodspeed, *History of the University of Chicago*, 439–42.

149. *Extracts from a Letter by La Verne Noyes*.

150. Iowa State University, "History of Iowa State."

151. "Spelter Market Affects Windmills and Towers," 13; Baker, *Field Guide to American Windmills*, 106.

152. Elgin Wind Power and Pump Company to J. H. Wormly, Ransom, IL, May 25, 1915.

153. Baker Manufacturing Co., *Gentlemen:—We Regret*.

154. Aermotor Company, *Reduction in Prices on Painted Goods*.

155. US Department of Commerce, Bureau of Foreign and Domestic Commerce, *Statistical Abstract, 1921*, 247.

CHAPTER 4

1. Dickerson, "Care of Windmills," 126, 128; Baker, "Any Squeak or Grind," 2.

2. Dickerson, "Care of Windmills."

3. Baker, *Field Guide to American Windmills*, 41.

4. Baker, "Any Squeak or Grind," 2.

5. Ibid.

6. Aermotor Company, *Eleventh Annual Price List, October 1, 1899*, 63.

7. Ibid.

8. Ibid.

9. Baker, "Any Squeak or Grind," 4.

10. Ibid.

11. Baker, "Product History," 7.

12. Aermotor Company, *Eleventh Annual Descriptive Catalogue*, 10.

13. Ibid.

14. Aermotor Company, *Thirteenth Annual Descriptive Catalogue*, 8.

15. Challenge Wind Mill and Feed Mill Co., *Dandy Steel Wind Mill*; Baker, "Product History," 5.

16. Aermotor Company, *To Aermotor Agents. Bulletin No. 2*, 2.

17. Patent No. 667,615, Stephen E. Burke, US Department of Commerce, Patent Office.

18. "New Wind Mill Oiler," 36.

19. Burke-Bollmeyer Oiler Company, *You Can't Afford To Be Without It*.

20. Burke-Bollmeyer Oiler Company, "Oil Your Windmill from the Ground," 130; Burke-Bollmeyer Oiler Company, "Oil Your Windmill from the Ground by Pulling a Wire," 33.

21. Flint and Walling Manufacturing Company, *Triumph of Inventive Ingenuity*.

22. Patent No. 695,443, Stephen E. Burke, US Department of Commerce, Patent Office.

23. "Will Manufacture Wind Mills," 17; "New Wind Mill Company," 25.

24. Burke-Bollmeyer Manufacturing Company, *Red King Windmills*, 8–9.

25. Ibid., 6, 8.

26. Patent Nos. 749,526 and 761,143, Stephen E. Burke, US Department of Commerce, Patent Office; "Change in Burke-Bollmeyer Company," 16; "Change of Name," 22.

27. "Valuable Improvement on Wind Mills," 39; Red Cross Manufacturing Company, "Red Cross Wind Mills"; Red Cross Manufacturing Company, *Red Cross Line of Wind Mills*, 7–24.

28. Sears, Roebuck and Company, *Agricultural Implements*, 704.

29. Patent No. 755,802, Gottlieb Schneider, US Department of Commerce, Patent Office.

30. Patent No. 1,058,544, Samuel E. Burke, US Department of Commerce, Patent Office.

31. Patent No. 819,943, Gilbert B. Snow, US Department of Commerce, Patent Office; Elgin Wind Power and Pump Company, *1882 Up-to-Date Windmills 1906*, 14.

32. Elgin Wind Power and Pump Company, *Catalog Number Twenty-One*, 11.

33. Baker, *Field Guide to American Windmills*, 43–44.

34. Baldwin, e-mail message to author; "Class of 1904," 1; "Seniors," 253.

35. "Cornell Alumni Notes," 489.

36. Patent Nos. 821,126; 836,465; 872,758; 870,053; and 933,493, Daniel R. Scholes, US Department of Commerce, Patent Office.

37. Patent No. 1,101,211, Daniel R. Scholes, ibid.

38. Patent Nos. 1,143,324 and 1,155,518, Daniel R. Scholes, ibid.

39. Patent Nos. 1,141,356; 1,163,680; 1,163,681; and 1,171,631, La Verne Noyes, ibid.

40. Patent No. 1,151,815, La Verne Noyes, ibid.; Aermotor Company, *The Oil Lifter*.

41. Patent Nos. 1,163,682 and 1,188,093, La Verne Noyes, US Department of Commerce, Patent Office.

42. Aermotor Company, "To the Trade," 30; Aermotor Company, "Auto-Oiled Aermotor," 10.

43. Patent No. 13,171/14, Daniel R. Scholes, and No. 13,217/14, La Verne W. Noyes, Australia Department of Patents.

44. Aermotor Company, *Auto-Oiled Aermotor with Duplicate Gears*.

45. Ibid.

46. Aermotor Company, *Great Line of Towers*; Baker, "Aermotor Windmill Towers," 9.

47. Aermotor Company, *Auto-Oiled Aermotor*.

48. Aermotor Company, "To the Trade."

49. Ibid.

50. Ibid.; Aermotor Company, *To Put the Auto-Oiled Aermotor on an Old Aermotor*.

51. Aermotor Company, *Auto-Oiled Aermotor with Duplicate Gears*; Aermotor Com-

pany, *Aermotor Price List*[,] *February 1, 1915*, 10.; Baker, "Interchangeable Wheels and Vanes," 5–6.

52. Aermotor Company, *Reduction in Prices on Painted Goods*.

53. Aermotor Company, *Auto-Oiled Aermotor with Duplicate Gears*; Aermotor Company, *Instructions for Assembling the Auto-Oiled Aermotor*.

54. Aermotor Company, *Price List of Parts*, 2; Aermotor Company, *Aermotor Price List*[,] *February 15, 1917*, 1.

55. Baldwin, e-mail message to author.

56. Day, *Iowa State University*, 15–16.

57. "La Verne Noyes Gives $2,500,000 for Education"; "$2,500,000 to Teach Fighters at U. of C."; "$2,500,000 to Aid Soldiers"; "Splendid Act of Citizenship"; "Gift of a Grateful Patriot."

58. Aermotor Company, "Annual Stockholders Meeting" (January 18, 1916), 50–51.

59. Aermotor Company, "Annual Stockholders Meeting" (January 21, 1919), 59–60.

60. "La Verne W. Noyes," *Chicago Journal*.

61. "La Verne W. Noyes," *Chicago Evening Post*.

62. "Funeral Plans Private"; "Noyes Bequeaths Part of Wealth"; "La Verne Noyes Dies."

63. "Biographer Pays Final Tribute to L. W. Noyes."

64. "Noyes Bequeaths Part of Wealth."

65. "Noyes Wills Fortune to Educate Heros."

66. Last Will and Testament, 1–3, 11–14.

67. Ibid., 5.

68. Ibid., 1.

69. Ibid., 3.

70. Ibid., 5.

71. Aermotor Company, "Special Directors Meeting" (September 15, 1919), 61–64; Aermotor Company, *Stock Ledger* (September 8, 1919), 105.

72. Aermotor Company, "Special Director's Meeting" (September 17, 1919 and January 27, 1921), 64–65, 70.

73. Ibid., 67.

74. "L. W. Noyes Home"; "Chas. S. Peterson Buys Noyes Home"; "C. S. Peterson Purchases La Verne W. Noyes Home"; "C. S. Peterson Buys La Verne Noyes Home"; Drury, "Willow Park District."

75. Fogle, "C. S. Peterson."

76. "Noyes Scholarships Steadily Increase."

77. Ibid.; "Factory Prosperity Gives Scholarships."

78. "Noyes Scholarships Steadily Increase"; "Noyes Estate to Train 400 Vets a Year"; "Noyes Estate Scholarships to 400 Veterans"; "Veterans Will Receive College Scholarships"; "400 War Veterans Get Scholarships"; "400 Veterans Are Awarded Scholarships"; "Noyes Bequests to War Veterans"; "Scholarships for 400 Vets"; Associated Press, "Scholarships Given for Service"; "Lake Forest Gets 15 Scholarships."

79. Associated Press, "To Benefit from Tuition Scholarships"; Associated Press, "Women Who Served in World War"; Associated Press, "Service Women Given Tuition Scholarships"; "Women Veterans Get Free Tuition"; Associated Press, "Women of War Service Benefit from an Estate"; Associated Press, "Nurses Will Receive Tuition Scholarships"; Associated Press, "Scholarships for World War Women"; Associated Press, "Plan to Aid Women Vets"; Associated Press, "Ex-Service Girls"; "Scholarships Are Offered Women"; Associated Press, "Tuition Scholarships Offered War Women"; Associated Press, "Women Who Worked during War Helped"; Associated Press, "Women in War Get Scholarship"; Associated Press, "Women War Workers Get Scholarships."

80. Aermotor Company, "Special Directors Meeting" (December 20, 1921; December 19, 1922; December 18, 1923; December 16, 1924; December 14, 1925; December 20, 1926; December 20, 1927; and December 25, 1928), 73, 76, 79, 86, 90–91, 93, 96, 100.

81. "Annual Directors Meeting" (January 20, 1925), 87–88.

82. "Special Directors Meeting" (April 14, 1925), 88–89.

83. "Aermotor Buys Plant from U. of C."

84. "Victory Is Scored for Clean Building," 13–14; "Bomb Shatters Yuletide Calm."

85. Patent No. 1,341,080, Daniel R. Scholes, US Department of Commerce, Patent Office; Aermotor Company, *Auto-Oiled Aermotor*.

86. Aermotor Company, *Auto-Oiled Aermotor*.

87. Patent No. 1,545,610, Daniel R. Scholes, US Department of Commerce Patent Office, Aermotor Company, *Real Self-Oiling Windmill*; Aermotor Company, *This Is the Windmill*.

88. Patent Nos. 1,395,601; 1,529,088; and 1,545,611, Daniel R. Scholes, US Department of Commerce Patent Office; Aermotor Company, *Improved Self-Oiling Windmill with Double Gears*.

89. Aermotor Company, *Improved Auto-Oiled Aermotor with Adjustable Stroke*.

90. Patent No. 1,427,914, Daniel R. Scholes, US Department of Commerce, Patent Office.

91. Patent No. 1,529,089, Daniel R. Scholes, ibid.

92. Aermotor Company, *Be Thrifty—Pump with Wind*.

93. Aermotor Company, *Improved Auto-Oiled Aermotor with Adjustable Stroke*; Aermotor Company, *Storm-Defying Auto-Oiled Aermotor*.

94. Aermotor Company, *Announcing the 6-Foot Auto-Oiled Aermotor*.

95. Ibid.

96. Ibid.

97. Patent No. 1,545,609, Daniel R. Scholes, US Department of Commerce, Patent Office.

98. Yerian, e-mail message to author.

99. Ibid.

100. Aermotor Company, *To Aermotor Dealers* (1922), and *To Aermotor Dealers* (1924).

101. Aermotor Company, *Aermotor Price List*[,] *March 14, 1927*, 1.

102. Aermotor Company, *4 Times around the World*.

103. US Department of Commerce, Bureau of the Census, *Abstract of the Census of Manufactures*, 270, 738; US Department of Commerce, Bureau of Foreign and Domestic Commerce, *Statistical Abstract, 1921*, 247; Webb, *Great Plains*, 337.

104. Patent No. 827,611, William P. Brett, US Department of Commerce, Patent Office; "Windmill Suit Settled," 9; Dempster to G. F. Kregel.

105. Baker, "Close Look," 5; Eide to T. Lindsay Baker, March 14, 1994; May 20, 1996; and December 10, 2004.

106. Aermotor Company, *Instructions for Oiling Auto-Oiled Aermotor*; Aermotor Company, *Oiling System*; Baker, "Annual Lubrication," 7–8.

107. Gillis, *Windpower*, 27; *Farm Light and Power Year Book*, 65.

108. "Sources of Energy in Nature," 321.

109. "Electric Light"; "Mr. Brush's Windmill Dynamo," 389; "Brush, Charles Francis."

110. Nissen, "Visit to the Poul la Cour Musuem."

111. "Wind-Driven Generators for Farming," 490; "Electrical Value of Wind Power," 394–95; Righter, *Wind Energy*, 76–77.

112. Day, "Winds of the United States."

113. Bates, "Farm Electric Lighting by Wind Power," 262.

114. Baker, "Wind Electric News," 9–10; Richter, *Wind Energy*, 77–80.

115. Patent No. 1,022,205, La Verne W. Noyes, US Department of Commerce, Patent Office.

116. "Noyes, La Verne," 156; "Future of the Windmill," 309.

117. Patent No. 1,514,305, Daniel R. Scholes, US Department of Commerce, Patent Office.

118. Ibid.

119. Aermotor Company, *Electric Aermotor*.

120. Ibid.

121. Gillis, *Windpower*, 27–31; *Farm Light and Power Year Book*, 294–95; Toepfer, *Hybrid Electric Home*, 41–53.

122. Gillis, *Windpower*, 27; *Farm Light and Power Year Book*, 42–44; "Aerodynamic Wind Mills," 525; Jacobs, "Wind Driven Electric Generating Plant."

123. Toepfer, *Hybrid Electric Home*, 55–69; Aerodyne Company, *Electricity from Wind!*; Herbert E. Bucklen Corporation, *Power and Light from the Free Wind*; Herbert E. Bucklen Corporation, *Where Service Is a Problem*; Wind-Power Manufacturing Company, *WinPower*; Wincharger Corporation, *Every Farm Home*; Wincharger Corporation, *Light Your Farm*.

124. Aermotor Company, "Special Directors Meeting" (February 15, 1924), 82–83.

125. Aermotor Company, *Aermotor Galvanized Steel Bell Towers*; Aermotor Company, *Aermotor Observation Towers*.

126. Aermotor Company, "Special Directors Meeting" (July 7, 1924), 84.

127. US Department of Commerce, Coast and Geodetic Survey, *Bilby Steel Tower*, 1–27.

128. Leigh, "Bilby Towers."

129. Ibid.; Crattie, "Mr. Bilby's Elegant Assembly."

130. "In Memoriam[,] Thomas Osborn [*sic*] Perry," 520; "Thomas Osborne Perry, Long and Effective Career," 14.

131. Patent Nos. 33,010 and 655,270, Thomas O. Perry, US Department of Commerce, Patent Office.

132. Patent No. 772,052, Thomas O. Perry, ibid.

133. Patent No. 717,916, Thomas O. Perry, ibid.

134. Patent Nos. 901,555 and 928,326, Thomas O. Perry, ibid.; Perry survey; "Electro-Pneumatic Water-Supply System," 744–45; "Perry Aquapneumatics," 126; Huppertz, "Final Solution of the Water Supply System," 64–68; United Pump & Power Co., *Perry Water System*; United Pump & Power Co., *Fresh Water on the Farm*; Putnam, *Gasoline Engine*, 423–24.

135. Thomas Osborne Perry, letter to editor of the *Michigan Alumnus*.

136. Ibid.; "'Oldest Airman' Takes Ride."

137. Patent Nos. 1,272,846; 1,345,101; and 1,524,309, Thomas O. Perry, US Department of Commerce, Patent Office.

138. "Helicopter, Thomas O. Perry."

139. "In Memoriam[,] Thomas Osborn [*sic*] Perry."

140. Perry, *Life in the Universe*.

141. Albert W. Palmer, preface to Perry, *Life in the Universe*, i.

142. Thomas O. Perry, letter to the editor, *Chicago Evening Post*, August 1, 1919.

143. Mary E. Perry to Julius Rosenwald.

144. Mary E. Perry to Leo F. Wormser, March 5, 1927.

145. Ibid.

146. Mary E. Perry to Leo F. Wormser, April 15, 1927, and July 7, 1930.

147. Rosengard, e-mail message to author.

148. Ibid.

149. Keeney, Gleim Family Tree.

CHAPTER 5

1. Wallace, "The Year In Agriculture," 6–8; "Farm Products Prices Show No Improvement," 14; "Poultry and Egg Values for 1921 Slump," 421.

2. Aermotor Company, *Aermotor Price List*[,] *February 15, 1917*, 1; Aermotor Company, *Aermotor Price List*[,] *April 4, 1921*, 9.

3. Aermotor Company, *Aermotor Price List*[,] *April 16, 1923*, 1; Aermotor Company, *Aermotor Price List*[,] *March 14, 1927*, 1.

4. Baker, "New Deal Specials," 3–5.

5. Ibid.; US Department of Commerce, Bureau of Foreign and Domestic Commerce, *Statistical Abstract, 1922*, 219; and *Statistical Abstract, 1931*, 836.

6. Aermotor Company, *Aermotor Price List*[,] *Oct. 2, 1933*, 1; Aermotor Company, *Aermotor Windmills and Water Systems*[,] *Price List* (Kansas City), 2.

7. Aermotor Company, "Special Directors Meeting" (November 14, 1930), 105; and "Special Directors Meeting (December 16, 1930), 106.

8. "Special Directors Meeting" (December 21, 1931), 108; and "Special Directors Meeting" (December 31, 1932), 109.

9. "Special Directors Meeting" (December 31, 1932), 112; and "Special Directors Meeting" (April 6, 1933), 116.

10. Aermotor Company, "President's Reemployment Agreement."

11. Aermotor Company and National Association of Farm Equipment Manufacturers, "Petition to the Administrator for N.R.A. Consent."

12. Aermotor Company, L[ewis] C. Walker to US Department of Foreign and Domestic Commerce.

13. US National Archives & Records Administration, "National Industrial Recovery Act (1933)."

14. Aermotor Company, "Special Directors Meeting" (December 21, 1934).

15. Aermotor Company, "Special Directors Meeting" (January 29, 1935).

16. Aermotor Company, "Special Directors Meeting" (December 17, 1935).

17. Aermotor Company, *20-Foot Aermotor Auto-Oiled*; Baker, "Aermotor Windmill Towers," 13.

18. Patent Nos. 1,740,627; 1,749,892; 2,099,036; and 2,115,286, Daniel R. Scholes, US Department of Commerce, Patent Office.

19. "Newest Windmill[,] an Improved Aermotor"; Aermotor Company, *Aermotor Price List*[,] *Oct. 2, 1933*, 1; Aermotor Company, *New Improved, Model 702*; Aermotor Company, *Instructions for Oiling 702 Model*; Aermotor Company, *Great Strength of the Aermotor Wheel*; Aermotor Company, *Instructions for Assembling*; Aermotor Company, *"Runs in Less Wind"*; Aermotor Company, *You Can Cut Your Pumping Cost*; Aermotor Company, *Add Pumping Savings to Your Profits!*; Aermotor Company, *Pump with Wind*; Aermotor Company, *Improved Aermotor*; Yerian, e-mail message to author.

20. Aermotor Company, *Chicago, September 28, 1933*.

21. Ibid.

22. *Aermotor Price List*[,] *Oct. 2, 1933*, 2.

23. Ibid., 3; Aermotor Company, *Crated Motors*.

24. Aermotor Company, *Pan American Airways*; Thomis, "Windmills Built Here."

25. Aermotor Company, *Pan American Airways*.

26. US Department of Agriculture, Rural Electrification Administration, *Brief History of*, A1–A2, C2; Schurr et al., *Electricity in the American Economy*, 234; Gillis, *Windpower*, 31.

27. Hurt, "Irrigation of the West," 63.

28. Aermotor Company, *Aermotor Electric Pumps for Shallow or Deep Wells*, 2–24; Aermotor Company, *Aermotor Electric Pumps for Deep or Shallow Wells*, 1–26.

29. Patent No. 1,901,061, Daniel R. Scholes, US Department of Commerce, Patent Office; Aermotor Company, *Facts about Aermotor Water Systems*.

30. Aermotor Company, *Aermotor Electric Water Systems*[,] *Modern Water Systems*, 4–5; Aermotor Company, *Information Blank for Electric Pumps*.

31. Aermotor Company, *How to Choose Pumping Equipment*, 2–3, 20–23.

32. Aermotor Company, *Aermotor Electric Water Systems*[,] *Modern Water Systems*, 3; Aermotor Company, *Aermotor Electric Water Systems*[,] *All Prices F.O.B.*, 2–3.

33. Aermotor Company, *Aermotor Windmills and Water Systems Price List* (Des Moines), 22–38; Aermotor Company, *All Prices F.O.B. Chicago*[,] *Aermotor Electric Water Systems Price List 52-A*, 1–13.

34. Aermotor Company, *Put the Wind to Work with Aermotor*.

35. Aermotor Company, "Meeting of Shareholders" and "Special Directors Meeting" (September 15, 1936).

36. Aermotor Company, "Regular Directors Meeting" and "Special Shareholders Meeting" (October 6, 1936).

37. Aermotor Company, "Special Directors Meeting" (December 15, 1936); and "Special Directors Meeting" (December 22, 1936).

38. Aermotor Company, "Aermotor Company[,] Minutes of a Special Meeting," 149; Aermotor Company, "Agreement" (October 28, 1937), 153–56; Aermotor Company, "Aermotor Company[,] An Illinois Corporation[,] Waiver of Notice," 157; "Aermotor Company[,] An Illinois Corporation[,] Minutes of a Special Meeting of the Stockholders," 157.

39. Aermotor Company, "Regular Directors Meeting" (December 12, 1939), 175–76.

40. Aermotor Company, "Special Directors Meeting" (March 25, 1938), 162–63; "Special Directors Meeting" (August 18, 1938), 164–65; "Special Directors Meeting" (September 19, 1938), 165; "Special Directors Meeting" (November 22, 1938), 166–67; "Special Directors Meeting" (December 13, 1938), 167; "Special Directors Meeting" (April 10, 1939), 172; "Special Directors Meeting" (June 25, 1940), 178–79; "Special Directors Meeting" (August 16, 1940), "Declaration of Establishment of Protective Easements," 178 ½; "Special Directors Meeting" (August 30, 1940), 179; "Special Directors Meeting" (September 27, 1940), 181; "Special Directors Meeting" (February 24, 1941), 185; "Special Directors Meeting" (March 10, 1941), 185; "Directors Meeting" (June 10, 1941), 187; "Special Directors Meeting" (September 22, 1941), 188; "Special Directors Meeting" (December 4, 1941), 189; "Special Directors Meeting" (December 9, 1941), 189–90.

41. Aermotor Company, "Special Directors Meeting" (December 13, 1938), 167–68; "Special Directors Meeting" (December 19, 1939), 175; "Special Directors Meeting" (December 17, 1940), 179; "Special Directors Meeting" (December 16, 1941), 190.

42. Aermotor Company, "Special Directors Meeting" (January 14, 1941), 180.

43. Hamilton and Jepson, *Stock-Water Developments*, 18; Rohwer and Lewis, *Small Irrigation Pumping Plants*, 21–22.

CHAPTER 6

1. Aermotor Company, *Aermotor Galvanized Steel Towers*.

2. Isvik, interview.

3. Isvik, "Talk on Windmills."

4. Baker, "'For the Duration,'" 4–5.

5. Aermotor Company, "Annual Directors Meeting" (January 21, 1943), 195–96.

6. Aermotor Company, "To Aermotor Dealers," December 18, 1942.

7. Baker, *Field Guide to American Windmills*, 108; Butler Company, "General Letter."

8. Redborg to William McCook.

9. Baker, *Field Guide to American Windmills*, 180.

10. Gustafson and Schielke, *Historic Batavia*, 84.

11. *Baker Manufacturing Co.*, 21.

12. Aermotor Company, "To Aermotor Dealers," July 30, 1945.

13. Aermotor Company, "Dear Mr. George."

14. Isvik, interview.

15. Ibid.

16. Anderson to author, April 28, 2012. Telegraphic codes were devised to avoid confusion in orders and to shorten and consequently save expenses in sending telegrams.

17. Isvik, interview.

18. Anderson, interview, June 11, 1997.

19. Isvik, "Talk on Windmills."

20. Anderson to author, April 28, 2012.

21. Ibid.

22. Ibid.

23. Aermotor Company, *This National Advertising*.

24. Carmichael, "Water from the Wind," 70.

25. Aermotor Company, *Aermotor Price List No. 59*, 1.

26. Carmichael, "Water from the Wind."

27. Aermotor Company, *Announcement[,] New Aermotor Branch House.* The announcement, dated November 11, 1953, made public the information that the Aermotor Company had just opened a new building with office and warehouse at 1839 Nagle Street in Dallas.

28. Orr, "Day by Day on the Farm."

29. Anderson to author, October 10, 2012.

30. Aermotor Company, *Wide Spread Aermotor Towers.*

31. Aermotor Company, *Aermotor Windmills and Towers*, 8.

32. Anderson to author, April 16, 2012.

33. Aermotor Company, *Memorandum to Salesmen.*

34. Aermotor Company, *Notice to All Branches.*

35. Aermotor Company, *Oil Hole Relocated.* Dated June 8, 1955, the announcement to branches explains how to make oil-leak repairs in earlier 702 mills.

36. Isvik, interview.

37. Aermotor Company, *Aermotor Towers Mean Extra Sales.*

38. Aermotor Company, *Aermotor Special Towers*, folder cover.

39. "Obituaries[,] Frederick E. Smith."

40. Aermotor Company, "Annual Meeting of Shareholders" (January 16, 1945); and "Annual Directors Meeting" (January 16, 1945), 215–16.

41. Aermotor Company, "Directors Meeting" (June 11, 1946), 227.

42. Aermotor Company, "Directors Meeting" (January 28, 1947), 233–34.

43. Aermotor Company, "Special Meeting of Shareholders" (July 21, 1947), 236–37; and "Directors Meeting" (July 25, 1947), 237.

44. Aermotor Company, "Special Directors Meeting" (October 27, 1947), 240.

45. Aermotor Company, "Special Directors Meeting" (May 18, 1948), 244.

46. Kleinke, interview.

47. Anderson to author, April 12, 2012.

48. Crist, interview.

49. "Machinists out in Many Cities."

50. "Workers of Two Plants Reject, Indorse Unions."

51. "CIO Picket Dies."

52. Crist, interview.

53. Anderson to author, April 13, 2012.

54. Aermotor Company, *How to Choose Your Water Pumping System*, 4–7; Aermo-

tor Company, *Aermotor Convertible Jet Pump*, 2–5; Aermotor Company, *Another First for Aermotor*; Aermotor Company, *Aermotor Running Water*; Wegehoft, *Significant Dates in Aermotor History*, 4.

55. Anderson to author, May 10, 2012.

56. Aermotor Company, *Aermotor Special Towers*, 2.

57. Aermotor Company, "Special Directors Meeting" (June 23, 1953), 272; "People and Events."

58. Aermotor Company, "Special Directors' Meeting" (April 10, 1953), 271; "Special Directors' Meeting" (April 2, 1954), 277; "Special Directors Meeting" (January 26, 1955), 281; "Special Directors Meeting" (March 1, 1955), 281; "Special Directors Meeting" (March 18, 1955), 281; "Special Directors' Meeting" (July 26, 1955), 282; "Special Directors' Meeting" (February 15, 1956), 284; "Special Directors' Meeting" (December 3, 1956), 285; Estate of La Verne Noyes to Board of Directors, Aermotor Company, 285; "68th Annual Meeting of Shareholders" (January 15, 1957), 286; "Special Directors' Meeting" (March 4, 1957), 286; "Regular Directors' Meeting" (December 3, 1957), 289.

59. Aermotor Company, "Sixty-Ninth Annual Meeting of Shareholders" (January 28, 1958), 290.

60. Aermotor Company, "Annual Directors' Meeting" (January 28, 1958), 290.

61. Ibid.

62. Aermotor Company, "Regular Directors' Meeting" (March 6, 1958), 290.

63. Aermotor Company, "Regular Directors' Meeting" (April 15, 1958), 290.

64. Aermotor Company, "Revised Minutes," 290.

65. "Detroit Firm Has Acquired Aermotor"; Motor Products Corporation, *44th Annual Report*, [1]–3; Anderson, interview.

66. Aermotor Company, *Stock Ledger* (June 2, 1958, and June 3, 1958), 110.

67. Baldwin, e-mail message to author.

68. "D. R. Scholes Funeral Set at 11 Today."

69. Motor Products Corporation, *44th Annual Report*, [1]–3.

70. Motor Products Corporation, *45th Annual Report*, 10.

71. Ibid., 11; Aermotor Division, Motor Products Corporation, *How to Choose Your Aermotor Water System*.

72. Wegehoft, *Significant Dates in Aermotor History*, 4.

73. Motor Products Corporation, *45th Annual Report*, 11.

74. Wegehoft, *Significant Dates in Aermotor History*, 4.

75. Nautec Corporation, *Annual Report*[,] *Fiscal Year Ended June 30, 1961*, 10.

76. Nautec Corporation, *Annual Report*[,] *Fiscal Year Ended June 30, 1962*, 7.

77. Ibid., 11; Aermotor Division, Nautec Corporation, *Aermotor Everpure Chlorination-Dechlorination Equipment*.

78. Nautec Corporation, *Annual Report*[,] *Fiscal Year Ended June 30, 1962*, 7.

79. Wegehoft, *Significant Dates in Aermotor History*, 5.

80. Nautec Corporation, *Annual Report*[,] *Fiscal Year Ended June 30, 1961*, 11.

81. Baker, *Field Guide to American Windmills*, 108.

82. Nautec Corporation, *Annual Report*[,] *Fiscal Year Ended June 30, 1963*, 2, 5; "Braden Winch Plant Important Cog."

83. Nautec Corporation, *Annual Report*[,] *Fiscal Year Ended June 30, 1964*, 9.

84. Aermotor Division, Nautec Corporation, *Announcing*; Anderson to author, April 16, 2012.

85. Tucker, interview.

86. Anderson to author, April 14, 2012.

87. "Message from Bob Schuetz," 1.

88. Nautec Corporation, *Annual Report*[,] *Fiscal Year Ended June 30, 1965*, 3–4, 11; Braden-Aermotor Corporation, *Braden-Aermotor Corporation*[,] *Formerly Braden Winch*; Braden-Aermotor Corporation, *Annual Report*[,] *Fiscal Year Ended June 30, 1966*, unpaged.

89. "Company Involved in 2-Step Lease Plan."

90. Braden-Aermotor Corporation, *Annual Report*[,] *Fiscal Year Ended June 30, 1966*, unpaged.

91. Ibid.

92. Isvik, interview.

93. Anderson to author, April 24, 2012.

94. Aermotor Division, Braden-Aermotor Corporation, *Windmills . . . Lowest-Priced Pumping Power*.

95. Aermotor Division, Braden-Aermotor Corporation, *Aermotor Repairs and Replacement Units*, 1–2.

96. Monk to E. L. Weid Hardware.

97. W. D. Moore & Co., "History of the Yellowtail Purchase," 5.

98. Andrag, "Concise History of P Andrag & Sons," 4.

99. Walter, "The 'Metters Ltd' Nuoil Windmills," 2.

100. Moore to William McCook; John Danks & Son Pty. Ltd., *The "Coo-ee."*

101. W. D. Moore & Co., "History of the Yellowtail Purchase," 2–3; Moore and Walter, "W. D. Moore & Co.," 4; Walter, "Imported Windmills," 7.

102. Moore to William McCook; McCook to T. Lindsay Baker; W. D. Moore & Company, *Cheapest Power for Pumping Water*.

103. Andrag, "Concise History of P Andrag & Sons," 1–4.

104. Andrag to Martin Andrag.

105. Ibid.

106. Andrag, "Concise History of P Andrag & Sons," 1–4; Walton and Pretorius, *Windpumps in South Africa*, 26.

107. "Braden Winch Plant Important Cog."

108. Braden-Aermotor Corporation, *Annual Report*[,] *for the Fiscal Year Ended June 30, 1967*, unpaged.

109. Wegehoft, *Significant Dates in Aermotor History*, 5; Braden-Aermotor Corporation, *Annual Report 1968*, unpaged; Jo Krieger Daniels, "Oklahoma Firm's Windmills."

110. Braden-Aermotor Corporation, *Annual Report*[,] *for the Fiscal Year Ended June 30, 1967*, unpaged; Aermotor Division, Braden-Aermotor Corporation, *Aermotor Galvanized Steel Towers*, looseleaf notebook page; Aermotor Division, Braden-Aermotor Corporation, *Aermotor Antenna Towers*, looseleaf notebook page; Aermotor Division, Braden-Aermotor Corporation, *Aermotor Observation Tower Price List*.

111. Braden-Aermotor Corporation, *Annual Report 1968*[,] *for the Fiscal Year Ended June 30, 1968*; Sanz to Whom It May Concern; "Brief History of FIASA."

112. Isvik, "Report on 8 Ft. Argentine Aermotor Mill."

113. Ibid.

114. Anderson to author, October 10, 2012.

115. Marks to author; Anderson to Aero Manufacturing; Marks to Aermotor; Anderson to Howard C. Marks; Aermotor Division, Braden-Aermotor Corporation, *General Letter No. 75–67*; Aermotor Division, Braden-Aermotor Corporation, *It's New!*

116. Braden Industries, Inc., *Annual Report 1969*, 9.

117. Ibid.

118. Tucker, interview.

119. Braden Winch, Division of Braden Industries, Inc., "'New' Braden Industries Philosophy," 3.

120. Ibid.

121. Nigh, interview.

122. Braden Winch, Division of Braden Industries, Inc., "'New' Braden Industries Philosophy."

CHAPTER 7

1. Fetters, "Windmills," 29.

2. McAdams, "Disappearing Windmill," 53.

3. Braden Industries, Inc., *Quarterly Report*.

4. Braden Industries, Inc., "We're Gaining Ground Again," unpaged.

5. Wegehoft, *Significant Dates in Aermotor History*, 5.

6. Anderson, interview, September 23, 2013.

7. Aermotor Division, Braden Industries, Inc., "Pride of Quality," unpaged.

8. Newman, *Conway*, 7; Faulkner County Historical Society, "Welcome to Conway Arkansas."

9. Anderson, interview, September 23, 2013.

10. Aermotor Division, Braden Industries, Inc., "Pride of Quality."

11. Ibid.

12. Ibid.

13. Scott, e-mail message to author.

14. Aermotor Division, Braden Industries, Inc., *Aermotor 1975 Sales Meeting*; *Aermotor 1976 Sales Meeting*; Aermotor Division, Braden Industries, Inc. "1977 Sales Meeting Packed with Information," unpaged.

15. Fetters, "Viewpoint," unpaged; Anderson to author, March 14, 2013; Gillis, "A Father's Influence," 8–12.

16. Atkinson, e-mail message to author, March 23, 2012.

17. Aermotor Division, Braden Industries, Inc., *Aermotor 1976 Sales Meeting*; Aermotor Division, Braden Industries, Inc., "Branch Openings," unpaged.

18. Valley Industries, Inc., *1972 Annual Report*, 4, 8. Counter to Valley Industries' 1972 annual report, the company purchased its first one hundred submersible pumps from manufacturer Flint & Walling in Kendallville, IN, for testing before manufacturing its own line of these pumps the following year, according to David P. Suey Sr., who joined Valley as executive vice president of the pump division in 1973.

19. Ibid., 9.

20. Valley Industries, Inc., *1973 Annual Report*, 5; Valley Industries, Inc., "'72 in Review," 1; Valley Industries, Inc., "David Suey Heads New Pump Division," 1; Valley Industries, Inc., "Submersible Pumps," 3.

21. Valley Industries, Inc., *1974 Annual Report*, 2, 14.

22. Valley Industries, Inc., *Data Book*, 14.

23. Valley Industries, Inc., *1975 Annual Report*, 5–6; Valley Industries, Inc., "New Developments," 4; Valley Industries, Inc., "New Places," 1–2; Valley Industries, Inc., "More News from Texas," 1.

24. Valley Industries, Inc., *Annual Report 1976*, 2, 8; Schafers, "Valley Industries Sees Upturn Ahead"; Valley Industries, Inc., "Valley Industries Acquires Braden Aermotor Division," 1; Valley Industries, Inc., "More about Valley's New Braden Aermotor Division," 1–2.

25. Valley Industries, Inc., *Aermotor Engineering Program*, unpaged.

26. Valley Industries, Inc., *1977 Annual Report*, 4.

27. Valley Industries, Inc., *1978 Annual Report*, 4.

28. Ibid.

29. Valley Industries, Inc., *1979 Annual Report*, 1, 4, 6; Valley Industries, Inc., "Roaring of Motors," 1–2; Valley Industries, Inc., "An Overall View," 1.

30. Valley Industries, Inc., *Annual Report 1976*, 8.

31. Valley Industries, Inc., *1979 Annual Report*, 1.

32. Valley Industries, Inc., "Aermotor Opens Omaha Branch," 1–2.

33. Valley Industries, Inc., *Annual Report 1976*, 8.

34. Henry Clews, "Electric Power from the Wind," 13–14; Plowboy, "Meet the Man Who Invented the Modern Windplant"; Vale and Vale, *Autonomous House*, 80–81; Sullivan, *Wind Power*, 9–13, 88–92; Righter, *Wind Energy*, 161–69; Gillis, *Windpower*, 31–33.

35. Aermotor Company, *Electric Aermotor*.

36. Aermotor Division, Valley Industries, Inc., *Development of a 1kW High-Reliability Wind Turbine Generator*, pt. 1, 1, 5; pt. 2, 1–2, 4, 7–8, 24–37; and pt. 3, 1.

37. Brulle, *Feasibility Investigation*, vol. 1, *Executive Summary*, 1–7, 21; and vol. 2, *Technical Discussion*, 2–3, 12–20, 59–63; Moran, *Giromill Wind Tunnel Test*, 1, 82.

38. Brulle, *Engineering the Space Age*, 220.

39. Ibid., 221.

40. McConnell, *Giromill Overview*, 1–2, 10; Anderson et al., *Development of a 40 kW Giromill*, 1, 22–23, 229; Valley Industries, Inc. "Fabrication of Giromill Underway," 4; Valley Industries, Inc., "Giromill Readied for Preliminary Tests," 3.

41. Valley Industries, Inc., "Giromill Begins Performance Tests," 1.

42. Anderson et al., *Development of a 40 kW Giromill*, 109; Brulle, *Engineering the Space Age*, 223.

43. Valley Industries, Inc., *1978 Annual Report*, 4, and *1979 Annual Report*, 6; Valley Industries, Inc., "New Project 'in the Wind,'" 1; Valley Industries, Inc., "Future Plans for the Giromill," 4.

44. Brulle, *Engineering the Space Age*, 223–27; Kovarik, Pipher, and Hurst, *Wind Energy*, 27–30.

45. Anderson to author, October 10, 2012.

46. Anderson to Jim Fetters.

47. Aermotor Division, Valley Industries, Inc., *Performance Table*; Anderson to Whom It May Concern; Anderson to Bob Kain; Anderson to author, October 10, 2012. The historic record available to the author about the technical shortcomings of Aermotor developing a windmill with an eighteen-foot-diameter wind wheel was limited. According to a technical expert, the most notable flaw with wind wheels as they become larger is a diminishing return in their efficiency, in addition to the increased weight and cost factors of making them larger.

48. Atkinson, e-mail message to author, March 23, 2012; Aermotor Division, Braden Industries, Inc., *100,000 Windmills Received in Conway*.

49. Anderson to author, October 10, 2012.

50. Gillis, "A Father's Influence," 10; Anderson to author, April 14, 2012.

51. Aermotor Division, Valley Industries, Inc., *Installation, Operating and Maintenance Instructions*; Aermotor Division, Valley Industries, Inc., *Installation Instructions*.

52. Fetters, General Letter.

53. Atkinson, e-mail message to author, March 23, 2012.

54. Suey to author.

55. Valley Industries, Inc., *1980 Annual Report*, 4, 10.

56. Sanz, interview.

57. Wegehoft, *Significant Dates in Aermotor History*, 6.

58. Atkinson, e-mail message to author, March 23, 2012.

59. Kain to Aermotor Windmill Customers, 2.

60. Ibid., 1–2; Stefanides, "Self-Lubricating Bearings," 88–89.

61. Atkinson, e-mail message to author, March 23, 2012.

62. Valley Pump Group Aermotor-Weinman-Midland, *Valley Pump Group*, folder.

63. "Longtime Manufacturers Face Brighter Future", 36.

64. Wegehoft, *Significant Dates in Aermotor History*, 7.

65. Stroud, "Windmills Pump Valley Industries' Profits."

66. Aermotor Division, Valley Industries, Inc., *Aermotor Wind Energy Price List*, 2.

67. Stroud, "Windmills Pump Valley Industries' Profits."

68. Sanz to Whom It May Concern; Sanz, interview.

69. Atkinson, e-mail message to author, March 23, 2012.

70. Swanson, "Windmill Manufacturers Keep on Turning," 41; Atkinson, interview.

71. Kain to Aermotor Windmill Customers, 1–4.

72. Atkinson, e-mail message to author, March 23, 2012.

73. Suey to author.

74. Kain to Aermotor Windmill Customers, 1–4.

75. Benson to Aermotor Windmill Customers, 1–4.

76. Valley Industries, Inc., *Annual Report 1983*, 4, 6–7.

77. Valley Industries, Inc., *Annual Report 1984*, 2, 9–10, 17–18; Baker, "Complex Story," 9.

78. Wegehoft, interview; Atkinson, e-mail message to author, March 23, 2012; Sanz, interview.

79. Swanson, "Windmill Manufacturers Keep on Turning," 40; Owens, "Angelo Becomes Windmill Capital"; Schovajsa, interview; Wegehoft, interview.

80. Schovajsa, interview.

81. Owens, "Angelo Becomes Windmill Capital."

82. Pasztor, "Windmills Breeze into Oblivion."

83. Schovajsa, interview.

84. Ibid.

85. Ibid.

86. Aermotor Windmill Corporation, *Aermotor Windmills*.

87. Martin, "The Business of Breeze"; Martin, "Aermotor Has Long History."

88. Heller-Aller Company, *December 1983*[,] *Dear Dealer*; Eide to T. Lindsay Baker, January 15, 1985; Kelley, interview.

89. Essex Associates, Inc., *FIASA Moves One Step Closer*; Shippard, interview; Grant, interview; Baker, *North American Windmill Manufacturers' Trade Literature*, 244–45.

90. Atkinson, e-mail message to author, March 23, 2012; Durham, interview.

91. Lowenstein to T. Lindsay Baker; Molinos de Viento, S.A., *6 Tamaros Diferentes*.

92. Felizardo Elizondo Guajardo, S.A. de C.V., *Aermotor*; Second Wind Windmill Service, *The Wind Engine 702*.

93. Atkinson, e-mail message to author, February 6, 2012.

94. Thomerson, interview.

95. LoLordo, "Fashioning Life among Windmills"; Swindle, "Spinning the Revival of an American Tradition"; Ritz, "A Different Kind of American Music"; Verheul, e-mail message to author, January 12, 2012.

96. Verheul, e-mail message to author, January 12, 2012.

97. Ibid.

98. Verheul, e-mail message to author, January 26, 2012.

99. Ibid.

100. Ibid.

101. Flippin, "Aermotor Owner Secure in Future."

102. Gillis, "U.S./Africa Trade Slow to Ripen," 44–46.

103. Flippin, "Aermotor Owner Secure in Future."

CHAPTER 8

1. Share, "Windmill on Nebraska Plates."
2. Jones, interview.
3. "Ranchers in West Seeking to Restore Use of Windmills"; "Course in Windmills Is Offered."
4. Runyan, e-mail message to author, October 9, 2012; New Mexico State University, College of Agricultural, Consumer, and Environmental Sciences, "NMSU Windmill Technology Certification Workshop"; Bannister, "NMSU to Host Workshop."
5. Runyan, e-mail message to author, April 29, 2014.
6. Iron Man Windmill Co., *Perfect Solution*; Durham, e-mail message to author.
7. Conlon, e-mail message to author.
8. Clark, "Performance Comparison," 147–49; Clark, interview.
9. Clark and McCarty, "Variable Stroke Pumping for Mechanical Windmills," 217–21; Clark, interview.
10. US Department of Agriculture, Forest Service Equipment Development Center, *Range Water Pumping Systems*, 2–3.
11. Morrow, e-mail message to author.
12. Headings, e-mail message to author; US Department of Agriculture, Natural Resources Conservation Service, "Environmental Quality Incentives Program" and "Farm Bill 2008 Fact Sheet"; Galbraith, "Windmilling."
13. Morrow, e-mail message to author; Clark, interview; Clark, *Small Wind*, 147.
14. Morrow, e-mail message to author.
15. O'Brock Windmill Distributors, *"Harness the Wind,"* 34; O'Brock Windmill Distributors, *E-Z Air Windmill Compressor*; Brock, e-mail message to author; Brock, interview.
16. Gillis, *Windpower*, 27.
17. Fernandez-Bueno, interview.
18. Ibid.
19. Ibid.
20. Potomac Wind Energy, *American Classic Wind Power*, 4.
21. Fernandez-Bueno, interview.
22. Ibid.
23. Whelan to author.
24. Baker, "Mid-America Windmill Museum Opens," 9; Baker, "American Wind Power Center Opens," 5–6; Redmond, "Gone with the Wind," 96–102; Kash, "Blowing in the Wind"; Galbraith, "Museum Shows History and Power of Wind Energy"; Baker, "Windmill Museums as Destinations"; Mid-America Windmill Museum, *Mid-American Windmill Museum*; American Wind Power Center, *American Wind Power Center*; Southwestern Pioneer Windmill Association, *Southwestern Pioneer Windmill Association*.
25. Baker, "Windmill Museums as Destinations," 7–8.
26. International Molinological Society, "About TIMS."
27. Baker, *Guide to United States Patents*, 9.
28. "Restoration of Wind-Powered Water Pumps Scheme"; "Eighteen Wind Powered Water Pumps Restored in Gozo."
29. Kjerfve et al., "Hydrology and Salt Balance," 705; Simpson, "Salinas"; Simpson, e-mail message to author.

CHAPTER 9

1. Verheul, e-mail message to author, February 10, 2012.
2. Bracher, interview.
3. Ibid.

4. Ibid.

5. Gillis, "Lean and Clean," 10.

6. Aermotor Windmill Company, "Aermotor Windmill Names Bracher President."

7. Verheul, e-mail message to author, February 10, 2012.

8. Bracher, interview.

9. Verheul, e-mail message to author, February 10, 2012.

10. Bracher, interview.

11. Zwiebel, interview.

12. Ibid.

13. Ibid.

14. Ibid.

15. Ibid.

16. Morrow, interview.

17. Zwiebel, interview.

18. Ibid.

19. Bracher, interview.

20. Associated Press, "Texas Drought Endangers Animals"; Campbell, "Texas Weather"; Lackey, "Windmill Almanac 2012"; Smith and Yang, *8 Industries That Will Lose*, 1–4; Lackey, "Windmill Country."

21. "Farewell to Windmills," 3; Clark and Vick, "Livestock Watering," 239–40.

22. Smith, "Solar Systems Grow"; Cantu, "Solar Pumps vs. Windmills," 28–29; Blakeley, interview; Runyan, e-mail message to author, March 10, 2014; Grundfos Holding A/S, "SQFlex"; Grundfos Holding A/S, "SQFlex at the Museum of Modern Art."

23. Runyan, e-mail message to author, March 10, 2014.

24. Comis, "Wind Power Where You Want It," 4–7; Clark, "Wind-Electric Water Pumping Systems," 1136–37, 1139.

25. Gipe, "Windmills," 64; Bergey, "Wind-Electric Pumping Systems for Communities"; Ziter, "Electric Wind Pumping," 15.

26. Dewey, "Aermotor Windmills."

27. Morrow, interview.

28. United Nations, *Millennium Development Goals*, 59; Gross, van Wijk, and Mukherjee, *Linking Sustainability*, 16; Deen, "Women Spend 40 Billion Hours Collecting Water"; United Nations, "Global Issues"; Water.org, "Water Facts."

29. Bracher, interview.

30. Dewey, "Aermotor Windmills."

31. "Wind Energy Used in Iraqi Water Pumps"; "Agribusiness Development Team in Afghanistan"; Runyan, e-mail message to author, October 9, 2012.

32. Bracher, interview.

33. Morrow, interview.

34. Ibid.

35. Dunker, "Dempster Had a Long History"; "Wallace Davis Hoped to Turn Company Around"; Dunker, "Debts Unpaid"; Dunker, "'The More We Went There, the Worse It Got'"; Dunker, "Money for Nothing"; Dunker, "Tilting at Windmills"; Dunker, "Davis Settles with Dempster Employees"; Dunker, "City Council Rejects Dempsters Loan Application"; Dunker, "Dempsters LLC Not Registered"; Dunker, "Past Likely Haunting Current Dempsters' President"; Dempster Industries LLC, "About Us."

36. Jack Nitz & Associates, "Complete Manufacturing Division & Relocation Dispersal Auction"; Koperski, "Dempsters Prepares to Auction Equipment."

37. Morrow, interview.

38. Zamudio, "San Angelo's Aermotor"; "Water Pumping Windmills."

GLOSSARY

Due to the variety of local, regional, technical, and corporate terminology that has evolved in 150 years of windmill manufacture and use, this glossary provides only a summary of key words and definitions related to this history and the Aermotor windmill. Sources include Aermotor trade literature and definitions compiled from the *Windmillers' Gazette*.

TYPES OF WINDMILLS

Back-geared windmill   Windmill mechanism designed so that more than one revolution of the wheel was required to produce a single pump stroke; from the outset Aermotor produced a back-geared windmill with a variable stroke.

Combination outfit   Windmill installed on top of a stand-alone tower supporting a large wooden or steel water tank with a cover. Water was piped up from the ground into the tank by the action of the windmill's pump rod and pole. Hydrostatic (gravity) pressure was used to deliver water from the tank to its intended use.

Direct-stroke windmill   Windmill with a mechanism that resulted in one complete pump stroke with each revolution of the wind wheel, a typical setup for most early commercially available and homemade water-pumping windmills.

Homemade windmill   Wind-driven device often made with wooden paddles and shaft supports and held together with various recycled materials, such as leather and nails, or produced locally by blacksmiths and carpenters; for pumping water or operating small farm machines. Developed by frugal farmers, these windmills coalesced around several basic designs and were given colloquial names such as "ground tumbler," "battle-axe," and "merry-go-round."

Oil-bath windmill   Windmill with moving mechanical parts that are partially submersed in oil for continuous lubrication during operation; also called "self-oiling."

Open-gear windmill   Back-geared windmill in which the gears were exposed and open to the weather, requiring weekly lubrication. This was the common construction of the earliest back-geared US water-pumping windmills.

Outfit   Combination windmill and tower.

Power windmill   Windmill geared to rotate a shaft that was then geared near ground-level to operate light machinery such as feed grinders and wood saws.

Pumping windmill   Windmill that drove a reciprocating shaft for a piston-type water pump.

Railroad outfit   Windmill system designed to provide water for steam locomotives, which consisted of a large-diameter windmill on its own tower; a large wooden or steel water tank tower next to the windmill received the water and fed it via a spout into locomotive tenders.

Sectional-wheel windmill   Windmill with a governing device to protect it from turning too quickly in high winds; individual sections of the wind wheel pivoted inward with increasing wheel speed or wind pressure to reduce the surface area exposed to the wind. Sectional-wheel windmills may or may not have had vanes.

Self-regulating windmill   Windmill that automatically swiveled to face the wind and at the same time employed a governing mechanism to protect it from self-destruction due to centrifugal forces during high winds.

Suburban outfit   Windmill water system that consisted of a windmill on a tower and a wooden or steel water tank mounted within the tower legs; it used hydrostatic pressure to deliver water for household and agricultural purposes.

Vaneless windmill   Windmill with no vane to direct its wheel into the wind, but instead operated downwind, using a wheel on the leeward side of the tower. These windmills may or may not have incorporated a counterbalance weight to keep them balanced on top of the towers.

AERMOTOR WINDMILL PARTS

Bearings   A device that allowed two mechanical parts to move in proximity with minimal friction. Aermotor oil-bath windmills had two removable bearings, either made of babbitt (for Models 502, 602, and 702) or Garlock (for Model 802), that permitted wind wheel shafts to rotate freely in the mainframe castings; also known as boxings and journals. The 502s and 602s had nonremovable bearings poured into the castings.

Furling device   A mechanism that turned the windmill on or off. Aermotor had an external furling device, whereas it was more common among other makes of mills to use a pull-out chain that ran inside the mast pipe.

Gears   Oil-bath Aermotor windmills had four gears; two small pinion gears were attached to the horizontal wind wheel shaft and transferred their circular motion to two large crank gears, which drove the pitman arms and reciprocal pump rod in a ratio of about three wind wheel revolutions to one pump stroke.

Helmet   A galvanized sheet-metal covering over the motor that kept out moisture and dirt and also prevented oil from splashing out of the mainframe; also known as a hood.

Hub   Cast iron cylinder in front of the motor to which the steel arms of the wind wheel blade sections connected; attached to a steel shaft that drives the gears of the motor.

Mainframe   Main casting of the windmill, which supported its parts and held lubrication; also known as the main casting.

Mast pipe   A metal pipe attached through the base of the mainframe and secured in the center of the tower top. For Aermotor, this was a stationary upright member on which its windmills swiveled to face changing wind directions. Other makes had the mast pipes attached to the underside of the working mill head, where it rotated as the mill yawed.

Motor   All the working parts of a windmill, including the hub and gears, which translated the circular motion of the wind wheel to reciprocating movement of a pump rod or rotary motion of a vertical shaft; also known as the head or engine.

Pitmans   Steel arms attached to two large gears in the mainframe, which helped facilitate the conversion from circular motion of the wind wheel and shaft to reciprocating motion to drive the pump rod. The pitman arms of the oil-bath Aermotor offered both long- and short-stroke capability, with the short-stroke position increasing the pumping depth by one-third but decreasing the pumping capacity by 25 percent.

Pump rod   A steel rod that fastened to the moving parts in the motor and extended downward in the tower to transmit the reciprocating motion to a swivel connector attached to the wooden pump poles that passed farther down to the pump in the well.

Vane Sheet   A flat piece of galvanized sheet metal attached to a steel tailbone and governor spring that directed the wheel into the wind, helped control operating speed, and counterbalanced the weight of the wheel.

Wind wheel   A set of eighteen galvanized, contoured, trapezoidal blades or sails that were held together and evenly spaced by steel outer and inner rims; a series of steel arms secured the blade sections to the hub. Aermotor maintained eighteen sails for all its windmill sizes, whereas other makers often varied the number of sails with different wind wheel sizes.

WELLS AND PUMPS

Bored well   A relatively small-diameter hole placed into the ground by special drilling machinery for the purpose of entering water-bearing strata and giving access to it from the surface; also referred to as a "drilled well."

Casing   The pipe used to line a bored well to prevent the sides from caving in.

Driven well   Well created by manually driving special boring tools into the ground to reach water at relatively shallow depths.

Drop pipe   Pipe that hung in the well to which the cylinder was connected and encased the sucker rod, which drove the pump and channeled the water to the surface.

Hand pump   A hand-activated device placed over an open or bored well, which through varied mechanical means lifted water to the surface; water exited a spout and was collected in buckets for household purposes, livestock watering, and garden irrigation.

Head   The vertical distance, usually measured in feet, from the lowest level at which water held in a well to the highest point to which it was elevated; the computation of the head often included the additional factor of the friction of the water within the pipes or pump.

Open well   A hand-dug well usually several feet in diameter and excavated to various depths depending on where the water-bearing strata was reached; this hole in the ground was generally lined with bricks, stone, timber, or concrete to prevent the walls from caving.

Pumping cylinder   The cylindrical chamber that contained the moving parts of a simple single-stroke-action piston pump typically used in wells under windmills; also known as a cylinder.

Stuffing box   A brass cap over the top of the drop pipe of the well that was used to create pressure with conduits leading from the well to points higher in elevation, which thus allowed the pumping cylinder in the well to force water higher than the ground surface.

Sucker rod   A rod made of wood, steel, or fiberglass and produced in standard or various lengths; when screwed together, they reached from the surface down to the plunger valve in the cylinder.

AERMOTOR TOWERS

Anchor posts   Metal or wooden posts that extended typically five to six feet underground on which the corner posts of the windmill tower were fastened.

Corner posts   Individual legs or posts of a windmill tower.

Four-post tower   Windmill tower that consisted of four legs or posts, the most common style in North America.

Looped ladder   A series of steel rod steps that were bolted top to bottom to alternate sides of a single tower leg. It was introduced to Aermotor towers in 1899.

Platform   A wooden or steel table-like platform attached near the top of a tower that offered footing during maintenance and repair of a windmill.

Pump pole guides   Attachments in the center of the windmill tower that kept the wooden pump pole in line with the pump in the well so that the sucker rods did not become unscrewed.

Stub tower   A short steel tower mounted beneath the windmill that adapted the windmill for installation on either a wood or a steel tower.

Suburban tower   A windmill tower designed to support at some elevation within its legs a wooden or steel tank to store water under hydrostatic pressure for domestic use.

Tilting tower   Steel or wooden windmill tower made with hinges near the center, which allowed for the windmill motor to be lowered to the ground for convenience in lubrication and other maintenance.

Tripod tower   A windmill tower supported by three legs; also known as a three-post tower.

Wide-spread tower   A windmill tower designed with its base extra wide to provide additional stability in high-wind areas; one side of the tower base could be left open of girts for laying out pipe and rod in large numbers for deep wells.

AERMOTOR WINDMILL ACCESSORIES

Crab   A steel geared device, which, combined with block and tackle and rope or steel cable and cranked by hand, allowed for the erection of a preassembled windmill tower.

Foot gear   A gear mechanism that translated the vertical movement of the upright shaft of a power windmill into horizontal rotary motion to operate machinery more conveniently.

Regulator   A device designed for use with windmills to turn them on or off according to the level of water in storage tanks or troughs; wires typically connected the regulating device with floats on the surface of the water; similar in principle to how a modern-day bathroom toilet operates.

Stock tank   A steel or wooden structure, usually circular or rectangular in shape, into which water was pumped by a windmill to provide a watering trough for livestock.

Triangles   Three-sided metal templates, operated in pairs and connected with two heavy wires, which transmitted the reciprocating motion of a windmill's pump rod to a pump from some distance away.

BIBLIOGRAPHY

ARCHIVAL MATERIALS

"Address, L. W. Noyes, '72." *The Student*. Ames, IA: Iowa State College, 1897. Special Collections Department, Iowa State University Library, Ames.

Aermotor Company. "68th Annual Meeting of Shareholders" (January 15, 1957). Record Book, 286. Private collection of Mark S. Welch, Fort Worth, TX. (This record book is a compilation of handwritten notes from the corporate directors' and shareholders' meetings from January 20, 1903, to May 26, 1958, as well as typed minutes from the meetings; other types of documents were often glued into the book's pages. Subsequently referenced as Record Book.)

———. "Aermotor Company[,] an Illinois Corporation[,] Minutes of a Special Meeting of the Stockholders," 1–3. Attached in Record Book, 157.

———. "Aermotor Company[,] an Illinois Corporation[,] Waiver of Notice of Special Meeting of Stockholders, 1. Attached in Record Book, 157.

———. "Aermotor Company[,] Minutes of a Special Meeting of the Directors of Aermotor Company, an Illinois Corporation" (October 28, 1937), 1–5. Attached in Record Book, 149.

———. "Agreement" (October 28, 1937), 1–7. Contained in Record Book, 153–56.

———. "Annual Directors Meeting" (January 20, 1925). Record Book, 87–88.

———. "Annual Directors Meeting" (January 21, 1943). Record Book, 195–96.

———. "Annual Directors Meeting" (January 16, 1945). Record Book, 215–16.

———. "Annual Directors' Meeting" (January 28, 1958). Record Book, 290.

———. "Annual Meeting of Shareholders" (January 16, 1945). Record Book, 214–15.

———. "Annual Stockholders Meeting" (January 18, 1916). Record Book, 50–51.

———. "Dear Mr. George." September 15, 1945. Typewritten letter. In possession of author.

———. "Directors Meeting" (June 10, 1941). Record Book, 187.

———. "Directors Meeting" (June 11, 1946). Record Book, 227.

———. "Directors Meeting" (January 28, 1947). Record Book, 233–34.

———. "Directors Meeting" (July 25, 1947). Record Book, 237.

———. (Kansas City, MO). To J. E. Bonebrake, Minco, I[ndian] T[erritory], October 23, 1895. Aermotor Company files, Windmill Manufacturers' Trade Literature Collection, Research Center, Panhandle-Plains Historical Museum, Canyon, TX.

———. L[ewis] C. Walker, president, to US Department of Foreign and Domestic Commerce, August 24, 1933. Record Book, 121.

———. "Meeting of Shareholders" and "Special Directors Meeting" (September 15, 1936). Record Book, 139–40.

———. and National Association of Farm Equipment Manufacturers. "Petition to the Administrator for N.R.A. Consent to the Substitution of Paragraphs, Article V. Section 2 and 3 of Code of Fair Competition for the Farm Equipment Industry for Paragraphs 2, 3, 6 and 7 of the President's Reemployment Agreement." (August 15, 1933). Contained in Record Book, 121–22.

——. No. 28[,] *Buffalo Branch, Agency Contract. 1895. The Aermotor Company with N. D. Bartlett, Sales Agent, Earlville, N.Y.* Chicago: Aermotor Co., 1895. Folder plus envelope. Private collection of Mark S. Welch, Fort Worth, TX.

——. "President's Reemployment Agreement (Authorized by Section 4a National Industrial Recovery Act)." (July 27, 1933). Attached in Record Book, 118.

——. "Regular Directors Meeting" (December 12, 1939). Record Book, 175–76.

——. "Regular Directors' Meeting" (December 3, 1957). Record Book, 289.

——. "Regular Directors' Meeting" (March 6, 1958). Record Book, 290.

——. "Regular Directors' Meeting" (April 15, 1958). Record Book, 290.

——. "Regular Directors Meeting" and "Special Shareholders Meeting" (October 6, 1936). Record Book, 141.

——. "Resolution Respecting Workman's Compensation Law" (January 16, 1912). Record Book, 40.

——. "Revised Minutes[,] Regular Directors' Meeting" (May 12, 1958). Record Book, 290.

——. "Sixty-Ninth Annual Meeting of Shareholders" (January 28, 1958). Record Book, 290.

——. "Special Directors Meeting, April 20, 1905, Authorizing Sale Chicago Heights Property." Record Book, 15–18.

——. "Special Directors Meeting" (September 15, 1919). Record Book, 61–64.

——. "Special Directors Meeting" (September 17, 1919, and January 27, 1921). Record Book, 64–65, 67, 70.

——. "Special Directors Meeting" (December 20, 1921; December 19, 1922; December 18, 1923; December 16, 1924; December 14, 1925; December 20, 1926; December 20, 1927; December 25, 1928). Record Book, 73, 76, 79, 86, 90–91, 93, 96, 100.

——. "Special Directors Meeting" (February 15, 1924). Record Book, 82–83.

——. "Special Directors Meeting" (July 7, 1924). Record Book, 84.

——. "Special Directors Meeting" (April 14, 1925). Record Book, 88–89.

——. "Special Directors Meeting" (November 14, 1930). Record Book, 105.

——. "Special Directors Meeting" (December 16, 1930). Record Book, 106.

——. "Special Directors Meeting" (December 21, 1931). Record Book, 108.

——. "Special Directors Meeting" (December 31, 1932). Record Book, 109, 112.

——. "Special Directors Meeting" (April 6, 1933). Record Book, 116.

——. "Special Directors Meeting" (December 21, 1934). Record Book, 128.

——. "Special Directors Meeting" (January 29, 1935). Record Book, 131–32.

——. "Special Directors Meeting" (December 17, 1935). Record Book, 134.

——. "Special Directors Meeting" (December 15, 1936). Record Book, 143.

——. "Special Directors Meeting" (December 22, 1936). Record Book, 144.

——. "Special Directors Meeting" (March 25, 1938). Record Book, 162–63.

——. "Special Directors Meeting" (August 18, 1938). Record Book, 164–65.

——. "Special Directors Meeting" (September 19, 1938). Record Book, 165.

——. "Special Directors Meeting" (November 22, 1938). Record Book, 166–67.

——. "Special Directors Meeting" (December 13, 1938). Record Book, 167–68.

——. "Special Directors Meeting" (December 19, 1939). Record Book, 175.

——. "Special Directors Meeting" (June 25, 1940). Record Book, 178–79.

——. "Special Directors Meeting" (August 16, 1940). "Declaration of Establishment of Protective Easements, Restrictions, Limitations, Conditions, and Covenants Affecting the Real Property and Its Use Known as 'Stoneleigh Park in Glen Ellys." 1–5. Attached to Record Book, 178½.

——. "Special Directors Meeting" (August 30, 1940). Record Book, 179.

——. "Special Directors Meeting" (September 27, 1940). Record Book, 181.

———. "Special Directors Meeting" (December 17, 1940). Record Book, 179.

———. "Special Directors Meeting" (January 14, 1941). Record Book, 180.

———. "Special Directors Meeting" (February 24, 1941). Record Book, 185.

———. "Special Directors Meeting" (March 10, 1941). Record Book, 185.

———. "Special Directors Meeting" (September 22, 1941). Record Book, 188.

———. "Special Directors Meeting" (December 4, 1941). Record Book, 189.

———. "Special Directors Meeting" (December 9, 1941). Record Book, 189–90.

———. "Special Directors Meeting" (December 16, 1941). Record Book, 190.

———. "Special Directors Meeting" (October 27, 1947). Record Book, 240.

———. "Special Directors Meeting" (May 18, 1948). Record Book, 244.

———. "Special Directors' Meeting" (April 10, 1953). Record Book, 271.

———. "Special Directors Meeting" (June 23, 1953). Record Book, 272.

———. "Special Directors' Meeting" (April 2, 1954). Record Book, 277.

———. "Special Directors Meeting" (January 26, 1955). Record Book, 281.

———. "Special Directors Meeting" (March 1, 1955). Record Book, 281.

———. "Special Directors Meeting" (March 18, 1955). Record Book, 281.

———. "Special Directors Meeting" (July 26, 1955). Record Book, 282.

———. "Special Directors Meeting" (February 15, 1956). Record Book, 284.

———. "Special Directors Meeting" (December 3, 1956). Record Book, 285.

———. "Special Directors' Meeting" (March 4, 1957). Record Book, 286.

———. "Special Meeting of Shareholders" (July 21, 1947). Record Book, 236–37.

———. Stock Ledger (May 7, 1890–June 27, 1893), 100–101. Private collection of Lee Raney, Frisco, TX. (Subsequently referenced as Stock Ledger.)

———. Stock Ledger (May 7, 1890–September 8, 1919), 105.

———. Stock Ledger (June 2, 1958, and June 3, 1958), 110.

———. "To Aermotor Dealers." December 18, 1942. Typewritten letter. In possession of author.

———. "To Aermotor Dealers." July 30, 1945. Typewritten letter. In *Windmillers' Gazette* files, Rio Vista, TX.

Aermotor Division, Braden Industries, Inc. *100,000 Windmill Received in Conway*. Conway, AR: Aermotor Division of Braden Industries, ca. 1976. Brochure. In *Windmillers' Gazette* files, Rio Vista, TX.

———. *Aermotor 1975 Sales Meeting*. Conway, AR: Aermotor Division of Braden Industries, 1975. Brochure. Private collection of Stanley A. Anderson, Skiatook, OK.

———. *Aermotor 1976 Sales Meeting*. Conway, AR: Aermotor Division of Braden Industries, 1976. Brochure. Private collection of Stanley A. Anderson, Skiatook, OK.

Aermotor Division, Valley Industries, Inc. *Performance Table 18ft Diameter Windmill*. Conway, AR: Aermotor Division of Valley Industries, 1978. One page. Typewritten. In archive of American Wind Power Center, Lubbock, TX.

Aermotor Windmill Company. "Aermotor Windmill Names Bracher President," August 24, 2005. Press release. In *Windmillers' Gazette* files, Rio Vista, TX.

Anderson, Stanley A., to Bob Kain (18-Foot Windmill—Pumping Total Lift Elevations), May 26, 1978. Typewritten letter. In archive of American Wind Power Center, Lubbock, TX.

———, to Jim Fetters (Windmill Course with Navajo Tribe May 3 to May 7, 1976), May 10, 1976. Typewritten letter. In private collection of Stanley A. Anderson, Skiatook, OK.

———. To Whom It May Concern (18 Foot Windmill), May 25, 1978. Typewritten letter. In archive of American Wind Power Center, Lubbock, TX.

———. Sales, Aermotor, Division of Braden-Aermotor Corporation, Broken Arrow, OK, to Aero Manufacturing, Lincoln, NE, October 3, 1967. Typewritten letter. In files of Aero Manufacturing Company, Geneva, NE.

———, Aermotor Division, Braden-Aermotor Corporation, Broken Arrow, OK, to Howard C. Marks, Aero Manufacturing Company, Lincoln, NE, October 18, 1967. Typewritten letter. In files of Aero Manufacturing Company, Geneva, NE.

———, Skiatook, OK, to author, April 12, 2012; April 13, 2012; April 14, 2012; April 16, 2012; April 28, 2012; May 10, 2012; October 10, 2012; and March 14, 2013.

Andrag, Hellmut, P. Andrag & Sons (Pty.) Ltd., Bellville, South Africa, to Martin Andrag, P. Andrag & Sons (Pty.) Ltd., Wiesbaden, South Africa, April 4, 1964. In files of Agrico, Bellville, South Africa, and translated from Afrikaans by Theo Andrag, May 5, 2012. Copy in possession of author.

Andrag, Theo, consultant, Agrico (formerly P. Andrag & Sons), Bellville, South Africa. "A Concise History of P Andrag & Sons and Agrico." E-mail message to author, February 5, 2012.

Atkinson, H. Beck, Angel Fire, NM, to author, February 6, 2012 and March 23, 2012.

Baldwin, Marcia Scholes, daughter of Daniel R. Scholes, Twin Lakes, WI, e-mail message to author, February 6, 2012.

Barbour, Henry J., manager advertising for Fairbanks, Morse & Co., Dealers & Mfrs.' Div., Beloit, WI, to Walter Prescott Webb, July 15, 1927. Walter Prescott Webb Papers, Briscoe Center for American History, University of Texas at Austin.

Benson, D. P., manager, Domestic Products, Valley Pump Group, Conway, AR, to Aermotor Windmill Customers (Subject: Design Changes and Improvements), April 23, 1984. Typewritten letter. Private collection of David P. Suey Sr., Beatrice, NE.

"Bibliographical Note." La Verne and Ida Noyes Collection, (RS 21/7/235), Special Collections Department, Iowa State University Library, Ames. Accessed May 7, 2013. http://www.lib.iastate.edu/arch/rgrp/21-7-235.html.

"Biographer Pays Final Tribute to L. W. Noyes." (Unidentified newspaper clipping, no date). "Biography, Noyes, La Verne W." file, Chicago Historical Society, Chicago.

"Biography. Noyes, Mrs. La Verne W. (Ida Elizabeth Smith)" file, Chicago Historical Society, Chicago.

Braden-Aermotor Corporation. *Annual Report*[,] *Fiscal Year Ended June 30, 1966*. Broken Arrow, OK: Braden-Aermotor Corp., 1966. In *Windmillers' Gazette* files, Rio Vista, TX.

———. *Annual Report*[,] *for the Fiscal Year Ended June 30, 1967*. Broken Arrow, OK: Braden-Aermotor Corp., 1967. In *Windmillers' Gazette* files, Rio Vista, TX.

———. *Annual Report 1968*[,] *for the Fiscal Year Ended June 30, 1968*. Broken Arrow, OK: Braden-Aermotor Corp., 1968. In *Windmillers' Gazette* files, Rio Vista, TX.

Braden Industries, Inc., *Annual Report 1969*. Broken Arrow, OK: Braden Industries, 1969. In *Windmillers' Gazette* files, Rio Vista, TX.

———. *Quarterly Report for the Three Months Ended March 28, 1970*[,] *To Our Shareholders*[,] *June 10, 1970*. Broken Arrow, OK: Braden Industries, 1970. Brochure. Private collection of Stanley A. Anderson, Skiatook, OK.

"Brief History of FIASA." Typescript provided to T. Lindsay Baker on December 9, 1985, by James Fetters, president, Essex Associates, Inc. In *Windmillers' Gazette* files, Rio Vista, TX.

Brigolin, Mike, collector and restorer of historic windmills, Columbus, MI, e-mail messages to author, September 26, 2013; October 28, 2013; and March 5, 2014.

Brouwers, Frans, editor, *Levende Molens*, Ekeren, Belgium, e-mail message to author, January 3, 2007.

The Butler Company, Butler, IN, "General Letter," January 15, 1943. Photocopy in *Windmillers' Gazette* files, Rio Vista, TX.

"Class of 1904." *Cornell Daily Sun* 24, no. 164 (May 16, 1904): 1. Division of Rare and Manuscript Collection, Cornell University, Ithaca, NY.

Conlon, Tom, president, Iron Man Windmill Co., Wuhan, China, e-mail message to author, March 21, 2012.

"Cornell Alumni Notes." *Cornell Alumni News* 7, no. 29 (April 26, 1905): 489. Division of Rare and Manuscript Collection, Cornell University, Ithaca, NY.

Dempster, C. B., president, Dempster Mill Manufacturing Co[mpany], Beatrice, Nebr[aska], to G. F. Kregel, Nebraska City, Nebr[aska], December 16, 1921. Typewritten letter. Kregel Windmill Company Papers, Nebraska State Historical Society, Lincoln.

Durham, Mark, chief operations officer, Gicon Pumps & Equipment, Abernathy, TX, e-mail message to author, April 18, 2014.

Dutrieu, Jules. *L'Eau à la Campagne partout et gratuitement par l'Aeromoteur 'Dutrieu'[,] Moulin-a-vent automatique perfectionné en acier galvanisé*. Wetteren, Belgium: Jules Dutrieu, [ca. 1910]. Handbill. Museum voor de Oudere Technieken, Grimbergen, Belgium.

Eide, A. Clyde, Lewis Center, OH, to T. Lindsay Baker, Rio Vista, TX, January 15, 1985; March 14, 1994; May 20, 1996; and December 10, 2004. In *Windmillers' Gazette* files, Rio Vista, TX.

Elgin Wind Power and Pump Company, Elgin, IL, to J. H. Wormly, Ransom, IL, May 25, 1915. In *Windmillers' Gazette* files, Rio Vista, TX.

Essex Associates, Inc., Dallas, TX. *FIASA Moves One Step Closer to the Marketplace*. August 4, 1992. Press release. In *Windmillers' Gazette* files, Rio Vista, TX.

Estate of La Verne Noyes, Chicago. To Board of Directors, Aermotor Company, Chicago, December 6, 1956. Typewritten letter. In Record Book, 285.

Extracts from a Letter by La Verne Noyes to His Wife, Read at the Luncheon Following the Laying of the Cornerstone of Ida Noyes Hall, April 17th, 1915. Chicago: N.p., 1915. Copy of letter placed in Ida Noyes Hall building cornerstone. Private collection of Mark S. Welch, Fort Worth, TX.

Fetters, Jim. General Letter No. 36-79 (Wind Energy Catalog), June 7, 1979. Private collection of Stanley A. Anderson, Skiatook, OK.

Ghazi, Ellouze, Tunisian windmill researcher, Tunis, Tunisia, e-mail message to author, September 17, 2012.

Harris, Coy F., executive director, American Wind Power Center, Lubbock, TX, e-mail message to author, January 29, 2013.

Headings, Troy, civil engineer, Amarillo Technical Office, Natural Resources Conservation Service, US Department of Agriculture, Amarillo, TX, e-mail message to author, January 8, 2014.

Isvik, Marvin, branch manager, Omaha, Nebraska, Aermotor Division, Braden-Aermotor Corporation. "Report on 8 Ft. Argentine Aermotor Mill." November 17, 1966. Letter to Braden-Aermotor Corporation. Archive of American Wind Power Center, Lubbock, TX.

———. "Talk on Windmills." Speech. Stockmans Inn, North Platte, NE, June 6, 2002. Photocopy in possession of author.

Jacobs, Marcellus L. "Wind Driven Electric Generating Plant—1931–1957" [ca. 1957]. Copy in private collection of Craig Toepfer, Chelsea, MI.

Kain, Robert H., director of engineering, Valley Pump Group, Conway, AR, to Aermotor Windmill Customers, (Subject: Design Changes and Improvements, Aermotor Windmills), August 13, 1982. Typewritten letter in possession of David P. Suey Sr., Beatrice, NE.

Keeney, Mary Erena. Gleim Family Tree. Ancestry.com. Accessed November 20, 2011. http://trees.ancestry.com/tree/17565134/person/18114260761.

"La Verne Noyes Family Record." Biography, Noyes, La Verne W. file, Chicago Historical Society, Chicago.

Last Will and Testament and Two Codicils thereto of La Verne Noyes[;] Date of Death, July 24, 1919. (Includes a seven-page, typed document providing a room-by-room inventory of the Noyes estate on Lake Shore Drive.) Copy in private collection of Mark S. Welch, Ft. Worth, Texas.

Lowenstein, George H., International Department, Molinos de Viento, S.A., Chihuahua, Mexico, to T. Lindsay Baker, October 7, 1985. In *Windmillers' Gazette* files, Rio Vista, TX.

Marks, Howard C., owner, Aero Mfg. Co. [Lincoln, NE], to Aermotor, Broken Arrow, OK, October 14, 1967. Typewritten letter in files of Aero Manufacturing Company, Geneva, NE.

Marks, Norman H., co-owner, Aero Manufacturing Co., Geneva, NE, to author, June 27, 2012.

McCook, William, Springvale, Victoria, Australia, to T. Lindsay Baker, Rio Vista, TX, October 17, 1989. In *Windmillers' Gazette* files, Rio Vista, TX.

"In Memoriam[,] Thomas Osborn [sic] Perry, '69, '72e." *Michigan Alumnus* 33, no. 23 (March 1927): 520. Bentley Historical Library, University of Michigan, Ann Arbor.

Moore, Geoffrey J., W. D. Moore & Company, O'Connor, Australia, to William McCook, Springvale, Australia, September 21, 1989. Photocopy. In *Windmillers' Gazette* files, Rio Vista, TX.

Moore, Kevin. "Early Aermotor Windmill Company History in California." Attachment, e-mail message to author, March 18, 2012.

W. D. Moore & Co., O'Connor, Australia. "History of the Yellowtail Purchase," ca. 1985. Attachment from Geoffrey J. Moore, president, W. D. Moore & Co., e-mail message to author, January 2, 2012.

Monk, C. L., office manager, Aermotor Div[ision] of Nautec Corp., Dallas, TX, to E. L. Weid Hardware, Cameron, TX, June 19, 1964. Typewritten letter. In Windmill Manufacturers' Trade Literature Collection, Historical Research Center, Panhandle-Plains Historical Museum, Canyon, TX.

Morrow, Michael Guy, president, Aermotor Windmill Company, San Angelo, TX, e-mail message to author, January 3, 2014.

Motor Products Corporation. *44th Annual Report[,] Motor Products Corporation Fiscal Year Ended June 30, 1959*. Detroit: Motor Products Corp., 1959. In *Windmillers' Gazette* files, Rio Vista, TX.

———. *45th Annual Report[,] Motor Products Corporation Fiscal Year Ended June 30, 1960*. Detroit: Motor Products Corp., 1960. In *Windmillers' Gazette* files, Rio Vista, TX.

Nautec Corporation. *Annual Report[,] Fiscal Year Ended June 30, 1961*. New York: Nautec Corp., 1961. In *Windmillers' Gazette* files, Rio Vista, TX.

———. *Annual Report[,] Fiscal Year Ended June 30, 1962*. New York: Nautec Corp., 1962. In *Windmillers' Gazette* files, Rio Vista, TX.

———. *Annual Report[,] Fiscal Year Ended June 30, 1963*. New York: Nautec Corp., 1963. In *Windmillers' Gazette* files, Rio Vista, TX.

———. *Annual Report[,] Fiscal Year Ended June 30, 1964*. New York: Nautec Corp., 1964. In *Windmillers' Gazette* files, Rio Vista, TX.

———. *Annual Report[,] Fiscal Year Ended June 30, 1965*. New York: Nautec Corp., 1965. In *Windmillers' Gazette* files, Rio Vista, TX.

New Mexico State University, College of Agricultural, Consumer, and Environmental Sciences. "NMSU Windmill Technology Certification Workshop." Accessed September 27, 2012. http://aces.nmsu.edu/ces/windmill/training.html.

Noyes, Ida. Ida Noyes Papers. Department of Special Collections, University of Chicago, Chicago.

Noyes, La Verne W. "The Manufacturer and the Farmer." *The Alumnus* (Iowa State College) 5, no. 4 (January 1910): 2–7. Special Collections Department, Iowa State University Library, Ames.

O'Brock, Ken, owner, O'Brock Windmill Distributors, North Benton, OH, e-mail message to author, February 18, 2012.

Perry, Mary E., wife of Thomas O. Perry, Oak Park, IL, to Julius Rosenwald, Chicago, March 5, 1927. Typewritten letter. Archive, Museum of Science and Industry, Chicago.

———, to Leo F. Wormser, Chicago, March 10, 1927, April 15, 1927, and July 7, 1930. Typewritten letters. Archive, Museum of Science and Industry, Chicago.

Perry, Thomas Osborne. Letter to editor of the *Michigan Alumnus*, September 11, 1911. Necrology File. Bentley Historical Library, University of Michigan, Ann Arbor.

———. To the University of Michigan, Ann Arbor. Record for General Catalogue of Alumni and Former Students. Survey. October 26, 1910. Bentley Historical Library, University of Michigan, Ann Arbor.

Redborg, C. E., Farm Equipment Sales, United States Wind Engine & Pump Company, Batavia, IL, to William McCook, Armadale, Melbourne, Victoria, Australia, October 26, 1944, Typewritten letter. William McCook Papers, Science Place, Museum of Victoria, Melbourne, Australia.

Rogier, Etienne, researcher and biographer of Amédée Durand, Toulouse, France, e-mail messages to author, June 26, 2012, and February 15, 2013.

Rosengard, Steve, assistant curator, Museum of Science and Industry, Chicago, e-mail message to author, February 18, 2013.

Runyan, Craig, owner, SolandAer, Williamsburg, NM, e-mail messages to author, October 9, 2012; March 10, 2014; and April 29, 2014.

Sanz, Victor, President, Fabrica de Implementos Agricoles, S. A., Buenos Aires, Argentina. To Whom It May Concern, March 17, 1981. Typewritten letter. Windmill Manufacturers' Trade Literature Collection, Historical Research Center, Panhandle-Plains Historical Museum, Canyon, TX.

Simpson, Luis, filmmaker, France, e-mail message to author, September 28, 2012.

Scott, Susan Wegehoft, daughter of Dick Wegehoft, chief engineer, Aermotor Division of Braden Industries, Inc., Conway, AR, e-mail message to author, October 10, 2013.

"Seniors." *Cornell Alumni News* 6, no. 32 (May 18, 1904): 253. Division of Rare and Manuscript Collection, Cornell University, Ithaca, NY.

Smith, Jeannette Husk. "Memories." February 10, 1964. Scrapbook no. 63. Batavia Historical Society, Batavia, IL.

Suey, David P., Sr., Beatrice, NE, to author, September 12, 2013.

Thornton, Mark V., researcher on the history of forest fire lookout towers, Groveland, CA, to T. Lindsay Baker, Canyon, TX, February 5, 1985. Typewritten letter. In *Windmillers' Gazette* files, Rio Vista, TX.

Tustin, William Isaac. "Recollections of Early Days in California." Manuscript. 1880. Bancroft Library, University of California, Berkley.

Valley Industries, Inc. *1972 Annual Report*. St. Louis: Valley Industries, 1972. Private collection of David P. Suey Sr., Beatrice, NE.

———. *1973 Annual Report*. St. Louis: Valley Industries, 1973. Private collection of David P. Suey Sr., Beatrice, NE.

———. *1974 Annual Report*. St. Louis: Valley Industries, 1974. Private collection of David P. Suey Sr., Beatrice, NE.

———. *1975 Annual Report*. St. Louis: Valley Industries, 1975. Private collection of David P. Suey Sr., Beatrice, NE.

——. *1977 Annual Report*. St. Louis: Valley Industries, 1977. Private collection of David P. Suey Sr., Beatrice, NE.

——. *1978 Annual Report*. St. Louis: Valley Industries, 1978. Private collection of David P. Suey Sr., Beatrice, NE.

——. *1979 Annual Report*. St. Louis: Valley Industries, 1979. Private collection of David P. Suey Sr., Beatrice, NE.

——. *1980 Annual Report*. St. Louis: Valley Industries, 1981. Private collection of David P. Suey Sr., Beatrice, NE.

——. *Aermotor Engineering Program*. St. Louis: Valley Industries, 1977. Binder, unpaged. Private collection of David P. Suey Sr., Beatrice, NE.

——. *Annual Report 1976*. St. Louis: Valley Industries, 1976. Private collection of David P. Suey Sr., Beatrice, NE.

——. *Annual Report 1983*. St. Louis: Valley Industries, 1984. Private collection of David P. Suey Sr., Beatrice, NE.

——. *Annual Report 1984*. St. Louis: Valley Industries, 1985. Private collection of David P. Suey Sr., Beatrice, NE.

——. *Data Book*. St. Louis: Valley Industries, 1975. Private collection of David P. Suey Sr., Beatrice, NE.

Van Sante-Baetens, R. *Aermotor: Tours et moulins à vent en acier galvanisé*[,] *Électricité, Eau*[,] *Mouture, Travaux du Bois, etc.* Wetteren, Belgium: R. Van Sante-Baetens, [ca. 1895]. Folder. Museum voor de Oudere Technieken, Grimbergen, Belgium.

Vana, David, Davana LLC, Fire Tower Restoration, Bloomingdale, NY, e-mail message to author, February 21, 2013.

Verheul, Kees, Port O'Connor, TX, e-mail messages to author, January 12, 2012; January 26, 2012; and February 12, 2012.

Wade, H. N., president of U.S. Wind Engine and Pump Company, Batavia, IL, to Walter Prescott Webb, July 5, 1927. Walter Prescott Webb Papers, Briscoe Center for American History, University of Texas, Austin.

Walter, Helen, editor of the *Windmill Journal* (Australia) and Australian windmill researcher, Morawa, Australia, e-mail messages to author, July 31, 2012; August 17, 2012; September 4, 2012; September 20, 2012; and October 15, 2012.

Ward, A[llen] T. S[hawnee] Ind[ian] M[anual] L[abor] School, [postmark Westport, MO], to Miss Elizabeth T. Ward, Mount Pleasant, Jefferson County, OH, September 1, 1850. A. T. Ward Collection, Manuscript Department, Kansas State Historical Society, Topeka.

Wegehoft, Dick. *Significant Dates in Aermotor History*. Conway, AR: Aermotor, a Division of Valley Industries, Inc., [ca. 1982]. Typewritten and photocopied. Private collection of Stanley A. Anderson, Skiatook, OK.

Whalen, Vivian R., Hartford, WI, to author, August 19, 2012.

Yerian, Neal, windmill restorer, seller, and owner of Windmill-Parts, Westfield, IN, e-mail message to author, December 23, 2011.

INTERVIEWS

Anderson, Stanley A., by T. Lindsay Baker, Shattuck, OK, June 11, 1997. Transcript in *Windmillers' Gazette* files, Rio Vista, TX.

——, Skiatook, OK, telephone interview by author, September 23, 2013.

Atkinson, H. Beck, vice-president, Essex Associates, by T. Lindsay Baker, Amarillo, TX, May 28, 1981. Transcript in *Windmillers' Gazette* files, Rio Vista, TX.

Blakely, Scott, owner, Pronghorn Pump & Repair, Evansville, WY, telephone interview by author, February 14, 2014.

Bracher, Bob, former president, Aermotor Windmill Company, by author, San Angelo, TX, July 27, 2012.

Clark, R. Nolan, by author, Lubbock, TX, November 8, 2013.

Crist, William, by T. Lindsay Baker, Batavia, IL, July 25, 1992. Transcript in *Windmillers' Gazette* files, Rio Vista, TX.

Durham, Mark, chief operations officer, Gicon Pumps & Equipment, by author, Abernathy, TX, November 8, 2013.

Fernandez-Bueno, Carlos, owner, Potomac Wind Energy, by author, Dickerson, MD, November 13, 2011.

Jones, Chuck, owner, Flint Hills and Pump Service, Benton, KS, by author, Batavia, IL, June 14, 2012.

Grant, Vance, Amarillo Pump and Supply Company, Inc., Lubbock, TX, telephone interview by T. Lindsay Baker, May 10, 1996. Typewritten notes in *Windmillers' Gazette* files, Rio Vista, TX.

Isvik, Marvin, by T. Lindsay Baker, Comstock, NE, June 4, 2001. Transcript in *Windmillers' Gazette* files, Rio Vista, TX.

Kelley, Max, president, Heller-Aller Company, Napoleon, OH, telephone interview by T. Lindsay Baker, April 20, 1987. Notes in *Windmillers' Gazette* files, Rio Vista, TX.

Kleinke, Clarence, by T. Lindsay Baker, Batavia, IL, July 25, 1992. Transcript in *Windmillers' Gazette* files, Rio Vista, TX.

Morrow, Michael Guy, president, Aermotor Windmill Company, by author, San Angelo, TX, July 27, 2012.

Nigh, Robert R., Jr., Tulsa, OK, telephone interview by author, December 10, 2012.

O'Brock, Ken, owner, O'Brock Windmill Distributors, by author, El Dorado, KS, June 4, 2013.

Popeck, Bob, by author, Batavia, IL, December 2, 2011.

Sanz, Juan Carlos, president, FIASA, by author, Abernathy, TX, November 7, 2013.

Schovajsa, Calvin, plant supervisor, Aermotor Windmill Company, San Angelo, TX, telephone interview by author, January 16, 2012.

Shippard, Burton, Amarillo Pump and Supply Company, Inc., Amarillo, TX, telephone interview by T. Lindsay Baker, May 10, 1996. Typewritten notes in *Windmillers' Gazette* files, Rio Vista, TX.

Thomerson, Al, owner, Topper Company, San Angelo, TX, telephone interview by author, October 9, 2012.

Tucker, Patricia, secretary to Wendell Dean, by author, Broken Arrow, OK, December 1, 2012.

Wegehoft, Dick, Public Relations, Mueller Pump, Conway, AR, telephone interview by T. Lindsay Baker, April 16, 1986. Transcript in *Windmiller's Gazette* files.

Zwiebel, Jesse, sales manager, Aermotor Windmill Company, by author, San Angelo, TX, July 27, 2012.

BOOKS AND PAPERS

Achilles, Rolf. *Made in Illinois: A Story of Illinois Manufacturing.* Chicago: Illinois Manufacturers' Association, 1993.

Allen, Lewis F. *American Cattle: Their History, Breeding and Management.* New York: Orange Judd and Company, 1868.

Ardrey, R. L. *American Agricultural Implements: A Review of Invention and Development in the Agricultural Implement Industry of the United States.* Chicago: privately printed, 1894.

Baichun Zhang. "Ancient Chinese Windmills." Paper presented at the Third International Symposium on History of Machines and Mechanisms. National Cheng Keng University, Tainan, Taiwan. November 11–14, 2008. Accessed February 19, 2014. http://link.springer.com/chapter/10.1007%2F978-1-4020-9485-9_15.

Baker, T. Lindsay. "The Export of Wind Engines Manufactured in North America." In *The International Molinological Society 10th International Symposium Transactions, Stratford, Virginia, USA[,] 16–24 September 2000*, 105–22. Orange, VA: TIMS America, 2002.

——. *A Field Guide to American Windmills*. Norman: University of Oklahoma Press, 1985.

——. *A Guide to United States Patents for Windmills and Wind Engines[,] 1793–1950*. Biblotheca Molinologica 18. Herts, UK: International Molinological Society, 2004.

——. "The Marketing of Wind Engines by Manufacturers in the United States." In *The International Molinological Society: 9th Symposium in the Technical University of Budapest Hungary*, 31–43. Budapest, Hungary: Magyar Molinológiai Társaság, 2004.

——. *North American Windmill Manufacturers' Trade Literature: A Descriptive Guide*. Norman: University of Oklahoma, 1998.

——. "Patents as a Key to Understanding Wind Power History in the United States." In *Transactions, 11th International Symposium of TIMS: Portugal, Amadora, 25th September–2nd October 2004*, 265–71. Belas, Portugal: Etnoideia, 2007.

Baker Manufacturing Co.[,] 125th Anniversary[,] 1873–1998[,] September, 1998. Edited by Ruth Ann Montgomery. Evansville, WI: Baker Manufacturing Co., 1998.

Bancroft, Hubert Howe. *History of California*. Vol. 7, *1860–1890*. San Francisco: The History Company, 1890.

Bauters, Paul. "The Oldest References to Windmills in Europe." In *The International Molinological Society Transactions of the Fifth Symposium, France, 1982, April 5–10*, 111–18. Saint-Maurice, France: Fédération Française des Amis de Moulines, 1984.

——, and Gerrit Pouw. *Van Zadelsteen tot Zetelkruier, 2000 jaar molens in Vlaanderen*. Ghent, Belgium: Provinciebestuur Oost-Vlaanderen, 2002.

Bergey, Michael L. S. "Wind-Electric Pumping Systems for Communities." Paper presented at the First International Symposium on Safe Drinking Water in Small Systems, Washington, DC, May 10–13, 1998. Accessed February 19, 2014. http://bergey.com/wind-school/articles/wind-electric-pumping-systems-for-communities-2.

Berman, Constance Hoffman. "The Preservation of Records for Thirteenth Century Windmills in Southern France." In *13th International Symposium on Molinology Papers*, 226–33. Hadsund, Denmark: Nordjyllands Historiske Museum, Hadsund, 2011.

Bodmer, George Rudolph. *Hydraulic Motors: Turbines and Pressure Engines: For the Use of Engineers, Manufacturers, and Students*. New York: D. Van Nostrand Company, 1897.

Brulle, Robert V. *Engineering the Space Age: A Rocket Scientist Remembers*. Maxwell Air Force Base, AL: Air University Press, 2008.

Chaney, Henry Allen. *University of Michigan: The Class of Sixty-Nine in 1887*. Detroit: Thomas Smith, 1887.

Clark, R. Nolan. "Performance Comparison of Two Multibladed Windmills." In *Proceedings of the 11th ASME Wind Energy Symposium, Houston, TX, January 1992*, Vol. 12, 147–49. New York: American Society of Mechanical Engineers, 1992.

——. *Small Wind: Planning and Building Successful Installations*. Waltham, MA: Academic Press, 2014.

——. "Wind-Electric Water Pumping Systems for Rural Domestic and Livestock Water." Paper presented at the 5th European Wind Energy Association Conference and Exhibition, Thessaloniki-Macedonia, Greece, October 10–14, 1994. In *Conference Pro-*

ceedings, vol. 2, *Oral Sessions*, ed. J. L. Tsipouridis, 1136–40. Thessaloniki-Macedonia, Greece: Hellenic Wind Energy Association, 1994.

——, and J. W. McCarty. "Variable Stroke Pumping for Mechanical Windmills." In *Proceedings of Windpower '90, Washington, DC, September 24–28, 1990*, 217–21. Washington, DC: American Wind Energy Association, 1990.

——, and Brian D. Vick. "Livestock Watering with Renewable Energy Systems." In *Agriculture as a Producer and Consumer of Energy*, ed. J. Outlaw, K. J. Collins, and J. A. Duffield, 232–42. Oxford, UK: CABI, 2005.

Clews, Henry. "Electric Power from the Wind." In *Producing Your Own Power: How to Make Nature's Energy Sources Work for You*, ed. Carol Hupping Stoner, 13–43. Emmaus, PA: Rodale Press, 1974.

Davidson, J. Brownlee, and Leon Wilson Chase. *Farm Machinery and Farm Motors*. New York: Orange Judd Company, 1910.

Day, H. Summerfield. *The Iowa State University Campus and Its Buildings, 1859–1979*. Ames, IA: Iowa State University, 1980. Accessed May 7, 2013. http://www.fpm.iastate.edu/maps/historic/ISU_Campus_and_its_buildings.pdf.

De Little, Rodney. *The Windmills of England*. Partridge Green, UK: Colwood Press Ltd., 1997.

Devyt, Chr. *Westvlaamse windmolens. Inventaris volgens de toestand op 1 januari 1965*. Brugge, Belgium: N.p., 1966.

Dickerman, Charles W. *How to Make the Farm Pay; or, The Farmer's Book of Practical Information on Agricultural, Stock Raising, Fruit Culture, Special Crops, Domestic Economy & Family Medicine*. Philadelphia: Zeigler, McCurdy & Co., 1870.

Freese, Stanley. *Windmills and Millwrighting*. London: Cambridge University Press, 1957.

Gillis, Christopher. *Windpower*. Atglen, PA: Schiffer Publishing, 2008.

Goodspeed, Thomas Wakefield. *A History of the University of Chicago*[,] *Founded by John D. Rockefeller*[,] *The First Quarter-Century*. Chicago: University of Chicago Press, 1972.

Goslin, Samuel B. *The Relative Advantages of Wind, Water, and Steam as Motive Powers, Compared with Each Other, and a Description of the Motors the Most Suited for Utilising Them*. 2nd ed. London: John Warner & Sons and M'Corquodale & Co., 1879.

Gregory, Roy. *The Industrial Windmill in Britain*. West Sussex, UK: Phillimore & Co., 2005.

Grille, M., and M. G. LeLarge. *L'Agriculture et les Machines Agricoles aux États-Unis*. Paris: E. Bernard et Cie, 1896.

Gross, Bruce, Christine van Wijk, and Nilanjana Mukherjee. *Linking Sustainability with Demand, Gender and Poverty*[:] *A Study in Community-Managed Water Supply Projects in 15 Countries*. Delft, Netherlands: IRC International Water and Sanitation Centre, 2001. Accessed January 27, 2014. http://www.wsp.org/sites/wsp.org/files/publications/global_plareport.pdf.

Gustafson, John A., and Jeffrey D. Schielke. *Historic Batavia, Illinois*. Batavia, IL: Batavia Historical Society, 1980.

Harverson, Michael. *Persian Windmills*. Biblioteca Molinologica 10. Sprang Capelle, Netherlands: International Molinological Society, 1991.

Hills, Richard L. *Power from Wind: A History of Windmill Technology*. Cambridge: Cambridge University Press, 1994.

Holly, H. Hundson. *Modern Dwellings in Town and Country: Adapted to American Wants and Climate with a Treatise on Furniture and Decoration*. New York: Harper & Brothers, 1878.

Hong-Sen Yang. *Reconstruction Designs of Lost Ancient Chinese Machinery*. History of Mechanism and Machine Science 3. Dordrecht, Netherlands: Springer, 2007.

In Memoriam[,] *Ida E. Smith Noyes*[,] *Born, April 16, 1853*[,] *Died, December 5, 1912*. New York: James T. White & Co., 1918.

Johnson, Frank G. *The Wind as a Motive Power; Especially as Adapted to Supply the Wants of the Farmer; as Well as Many Classes of Mechanics. Together with the Advantage of Drainage and Irrigation; Showing How One Thousand Dollars a Year Can Be Economized on a Small Farm by the Use of Wind and Water. Also, Giving in Detail a New and Cheap Plan for General Irrigation. Besides Containing a General Discussion on the Subject of Windmills, with Their Objectives Heretofore Used and a New Method of Constructing Them*. [New York], H. S. Taylor, 1856.

Johnson, Rossiter, ed. *A History of the World's Columbian Exposition*. Vol. 2. New York: D. Appleton & Company, 1897–1898.

Kovarik, Tom, Charles Pipher, and John Hurst. *Wind Energy*. Northbrook, IL: Domus Books, 1979.

McConnell, Robert D. *Giromill Overview*. Golden, CO: Solar Energy Research Institute, 1979.

Marks, William, and Charles Coleman. *The History of Wind-Power on Martha's Vineyard*. N.p.: National Association of Wind-Power Resources, Inc., 1981.

[Munro-Fraser, J. P.] *History of Contra Costa County, California*. San Francisco: W. A. Slocum & Co., 1882.

Needham, Joseph. *Science and Civilization in China*. Vol 4, *Physics and Physical Technology; pt. 2, Mechanical Engineering*. London: Cambridge University Press, 1965.

Newman, Ann. *Conway*. Images of America. Charleston, SC: Arcadia Publishing, 1999.

Nichols, C. S. *Iowa State College of Agriculture and Mechanic Arts, Directory of Graduates of the Division of Engineering*. Ames, IA: N.p., 1912.

Nijs, Wilfried, and Frans Brouwers. "Wieksystemen." Windmill course, *Levende Molens*, Aartselaar, Belgium, 2011–2012.

Noyes, Ida E.S. *Occasional Verses*[,] *Toasts and Sentiments*. Chicago: La Verne W. Noyes, 1912.

"Noyes, La Verne." *The National Cyclopedia of American Biography*. Vol. 17. New York: James T. White & Co., 1927.

[Noyes, La Verne W.] *Descendants of Reverend William Noyes, Born, England, 1568, In Direct Line to La Verne W. Noyes and Frances Adelia Noyes-Giffen*. Chicago: La Verne W. Noyes, 1900.

[Pennsylvania State Agricultural Society.] *Report on Wind Engine Test, Authorized by the Pennsylvania State Agricultural Society, and Made on the Exhibition Grounds, Sept. 18 to 27, 1884*. Philadelphia: N.p., 1884.

Pernoud, R. *Die Kreuzzuge in Augenseugenberichten*. Dusseldorf, Germany: Karl Rauch, 1961.

Perry, Thomas O. *Life in the Universe*. Preface by Albert W. Palmer. Chicago: Frank Perry Keeney, 1927.

Plowboy. "Meet the Man Who Invented the Modern Windplant: Marcellus Jacobs." In *The Mother Earth News Handbook of Homemade Power*, ed. Mother Earth News staff, 144–61. New York: Bantam Books, 1974.

Putnam, Xeno W. *The Gasoline Engine on the Farm*. New York: Norman W. Henley Publishing, 1913.

Quinn, William P. *The Saltworks of Historic Cape Cod: A Record of the Nineteenth Century Economic Boom in Barnstable County*. Orleans, MA: Parnassus, 1993.

Ramsower, Harry C. *Equipment for the Farm and the Farmstead*. Boston: Ginn and Company, 1917.

Reynolds, John. *Windmills & Watermills*. New York: Praeger, 1975.

Righter, Robert W. *Wind Energy in America: A History*. Norman: University of Oklahoma, 1996.

Robinson, Marilyn, and Jeffery D. Schielke. *John Gustafson's Historic Batavia*. Batavia, IL: Batavia Historical Society, 1998.

Schurr, Sam H., Calvin C. Burwell, Warren D. Devine Jr., and Sidney Sonenblum. *Electricity in the American Economy: Agent of Technological Progress*. Westport, CT: Greenwood Publishing, 1990.

Smith, Deonta, and David Yang. *8 Industries That Will Lose or Gain from Ongoing Drought[,] February 2013*. Santa Monica, CA: IBISWorld, 2013.

Sullivan, George. *Wind Power for Your Home*. New York: Cornerstone Library, 1978.

Toepfer, Craig. *The Hybrid Electric Home*. Atglen, PA: Schiffer Publishing, 2010.

Traill, H. D., ed. *Social England: A Record of the Progress of the People in Religion Laws Learning Arts Industry Commerce Science Literature and Manners from the Earlier Times to Present Day*. Vol. 4, *From the Accession of James I, to the Death of Anne*. London: Cassell and Company, 1895.

United Nations. *The Millennium Development Goals Report 2010*. New York: United Nations, 2010.

Vale, Brenda, and Robert Vale. *The Autonomous House: Design and Planning for Self-Sufficiency*. New York: Universe Books, 1975.

Wailes, Rex. *The English Windmill*. London: Routledge & Kegan Paul, 1954.

Walter, John, and Régis Girard. *The Éolienne Bollée: An Illustrated History of the Unique French Wind-Turbine, from 1868 to Today*. East Sussex, UK: Nevill Publishing & Design, 2009.

Walton, James, and André Pretorius. *Windpumps in South Africa[,] Wherever You Go, You See Them: Whenever You See Them, They Go*. Cape Town, South Africa: Human & Rousseau, 1998.

Webb, Walter Prescott. *The Great Plains*. Boston: Ginn and Company, 1931.

Wendel, C. H. *American Gasoline Engines Since 1872*. Ed. George H. Dammann. Sarasota, FL: Crestline Publishing Co., 1983.

Wilcox, Lucius M. *Irrigation Farming*. New York: Orange Judd Company, 1909.

Wilson, Woodrow T. *History of Crisfield and Surrounding Areas of Maryland's Eastern Shore*. Baltimore: Gateway Press, 1974.

Wolff, Alfred R. *The Windmill as a Prime Mover*. New York: John Wiley & Sons, 1885.

Wood, De Volson. "Windmills." *Johnson's Universal Cyclopedia*. Vol. 8. New York: D. Appleton and Company / A. J. Johnson Company, 1895.

Wulff, Hans E. *The Traditional Crafts of Persia: Their Development, Technology, and Influence on Eastern and Western Civilization*. Cambridge, MA: MIT Press, 1966.

ARTICLES

"400 Veterans Are Awarded Scholarships." *Cleveland News*, February 12, 1923.

"400 War Veterans Get Scholarships under Noyes' Will." *Beloit (WI) News*, February 10, 1923.

"1899—Chronological. Principal Occurrences in the Trade of the Past Year as Recorded in Farm Implement News." *Farm Implement News* 21, no. 2 (January 11, 1900): 50–55.

"$2,500,000 to Teach Fighters at U. of C." *Chicago Daily News*, July 25, 1918.

Aermotor Company. "Aermotor Pumping Devices Are Known and Used the World Over" (advertisement). *Implement and Vehicle Journal* (Dallas, TX) 14, no. 9 (May 8, 1909): 9.

———. "The Auto-Oiled Aermotor with Duplicate Gears Running in Oil" (advertisement). *Field and Farm* (Denver) 30, no. 1520 (March 20, 1915): 10.

——. "To the Trade[:] The Auto-Oiled Windmill with Duplicate Gears Running in Oil" (advertisement). *Farm Implement News* 36, no. 5 (February 4, 1915): 30.

——. "Twenty Aermotor Branch Houses for 1895." *Harper's Magazine Advertiser* (ca. 1895): 86.

"Aermotor Buys Plant from U. of C. for $625, 000." *Chicago Tribune*, April 25, 1925.

"Aermotor Company Go to Chicago Heights." *Farm Implement News* 21, no. 20 (May 17, 1900): 20.

Aermotor Division, Braden Industries, Inc. "1977 Sales Meeting Packed with Information." *To Aermotor Agents* (Conway, AR), [ca. 1977], unpaged. Company newsletter, which used as its cover the first page of *To Aermotor Agents, Bulletin No. 1, Oct. 16, 1893*.

——. "Branch Openings." *To Aermotor Agents* (Conway, AR), [ca. 1977], unpaged.

——. "Pride of Quality Is Aermotor Standard." *To Aermotor Agents* (Conway, AR), [ca. 1977], unpaged.

"The Aermotor Wind Mill." *Rural Californian*, August 1891, 488.

"The Aermotor Windmill[,] the Oiling Problem Solved" (advertisement). *The Review of the River Plate, a Weekly Journal of Commercial and General Views* (Buenos Aires, Argentina) 24, no. 881 (October 16, 1908): 963.

"Aerodynamic Wind Mills." *Scientific American* 140 (June 1929): 525.

"Agribusiness Development Team in Afghanistan." *San Angelo Standard-Times*, December 20, 2010.

Associated Press. "To Benefit from Tuition Scholarships." *Santa Cruz (CA) Sentinel*, February 10, 1923.

——. "Ex-Service Girls to Get Special Scholarship Gift." *Bisbee (AZ) Ore*, February 14, 1923.

——. "Nurses Will Receive Tuition Scholarships." *Little Rock Gazette*, February 11, 1923.

——. "Plan to Aid Women Vets." *Detroit Free Press*, February 13, 1923.

——. "Scholarships for World War Women." *Oshkosh (WI) Northwest*, February 10, 1923.

——. "Scholarships Given for Service in War." *Butte (MT) Miner*, March 4, 1923.

——. "Service Women Given Tuition Scholarships." *Twin Falls (ID) News*, February 10, 1923.

——. "Texas Drought Endangers Animals." *Frederick (MD) News-Post*, August 31, 2011.

——. "Tuition Scholarships Offered War Women." *Santa Barbara News*, February 15, 1923.

——. "Women in War Get Scholarship." *Peoria Journal*, February 20, 1923.

——. "Women War Workers Get Scholarships." *Elgin (IL) News*, February 21, 1923.

——. "Women Who Served in World War." *Hillsboro (TX) Mirror*, February 10, 1923.

——. "Women Who Worked During War Helped." *Savannah Press*, February 16, 1923.

Baker, T. Lindsay. "Aermotor Loop-Step Tower Ladders." *Windmillers' Gazette* 28, no. 2 (Spring 2009): 6–7.

——. "The 'Aermotor Man' and His Haying Tools." *Windmillers' Gazette* 12, no. 4 (Autumn 1993): 2–6.

——. "Aermotor 'Trussed Tripod' Steel Towers." *Windmillers' Gazette* 30, no. 4 (Autumn 2011): 8–9.

——. "Aermotor Windmill Towers." *Windmillers' Gazette* 11, no. 3 (Summer 1992): 2–14.

——. "American Wind Power Center Opens in Lubbock, Texas." *Windmillers' Gazette* 17, no. 4 (Autumn 1998): 5–6.

——. "Annual Lubrication for Self-Oiling Windmills." *Windmillers' Gazette* 4, no. 1 (Winter 1985): 7–8.

——. "Any Squeak or Grind: The Lubrication of American Windmills." *Windmillers' Gazette* 13, no. 3 (Summer 1994): 2–10.

———. "'As Old As the Hills and Just As Stationary'[,] The Battle between Windmills and Portable Gas Engines." *Windmillers' Gazette* 9, no. 1 (Winter 1990): 2–6.

———. "Battle Axe Homemade Windmills." *Windmillers' Gazette* 30, no. 4 (Autumn 2011): 2–7.

———. "A Close Look at the Challenge 27 Windmill." *Windmillers' Gazette* 23, no. 4 (Autumn 2004): 3–9.

———. "The Complex Story of Aermotor Corporate Ownership." *Windmillers' Gazette* 5, no. 1 (Winter 1986): 9.

———. "'For The Duration': American Windmills during World War II." *Windmillers' Gazette* 12, no. 3 (Summer 1993): 2–7.

———. "Forest Service Towers Made by the Aermotor Company: The National Standard for Fire Lookouts." *Windmillers' Gazette* 29, no. 2 (Spring 2010): 7–9.

———. "Freeze-Proofing Windmills in Moderate Climates." *Windmillers' Gazette* 4, no. 4 (Autumn 1985): 5–6.

———. "Galvanized Steel Tanks and Troughs for Use in Windmill Water Systems." *Windmillers' Gazette* 27, no. 4 (Autumn 2008): 2–7.

———. "How Windmill Companies Used Branch Houses and General Agencies to Conduct Business." *Windmillers' Gazette* 28, no. 4 (Autumn 2009): 5–7.

———. "Industrial Sabotage in 1893: The Incident at the World's Fair." *Windmillers' Gazette* 12, no. 3 (Summer 1993): 8–10.

———. "Interchangeable Wheels and Vanes for Open-Geared and Oil-Bath Steel Windmills." *Windmillers' Gazette* 18, no. 3 (Summer 1999): 5–10.

———. "Irrigating with Windmills on the Great Plains." *Great Plains Quarterly* 9, no. 4 (Fall 1989): 216–30.

———. "Large-Diameter Halladay Standard Windmills." *Windmillers' Gazette* 30, no. 2 (Spring 2011): 2–6.

———. "Mid-America Windmill Museum Opens at Kendallville, Indiana." *Windmillers' Gazette* 17, no. 3 (Summer 1998): 9.

———. "New Deal Specials[,] Bargain Windmills of the Depression Era." *Windmillers' Gazette* 4, no. 2 (Spring 1985): 3–5.

———. "An Overview of Horizontal Windmills." *Windmillers' Gazette* 18, no. 1 (Winter 1999): 2–7.

———. "Pioneer Metal Windmills of the Plains: The Kirkwood Iron Wind Engine." *Windmillers' Gazette* 14, no. 1 (Winter 1995): 2–6.

———. "Power Aermotor Windmills." *Windmillers' Gazette* 30, no. 3 (Summer 2011): 2–13.

———. "Power Windmills." *Windmillers' Gazette* 6, no. 1 (Winter 1987): 3–7.

———. "A Product History of Aermotor Windmills." *Windmillers' Gazette* 5, no. 1 (Winter 1986): 7–9.

———. "'Suburban Outfit' Windmill Towers: Water under Pressure Where You Want It." *Windmillers' Gazette* 25, no. 2 (Spring 2006): 2–5.

———. "Survey of Windmill Regulators: Part I." *Windmillers' Gazette* 14, no. 3 (Summer 1995): 2–10.

———. "Tank Heaters for Stock Watering." *Windmillers' Gazette* 10, no. 1 (Winter 1991): 5–7.

———. "Tilting Windmill Towers." *Windmillers' Gazette* 16, no. 1 (Winter 1997): 2–6.

———. "Turbine-Type Windmills of the Great Plains and Midwest." *Agricultural History* 54, no. 1 (1980): 38–51.

———. "Wind Electric News: The Papers of Oliver P. Fritchle." *Windmillers' Gazette* 10, no. 4 (Autumn 1991): 9–10.

———. "Windmill Museums as Destinations for Heritage Tourists." *Windmillers' Gazette* 31, no. 2 (Spring 2012): 2–9.

——. "Windmills and Railroad Water Systems." *Windmillers' Gazette* 28, no. 2 (Spring 2009): 2–5.

——. "Windmills and the Union Pacific Railroad." *Windmillers' Gazette* 6, no. 4 (Autumn 1987): 3–5.

——. "Windmills and Towers of the Temple Pump Company." *Windmillers' Gazette* 22, no. 1 (Winter 2003): 2–8.

——. "The Windmills That Get Down and Wallow on the Ground: The Jumbo/Paddle-Wheel/Ground Tumbler/Go-Devil Homemade Windmills." *Windmillers' Gazette* 30, no. 1 (Winter 2011): 2–6.

——. "Windmills with Variable-Pitch Blades." *Windmillers' Gazette* 8, no. 4 (Autumn 1989): 2–7.

Bannister, Justin. "NMSU to Host the 2013 Windmill Technology Certification Workshop." *New Mexico State University News Center*, May 8, 2013. Accessed January 1, 2014. http://newscenter.nmsu.edu/9432/nmsu-host-2013-windmill-technology-certification-workshop.

Bates, Putnam A. "Farm Electric Lighting by Wind Power." *Scientific American* 107 (September 28, 1912): 262.

Birkland, Sandy Jones. "Replica Russian Windmill Gifted to Fort Bliss: A Symbol of Peace Linking Two Nations." *Old Mill News* 40, no. 4 (Fall 2012): 15–17.

"Bomb Shatters Yuletide Calm on West Side." *Chicago Tribune*, December 25, 1927.

Braden Industries, Inc. "We're Gaining Ground Again." *Braden Action Line* (Broken Arrow, OK) 2, no. 7 (November 1970): unpaged.

Braden Winch, Division of Braden Industries, Inc. "The 'New' Braden Industries Philosophy & Technology[,] Excerpts from a Talk Given by James E. Beebe, President of Braden Industries, Inc. to the Tulsa Society of Investment Analysts—November 19, 1969." *Along the Braden Winchline* (Broken Arrow, OK) 9, no. 1 (1969): 2–3.

"Braden Winch Plant Important Cog in Prosperity of Broken Arrow." *Broken Arrow Ledger*, November 25, 1965.

B[rouwers], F[rans]. "Windmotoren." *Levende Molens* (Ekeren, Belgium) 34, no. 2 (February 2012): 15–21.

"Brush, Charles Francis." *American National Biography Online*. Ed. George Wise. February 2000. Accessed January 12, 2007. http://www.amb.org/articles/13/13-00214.html.

"Building Permits." *Chicago Daily Tribune*, December 22, 1893.

Burke-Bollmeyer Oiler Company. "Oil Your Wind Mill from the Ground" (advertisement). *Farm Implement News* 23, no. 2 (January 9, 1902): 130.

——. "Oil Your Windmill from the Ground by Pulling a Wire" (advertisement). *Farm Implement News* 23, no. 16 (April 17, 1902): 33.

"CIO Picket Dies on March at Struck Plant." *Chicago Tribune*, May 17, 1951.

Campbell, Steve. "Texas Weather: 'Drought Is Rearing Its Ugly Head Again.'" *Fort Worth Star-Telegram*, November 30, 2012.

Cantu, Lorrie Woodward. "Solar Pumps vs. Windmills: Seven Questions to Help You Pencil Out the Costs of Getting Water from Your Well by Wind or Solar." *The Cattleman* 99, no. 12 (May 2013): 28–29.

Carmichael, Joe M. "Water from the Wind." *The Cattleman* 36, no. 5 (October 1949): 23–25, 70.

"Change in Burke-Bollmeyer Company." *Farm Implement News* 25, no. 47 (November 24, 1904): 16.

"Change of Name." *Farm Implement News* 26, no. 49 (December 7, 1905): 22.

"Chas. S. Peterson Buys Noyes Home on Drive." *Chicago American*, August 9, 1921.

Church, Albert Cook. "The Padanaram Salt Works." *New England Magazine* 41, no. 2 (October 1909): 489–92.

Cline, Eric. "Landmark Windmill Finds New Home." *Progress-Index*, May 17, 2010. Accessed January 19, 2013. http://progress-index.com/news/landmark-windmill-finds-a-new-home-1.792194.

Collar, M. "The Iron Turbine Wind Engine" (advertisement). *Dodge City (KS) Times*, March 8, 1879.

Comis, Don. "Wind Power Where You Want It." *Agricultural Research* (Washington, DC) 42, no. 6 (June 1994): 4–7.

"Company Involved in 2-Step Lease Plan." *Chicago Tribune*, December 19, 1965.

"Conspired to Defraud. W. E. Hampton and August Holtgen Held for Felony Embezzlement." *San Francisco Call*, November 21, 1899.

Cook, Harold. "Water Stop." *Railroad Magazine* 66, no. 6 (October 1955): 12–23.

"Course in Windmills Is Offered by a University in New Mexico." *New York Times*, December 21, 1975.

Crattie, C. Barton. "Mr. Bilby's Elegant Assembly, Being Both Elegant and Sublime." *American Surveyor* 8, no. 4 (2011): unpaged. Accessed May 23, 2013. http://www.ameri-surv.com/PDF/TheAmericanSurveyor_Crattie-BilbyTowers_Vo18No4.pdf.

"C. S. Peterson Buys La Verne Noyes Home." *Chicago Evening Post*, August 13, 1921.

"C. S. Peterson Purchases La Verne W. Noyes Home." *Chicago Journal*, August 9, 1921.

Daniels, Jo Krieger. "Oklahoma Firm's Windmills, Pump[s] Improve Underdeveloped Areas." *Tulsa Sunday*, May 12, 1968.

Danker, Donald F. "Nebraska's Homemade Windmills." *American West* 3, no. 1 (Winter 1966): 13–19.

De Decker, Kris. "Wind Powered Factories: History (and Future) of Industrial Windmills." *Low-Tech Magazine* (October 8, 2009). Accessed January 18, 2013. http://www.lowtechmagazine.com/2009/10/history-of-industrial-windmills.html.

Deen, Thalif. "Women Spend 40 Billion Hours Collecting Water." *Inter Press Service News Agency*, August 31, 2012. Accessed January 27, 2014. http://www.ipsnews.net/2012/08/women-spend-40-billion-hours-collecting-water/.

Dempster Industries LLC, Beatrice, NE. "About Us." Accessed March 18, 2014. http://www.dempstersllc.com/demp_about_staff.php.

Denewet, Lieven. "Niet Nederland maar Vlaanderen was de bakermat! Belangrijke archiefvondst. De eerste vermelding van een poldermolen: de nog bestaande 'hoesse molen' bij Gent (1316)." *Molenecho's* 34, no. 3 (2006): 170–90.

"Detroit Firm Has Acquired Aermotor." *Chicago Tribune*, June 6, 1958.

Dewey, Mark. "Aermotor Windmills: 19th Century Company Seeking 21st Century Customers." *KUT News* (Austin, TX), May 28, 2013. Accessed February 9, 2014. http://kut.org/post/19th-century-windmill-company-seeks-21st-century-customers.

Dick, Everett. "Water: A Frontier Problem." *Nebraska History* 49, no. 3 (Autumn 1968): 215–45.

Dickerson, I. W. "Care of Windmills." *Wallace's Farmer* 43, no. 4 (January 25, 1918): 126, 128. Reprinted in *Windmillers' Gazette* 3, no. 1 (Winter 1984): 7–8.

Dole, F. L. "History and Development of the Windmill." *Export Implement Age* 14, no. 5 (August 1906): 27–28, 30.

"D. R. Scholes Funeral Set at 11 Today." *Chicago Tribune*, June 17, 1966.

Drury, Jack. "Willow Park District Will Open for Homes." *Chicago Herald and Examiner*, August 13, 1921.

Dunker, Chris. "City Council Rejects Dempsters Loan Application." *Beatrice (NE) Daily Sun*, November 18, 2013.

——. "Davis Settles with Dempster Employees." *Beatrice Daily Sun*, November 18, 2013.

——. "Debts Unpaid" (part 1 of 4), *Beatrice Daily Sun*, July 30, 2013.

——. "Dempster Had a Long History in Beatrice." *Beatrice Daily Sun*, July 29, 2013.

——. "Dempsters LLC Not Registered to Do Business in Neb.." *Beatrice Daily Sun*, November 19, 2013.

——. "Money for Nothing" (part 3 of 4), *Beatrice Daily Sun*, August 1, 2013.

——. "'The More We Went There, the Worse It Got'" (part 2 of 4), *Beatrice Daily Sun*, July 31, 2013.

——. "Past Likely Haunting Current Dempsters' President." *Beatrice Daily Sun*, November 20, 2013.

——. "Tilting at Windmills" (part 4 of 4), *Beatrice Daily Sun*, August 4, 2013.

Eastman, Philip. "Windmill Irrigation in Kansas." *Review of Reviews* 29, no. 2 (February 1902): 183–87.

"Editorial Notes." *Michigan University Magazine* 3, no. 9 (June 1869): 364–67.

Eide, A. Clyde. "Free as the Wind." *Nebraska History* 51, no. 1 (Spring 1970): 25–48.

"Eighteen Wind Powered Water Pumps Restored in Gozo." *Gozo News*, January 8, 2010. Accessed September 23, 2012. http://gozonews.com/11882/eighteen-wind-powered-water-pumps-restored-in-gozo/.

"The Electric Light[,] the Union Depot Illuminated by Fourteen Lamps, and the Business Community Rapidly Adopting the Electric Light Generally." *Cleveland Leader*, July 1, 1881.

"The Electrical Value of Wind Power." *Scientific American* 93 (November 18, 1905): 394–95.

"Electro-Pneumatic Water-Supply System." *Electrical World* (New York, NY) 56, no. 13 (September 29, 1910): 744–45.

"Factory Prosperity Gives Scholarships." *Columbus Dispatch*, February 22, 1923.

"Farewell to Windmills." *USA Today* 119, no. 2553 (June 1991): 3.

"Farm Products Prices Show No Improvement." *Extension Service News* (Ithaca, NY) 5, no. 2 (February 1922): 14.

Faulkner County Historical Society. "Welcome to Conway Arkansas." Accessed October 6, 2013. http://www.conwayar.org/history.asp.

Fetters, Jim. "Viewpoint." *To Aermotor Agents* (Conway, AR), [1977], unpaged.

——. "Windmills[,] Phenomena in the Atomic Age." *Water Well Journal* (Columbus, OH) 26, no. 2 (February 1972): 27–29.

Flippin, Perry. "Aermotor Owner Secure in Future." *San Angelo Standard-Times*, August 28, 2007.

Fogle, F. D. "C. S. Peterson Buys 16-Room Noyes Home." *Chicago Herald and Examiner*, August 9, 1921.

Forest Fire Lookout Association. "Lookout Resources." Accessed February 21, 2013. http://www.firelookout.org/resources.html.

Fort Ross Windmill. "Fort Ross Windmill." Accessed September 26, 2013. http://fortross-windmill.com/en/fort-ross-windmill/stanica-2/.

"Funeral Plans Private for La Verne W. Noyes." *Chicago Tribune*, July 25, 1919.

"The Future of the Windmill." *Scientific American* 108, no. 4 (April 5, 1913): 309.

Gagey, R. "Les moulins à vent. Leur travail. Leurs applications en Tunisie." *Bulletin Agricole de L'Algérie et de la Tunisie* 10, no. 5 (1904): 91–102.

Galbraith, Kate. "Museum Shows History and Power of Wind Energy." *New York Times*, July 20, 2011.

——. "Windmilling, a Dying Art, Hangs On in Texas." *Texas Tribune*, October 28, 2011. Accessed October 22, 2013. http://www.texastribune.org/2011/10/28/windmilling-dying-art-hangs-texas/.

"Gift of a Grateful Patriot." *Chicago Daily News*, July 27, 1918.

Gillis, Christopher. "A Father's Influence: The Story of Roberts Pump & Supply Company." *Windmillers' Gazette* 32, no. 3 (Summer 2013): 8–12.

———. "Lean and Clean." *American Shipper* (Jacksonville, FL) 51, no. 8 (August 2009): 9–12.

———. "Sea Breezes to Salt." *Windmillers' Gazette* 32, no. 2 (Spring 2013): 9–10.

———. "U.S./Africa Trade Slow to Ripen." *American Shipper* 45, no. 10 (October 2003): 44–46.

Gipe, Paul. "Windmills: Still Primed for Pumping Water." *Independent Energy* 19, no. 6 (July–August 1989): 62–64.

Grundfos Holding A/S. "SQFlex." Accessed March 17, 2014. http://www.grundfos.com /products/find-product/sqflex.html.

———. "SQFlex at the Museum of Modern Art." August 17, 2007. Accessed March 17, 2014. http://www.grundfos.com/about-us/news-and-press/news/sqflex-at-the-muse-umofmodernart.html.

Hayward Area Historical Society. "A Short History of Hayward." Accessed January 19, 2013. http://www.haywardareahistory.org/sup/docs/A%20Short%20History%20°f%20 Hayward.pdf.

"Helicopter, Thomas O. Perry." Smithsonian National Air and Space Museum, Washington, DC. Accessed December 1, 2011. http://www.nasm.si.edu/collections/artifact .cfm?id=A19390031000.

Hendrix, John M. "Windmill Monkeys." *The Cattleman* 25, no. 8 (January 1939): 51–52.

"Here and There at the Fair." *Farm Implement News* 14, no. 35 (August 31, 1893): 24.

"A High Windmill." *Scientific American* 70, no. 19 (May 12, 1894): 292–93.

Hill, James W. "The Wind-Mill of To-Day and Its Uses." *Journal of the Association of Engineering Societies* 3, no. 8 (June 1884): 160–67.

Huppertz, George C. "Final Solution of the Water Supply System." *American Carpenter and Builder* 12, no. 3 (December 1911): 64–68.

Hurt, R. Douglas. "Irrigation of the West." *Journal of the West* 30, no. 2 (April 1991): 63–77.

———. "Windcatchers and Eyecatchers: Technology Down on the Farm." *Timeline* 2, no. 2 (April–May 1985): 27–39.

"Improved Windmill." *Scientific American* 11, no. 30 (April 5, 1856): 236.

International Molinological Society. "About TIMS." Accessed January 5, 2014. http:// www.molinology.org/index.php?option=com_content&view=article&id=72&ltemid =480.

"Inventions Wanted in Texas." *Scientific American,* n.s., 3, no. 9 (August 25, 1860): 132.

Iowa State University. "History of Iowa State: Campus Buildings—For Whom It Is Named." Iowa State University, Ames. Accessed January 31, 2015. http://www.public .iastate.edu/~isu150/history/forwhom-buildings.html.

Jack Nitz & Associates. "Complete Manufacturing Division & Relocation Dispersal Auction for Dempsters LLC," Beatrice, NE, April 4, 2014. Accessed March 18, 2014. http:// ignite.auctionservices.com/print/auction/222710.

Jordan, Terry G. "Evolution of the American Windmill: A Study in Diffusion and Modification." *Pioneer America, the Journal of the Pioneer America Society* 5, no. 2 (July 1973): 3–12.

Kash, Steve. "Blowing in the Wind." *Terre Haute (IN) Tribune-Star*, June 5, 2011.

Kirk, Margaret. "The Windmill Man." *Western Livestock and the Westerner* (Denver) 37, no. 2 (September 1951): 16, 98–99.

Kjerfve, Björn, C.A.F. Schettini, Bastiaan Knoppers, Guilherme Lessa, and H. O. Ferreria. "Hydrology and Salt Balance in a Large Hypersaline Coastal Lagoon: Lagoa de Araruama,

Brazil." *Estuarine, Coastal and Shelf Science* 42 (1996): 705. Accessed January 2, 2014. http://geotest.tamu.edu/userfiles/167/146.pdf.

Koperski, Scott. "Dempsters Prepares to Auction Equipment." *Beatrice (NE) Daily Sun*, March 14, 2014.

Korsman, Hans. "De Noord-Hollandse achtkante binnenkruier." *Neerslag Magazine* (The Hague, Netherlands), N.d. Accessed March 5, 2013. http://www.neerslag-magazine.nl/magazine/artikel/544/.

Lackey, Jerry. "Windmill Almanac 2012: Two Years of Drought Has Plagued Farmers and Ranchers." *San Angelo Standard-Times*, December 9, 2012.

———. "Windmill Country: Dry Range Forces Ranchers to Burn Cactuses for Cattle." *San Angelo Standard-Times*, April 20, 2013.

"Lake Forest Gets 15 Scholarships." *Lake Forest (IL) Register*, March 16, 1923.

"La Verne Noyes Dies after Long Fight for Life." *Chicago Tribune*, July 24, 1919.

"La Verne Noyes Gives $2,500,000 for Education." *Chicago Evening Post*, July 15, 1918.

"La Verne W. Noyes." *Chicago Evening Post*, July 25, 1919.

"La Verne W. Noyes." *Chicago Journal*, July 25, 1919.

"List of Exhibitors in Implement Annex." *Farm Implement News* 14, no. 25 (June 22, 1893): 18.

LoLordo, Ann. "Fashioning Life among Windmills; Couple: In Retirement, Kees and Jane Verheul Took on an Unlikely Business Project, Reviving a Once-Dying Business and Finding Fulfillment." *Baltimore Sun*, April 22, 2000.

"Longtime Manufacturers Face Brighter Future[,] Aermotor." *Ground Water Age* (Elmhurst, IL) 15, no. 12 (August 1981): 36–37.

Lusk, Sonia Marcel. "Water Sources in Early West Texas." *Permian Historical Annual* 20 (1980): 55–65.

"L. W. Noyes—The Aermotor Co." *Farm Implement News* 11, no. 8 (August 1890): 43.

"L. W. Noyes Home on Drive Is Sold to C. S. Peterson." *Chicago Tribune*, August 9, 1921.

McAdams, Ken. "The Disappearing Windmill." *Ford Times* 67, no. 8 (August 1974): 48–53.

"Machinists Out in Many Cities. Struggle for a Nine-Hour Day with Old Pay Is Begun in Earnest." *Chicago Daily Tribune*, May 21, 1901.

"A Magnificent Show: Wind Mills and Appurtenances." *Farm Implement News* 14, no. 41 (October 12, 1893): 19.

Manning, Roger S. "The Windmill in California." *Journal of the West* 14, no. 3 (July 1975): 33–39.

"Manufacturers Meet Jan. 24. National Association to Hold a Three Days' Session in Cincinnati, Ohio." *Chicago Daily Tribune*, January 23, 1899.

Martin, Todd. "Aermotor Has Long History." *San Angelo Standard-Times*, February 11, 1996.

———. "The Business of Breeze." *San Angelo Standard-Times*, February 11, 1996.

"Meeting of Wind Mill Manufacturers." *Farm Implement News* 16, no. 50 (December 12, 1895): 20.

"Message from Bob Schuetz." *Braden Employees Journal* (Broken Arrow, OK) 1, no. 2 (May 1965): 1.

Moore, Geoffrey, and Malcolm Walter. "W. D. Moore & Co." *Windmill Journal* (Australia) 2, no. 3 (June 2003): 2–5.

"Mr. Brush's Windmill Dynamo." *Scientific American* 63 (December 20, 1890): 389.

"New Incorporations." *Chicago Daily Tribune*, July 1, 1893.

"New Wind Mill Company." *Farm Implement News* 23, no. 35 (August 28, 1902): 25.

"New Wind Mill Oiler." *Farm Implement News* 22, no. 26 (September 5, 1901): 36.

"The Newest Windmill[,] an Improved Aermotor" (advertisement). *Nebraska Farmer*, August 19, 1933, n.p.

Nissen, Povl-Otto, chairman of the Poul la Cour Museum's Friends. "A Visit to the Poul la Cour Museum." Accessed February 1, 2015. http://www.poullacour.dk/engelsk/museet.htm.

Norberg, Bob. "Fort Ross Shows Off New Russian-Built Windmill." *Press Democrat* (Santa Rosa, CA), October 20, 2012.

"Noyes Bequeaths Part of Wealth to Public Good." *Chicago Herald and Examiner*, July 25, 1919.

"Noyes Bequests to War Veterans." *Mt. Carmel (IL) Register*, February 14, 1923.

"Noyes Estate Scholarships to 400 Veterans." *Chicago Tribune*, February 10, 1923.

"Noyes Estate to Train 400 Vets a Year." *Chicago Herald and Examiner*, February 10, 1923.

"Noyes Scholarships Steadily Increase." *Chicago Daily News*, January 24, 1923.

"Noyes Wills Fortune to Educate Heros." *Chicago Evening News*, July 28, 1919.

Ogden, Derek, and Anne Burke. "The Windmill at Flowerdew Hundred." *Old Mill News* 6, no. 1 (January 1978): 4–6.

"Obituaries[,] Frederick E. Smith." *Chicago Tribune*, April 26, 1940.

"'Oldest Airman' Takes Ride." *Chicago Examiner*, August 27, 1911.

Orr, Richard. "Day by Day on the Farm[,] Windmills Still Extant." *Chicago Tribune*, March 6, 1953.

Owens, Jim. "Angelo Becomes Windmill Capital." *San Angelo Standard-Times*, April 27, 1986.

"Paris Exposition: Not Many American Exhibitors, but a Representative Showing." *Farm Implement News* 21, no. 30 (July 26, 1900): 18–21.

"People and Events." *Chicago Tribune*, July 3, 1953.

"Perry Aquapneumatics." *Domestic Engineering and the Journal of Mechanical Contracting* (Chicago) 57, no. 5 (November 4, 1911): 126.

Perry, Thomas O. Letter to the editor, *Chicago Post*, August 1, 1919.

———. "Thomas Perry in Household of Gideon Perry, United States Census, 1860." FamilySearch.org, Accessed February 7, 2004. https://familysearch.org/pal:/MM9.1.1/MW69-PPW.

"Poultry and Egg Values for 1921 Slump to $943,200,000." *Weather, Crops and Markets* (US Department of Agriculture, Washington, DC) 1, no. 19 (May 13, 1922): 421.

"Ranchers in West Seeking to Restore Use of Windmills." *New York Times*, November 3, 1974.

"R. B." Review of "Utilité des moulins à vent aux Indes," by A. Chatterton. *Le Mois Scientifique et Industriel* (Paris) 5, no. 52 (December 25, 1903): 1043.

"Recent Realty Sales and Leases: Property Which Has Figured in Transfers in Cook County." *Chicago Daily Tribune*, November 3, 1895.

Red Cross Manufacturing Company. "Red Cross Wind Mills to the Front with Patent Automatic Oiler" (advertisement). *Farm Implement News* 24, no. 48 (November 26, 1903): 31.

Redmond, Mike. "Gone with the Wind." *Indianapolis Monthly* 28, no. 2 (October 2004): 96–102.

"Restoration of Wind-Powered Water Pumps Scheme." *Gozo News*, May 15, 2009. Accessed January 2, 2014. http://gozonews.com/9016/scheme-for-restoration-of-wind-powered-water-pumps-launched/.

Ritz, Jennifer. "A Different Kind of American Music." *Texas Techsan* (Lubbock, TX) 55, no. 3 (May/June 2002): 14–16.

Sageser, A. Bower. "Windmill Irrigation: The Great Plains, 1890–1910." *Water Well Journal* 26, no. 2 (February 1972): 32–33.

Sanford, Gregory. "Windmills Saved Thirsty West." *Pacific Northwesterner* 21, no. 1 (January 1977): 12–16.

Schafers, Ted. "Valley Industries Sees Upturn Ahead." *St. Louis Globe Democrat*, July 13, 1977.

"Scholarships Are Offered Women." *Anderson (IN) Bulletin*, February 15, 1923.

"Scholarships for 400 Vets." *Chico (CA) Enterprise*, February 10, 1923.

Share, John. "Windmill on Nebraska Plates Puts Dempster in Bad Temper." *Omaha World Herald*, January 9, 1989.

"Showed Their Colors, Chicagoans Display Flags in McKinley's Honor. Buildings Are Filled, Almost Every Window Bears a Flag or Picture." *Daily Inter Ocean* (Chicago), November 1, 1896.

Simpson, Luis. "Salinas e o ciclo de salem Cabo Frio[,] um filme de Luis Simpson." *Salinas-Ofilme Blogspot*. Accessed September 3, 2012. http://salinas-ofilme.blogspot.fr.

Smeaton, J. "An Experimental Enquiry Concerning the Natural Powers of Water and Wind to Turn Mills, and Other Machines, Depending on a Circular Motion." *Philosophical Transactions of the Royal Society* 51 (1759): 100–174. Accessed February 5, 2014. http://www.engr.psu.edu/MTAH/projects/pdf/smeaton1759.pdf.

Smith, Gayle. "Solar Systems Grow as Alternative to Windmills." *Wyoming Livestock Roundup* (Casper, WY), September 24, 2011. Accessed January 29, 2014. http://www.wylr.net/component/content/article/210-water/technology/2000-solar-systems-grow-as-alternative-to-windmills.

"The Sources of Energy in Nature." *Engineering* (London) 32 (1881): 321–22.

"Spelter Market Affects Windmills and Towers." *Farm Implement News* 36, no. 24 (June 17, 1915): 13.

"A Splendid Act of Citizenship." *Chicago Tribune*, July 26, 1918.

"Steel Wind Mills." *Farm Implement News* 13, no. 12 (March 24, 1892): 12–17.

Stefanides, E. J. "Self-Lubricating Bearings Upgrade Windmill Performance: Redesigned Drive Uses Composite Bearing at Points Which Previously Had Highest Maintenance Requirements." *Design News* 37, no. 22 (November 16, 1981): 88–89.

Stroud, Jerri. "Windmills Pump Valley Industries' Profits." *St. Louis Post-Dispatch*, May 24, 1982.

Swanson, Gloria J. "Windmill Manufacturers Keep on Turning." *Water Well Journal* 41 no. 7 (July 1987): 40–41.

Swenson, Russell G. "Wind Engines in Western Illinois." *Western Illinois Regional Studies* 7, no. 1 (Spring 1984): 61–79.

Swindle, Howard. "Spinning the Revival of an American Tradition." *Dallas Morning News*, August 24, 2001.

"Thomas Osborne Perry[,] Long and Effective Career of Distinguished Scientist and Inventor Is Ended in Death." *Oak Leaves* (Historical Society of Oak Park & River Forest, Oak Park, IL), January 29, 1927, 14.

Thomis, Wayne. "Windmills Built Here on Way to Ocean Air Bases." *Chicago Tribune*, March 7, 1935.

"The Tilting 'Aermotor.'" *Implement and Machinery Review* (London) 19, no. 219 (July 1, 1893): 16,810.

"Tustin's Improved Adjustable Windmill." *Mining and Scientific Press* (San Francisco) 16, no. 16 (April 18, 1868): 241.

"Un Moteur à Vent." *Gazette du Village* (Paris), no. 3 (January 17, 1864): 17. Accessed February 5, 2014. http://gallica.bnf.fr/ark:/12148/bpt6k1205354/f35.image.

Valley Industries, Inc. "Aermotor Opens Omaha Branch." *Pipeline* (St. Louis), January 1979, 1–2.

———. "Fabrication of Giromill Underway at Tallulah Plant." *Pipeline*, November 1979, 4.

———. "Future Plans for the Giromill." *Pipeline*, November 1979, 4.

———. "Giromill Begins Performance Tests." *Pipeline*, June 1980, 1.

——. "Giromill Readied for Preliminary Tests." *Pipeline*, January 1980, 3.

——. "New Project 'in the Wind' at Valley's Aermotor Division." *Pipeline*, November 1978, 1.

——. "'72 In Review." *VI Pipeline* (St. Louis) 3, no. 1 (January 1973): 1, 4.

——. "David Suey Heads New Pump Division." *VI Pipeline* 3, no. 2 (March 1973): 1, 4.

——. "More about Valley's New Braden Aermotor Division." *VI Pipeline* 7, no. 3 (March 1977): 1–2.

——. "More News from Texas." *VI Pipeline* 6, no. 3 (April/May 1976): 1.

——. "New Developments." *VI Pipeline* 5, no. 7 (October/November 1975): 4.

——. "New Places[,] New Faces[,] New Products." *VI Pipeline* 6, no. 2 (February/March 1976): 1–2.

——. "An Overall View." *VI Pipeline* 8, no. 3 (March/April 1978): 1.

——. "Roaring of Motors Is Desert Song in Libyan Project." *VI Pipeline* 7, no. 5 (May/June 1977): 1–3.

——. "Submersible Pumps." *VI Pipeline* 3, no. 8 (October 1973): 3.

——. "Valley Industries Acquires Braden Aermotor Division." *VI Pipeline* 7, no. 2 (February 1977): 1.

United Nations. "Global Issues: Water." Accessed January 27, 2014. http://www.un.org/en/globalissues/water/.

United Press International. "Wind Energy Used in Iraqi Water Pumps." March 6, 2008.

"A Valuable Improvement on Wind Mills." *Farm Implement News* 24, no. 22 (May 28, 1903): 39.

Van Duijnhoven, Serge. "Irrigating Time with the Kinderdijk Windmills." *UNESCO Courier* 53, no. 12 (December 2000): 23–25.

"Veterans Will Receive College Scholarships." *Sacramento Bee*, February 10, 1923.

"Victory Is Scored for Clean Building." *Chicago Commerce* 18, no. 1 (July 29, 1922): 13–14.

"Wallace Davis Hoped to Turn Company Around." *Beatrice (NE) Daily Sun*, July 29, 2013.

Walter, Helen. "Windmills Imported into Australia from the USA—Manufacturers A to Z." *Windmill Journal* (Morawa, Australia) 12, no. 1 (March 2013): 9–11.

Walter, Malcolm. "Imported Windmills—The Aermotor Company of Chicago, USA." *Windmill Journal* 2, no. 3 (June 2003): 6–10.

——. "Metters Ltd (The Years before the Master Nuoil Windmill) 1896–1930." *Windmill Journal* 1, no. 1 (February 2002): 3–8.

——. "The 'Metters Ltd' Nuoil Windmills." *Windmill Journal* 4, no. 4 (December 2005): 2–10.

"Want Laws to Check Bribery: Civic Federation Members Propose Amendments to Make Conviction of Money Giver Easier." *Chicago Daily Tribune*, November 2, 1901.

Water.org. "Water Facts." Accessed January 27, 2014. http://water.org/water-crisis/water-facts/water/.

"Water Pumping Windmills." *How It's Made*. Aired January 21, 2011, Science Channel.

Watkins, John R. "A Common Crystal." *Strand Magazine* (London) 17, no. 98 (February 1899): 174–78.

"Where Chicago Millionaires Play: Midlothian Largest Country Club in America." *Chicago Daily Tribune*, August 4, 1907.

"Where the Wheels Go Round: Grief Visits the Wind Mill Exhibitors—Their Request Turned Down." *Farm Implement News* 14, no. 36 (September 7, 1893): 32.

"Will Manufacture Wind Mills." *Farm Implement News* 23, no. 32 (August 7, 1902): 17.

"Wind-Driven Generators for Farming." *Scientific American* 90 (June 25, 1904): 490.

"Windmill Firm to Move Plant." *Chicago Daily Tribune*, May 12, 1900.

"A Windmill Letter." *Farm Implement News* 8, no. 2 (February 1887): 30.

"Wind Mill Manufacturers Meet at Chicago and Organize a National Association." *Farm Implement News* 16, no. 39 (September 26, 1895): 22.

"Windmills and Wind-Engines, Part III." *Farm Implement News* 8, no. 3 (March 1887): 12–15.

"Windmill Suit Settled." *Farm Implement News* 42, no. 32 (December 29, 1921): 9.

"Women of War Service Benefit from an Estate." *Hoquiam (WA) Washingtonian*, February 10, 1923.

"Women Veterans Get Free Tuition in Will of Noyes." *Houston Post*, February 10, 1923.

"Workers of Two Plants Reject, Indorse Unions." *Chicago Tribune*, March 22, 1951.

Zamudio, Justin. "San Angelo's Aermotor to Be Featured on National TV." *San Angelo Standard-Times*, January 19, 2011.

Ziter, Brett G. "Electric Wind Pumping for Meeting Off-Grid Community Water Demands." *Guelph Engineering Journal* 2 (2009): 14–23.

TRADE LITERATURE AND DIRECTORIES

CU-AH—F. Hal Higgins Agricultural History Collection, Special Collections, University Library, University of California, Davis

DLC—Library of Congress, Washington, DC

DSImce—Windmill Vertical Files, Division of Mechanical and Civil Engineering, National Museum of American History, Smithsonian Institution, Washington, DC

DeGE—Eleutherian Mills Historical Library, Greenville, DE

ICHi—Chicago Historical Society, Chicago

InKendF&W—Corporate Archives, Water Systems Division, Flint & Walling, Inc., Kendallville, IN

KArMus—Cherokee Strip Land Rush Museum, Arkansas City, KS

MWA—American Antiquarian Society, Worcester, MA

MdCCG—Personal research files in the possession of Christopher Gillis, Frederick, MD

MiDbEI—Library, Henry Ford Museum and Greenfield Village, Dearborn, MI

MsSM—Mitchell Memorial Library, Mississippi State University, Mississippi State

NbAMC—Aero Manufacturing Company, Geneva, NE

NbHiK—Kregel Windmill Company Papers, Library, Nebraska State Historical Society, Lincoln

OkHi—Archives and Manuscripts Division, Oklahoma Historical Society, Oklahoma City

OkShaMus—Shattuck Windmill Museum, Shattuck, OK

SFPL—San Francisco Public Library, San Francisco

TxCaP—Windmill Manufacturers' Trade Literature Collection, Research Center, Panhandle-Plains Historical Museum, Canyon, TX

TxLT-NWP—American Wind Power Center, Lubbock, TX

TxLT-SW—Southwest Collection, University Libraries, Texas Tech University, Lubbock

TxSaW—History Department, Witte Memorial Museum, San Antonio, TX

TxTLB—Personal research files in the possession of T. Lindsay Baker, editor and publisher, *Windmillers' Gazette*, Rio Vista, TX

WyU-AHC—American Heritage Center, University Libraries, University of Wyoming, Laramie

Aermotor Company. *4 Times around the World with One Oiling[,] 100,000 Miles without Stopping for Oil*. Chicago: Aermotor Co., [ca. 1923]. Handbill. TxTLB, TxLT-SW.

——. *7th Annual Catalogue, Aug., 1895.* Chicago: Aermotor Co., 1895. MdCCG.

——. *8 Ft. $25 40 Ft. Steel Galvanized Fixed Tower with Anchor Posts $25.* Chicago: Aermotor Co., [ca. 1892]. Handbill. CU-AH.

——. *8ft. Aermotor $25[,] 12ft. Aermotor $50[,] 16ft. Aermotor $100[,] The Aermotor World's Fair Exhibit.* Chicago: Aermotor Co., [ca. 1895]. Handbill. TxTLB.

——. *20-Foot Aermotor Auto-Oiled[,] a Powerful Windmill for Pumping Deep Wells or Raising Large Quantities of Water from Shallow Wells.* Publication R20. Chicago: Aermotor Co., [ca. 1933]. Folder. TxTLB.

——. *A $25.00 Present for One out of Every 25 Aermotor Agents.* Chicago: Aërmotor Co., 1895. Circular Letter. TxCaP.

——. *110 Miles of Steel Towers.* Chicago: Aermotor Co., [ca. 1903]. Broadside. Private collection.

——. *The Accompanying Cut Shows the Outfit Which Caused Such a Bitter-Contest among the Windmill Exhibitors at the World's Fair.* Form No. 365. Chicago: Aermotor Co., [ca. 1895]. Handbill. TxTLB.

——. *Add Pumping Savings to Your Profits!* [Chicago]: Aermotor Co., [ca. 1940]. Folder. TxCaP.

——. *The Aermotor.* Chicago: Aermotor Co., [ca. 1890]. Broadsheet advertisement. TxTLB.

——. *Aermotor 4-Wheel Steel Truck[,] Capacity 2 tons, Weight 175 lbs. for Factory or Warehouse.* Chicago: Aermotor Co., [ca. 1910]. Handbill. TxCaP.

——. (Buffalo, NY). *Aermotor Company[,] Buffalo, May, 1894, 6th Annual Catalogue.* Chicago: Aermotor Co., 1894. MiDbEl.

——. *Aermotor Convertible Jet Pump[,] for Shallow and Deep Wells.* Form 243. Chicago: Aermotor Co., [ca. 1951]. Brochure. Private collection.

——. *Aermotor Electric Pumps for Deep or Shallow Wells, April 3[,] 1933.* Chicago: Aermotor Co., 1933. MdCCG.

——. *Aermotor Electric Pumps for Shallow or Deep Wells, June 1[,] 1931.* Chicago: Aermotor Co., 1931. MdCCG.

——. *Aermotor Electric Water Systems[,] Modern Water Systems for Deep & Shallow Well Requirements[,] Price List, June 14, 1938.* Publication DEP. Chicago: Aermotor Co., 1938. TxCaP.

——. *Aermotor Electric Water Systems[,] All Prices F.O.B. Chicago[,] Subject to Change Without Notice[,] Price List August 14, 1939.* Publication CEP. Chicago: Aermotor Co., 1939. MdCCG.

——. *Aermotor Galvanized Steel Bell Towers.* Publication B.T.1. Chicago: Aermotor Co., [ca. 1925]. Folder. Private collection.

——. *Aermotor Galvanized Steel Towers[:] Beacon[,] Survey[,] Substation[,] Observation[,] Fire Control[,] Flood Lighting[,] Bilby Triangulation[,] Electric Transmission[,] Made by Aermotor Company[,] 2500 Roosevelt Road[,] Chicago, Illinois.* Chicago: Aermotor Co., [ca. 1941]. Folder. TxTLB.

——. *Aermotor Gasoline Pump Air Cooled[;] It Has No Water Jacket to Freeze and Burst.* Publication G.P.2. Chicago: Aermotor Co., [ca. 1910]. Folder. TxCaP.

——. *Aermotor General Purpose Gasoline Engines with the Fluted Cooler Which Cools.* Publication E.S.1. Chicago: Aermotor Co., [ca. 1915]. Handbill. WyU-AHC, TxCaP, MiDbEI, TxTLB.

——. *Aermotor Heavy Back-Geared Gasoline Pumping Engine.* Publication B.G.3. Chicago: Aermotor Co., [ca. 1910]. Handbill. Private collection.

——. *Aermotor Irrigation.* Chicago: Aermotor Co., [ca. 1900]. Catalogue. TxLT-NWP.

——. *Aermotor Observation Towers[,] Siren Towers[,] Tank Towers[,] Made by Aermotor Co.* Chicago: Aermotor Co., [ca. 1930]. Folder. Photocopy. TxTLB.

———. *Aermotor Price List[,] February 1, 1915[,] Superseding All Former Prices*. Chicago: Aermotor Co., 1915. TxCaP.

———. *Aermotor Price List[,] February 15, 1917[,] Superseding All Former Prices*. Chicago: Aermotor Co., 1917. TxCaP.

———. *Aermotor Price List[,] April 4, 1921[,] Superseding All Former Prices*. Des Moines, IA: Aermotor Co., 1921. TxLT-NWP.

———. *Aermotor Price List[,] April 16, 1923[,] Superseding All Former Prices*. Chicago: Aermotor Co., 1923. TxLT-NWP.

———. *Aermotor Price List[,] March 14, 1927[,] Superseding All Former Prices[,] Prices Subject to Change without Notice*. Des Moines, IA: Aermotor Co., 1927. Private collection.

———. *Aermotor Price List[,] Oct. 2, 1933[,] the Improved Aermotor[,] All Prices F.O.B. Des Moines and Subject to Change without Notice*. Des Moines, IA: Aermotor Co., 1933. TxCaP.

———. *Aermotor Price List No. 59[,] Applying to Windmills and Towers[,] Pumps and Cylinders[,] Electric Pumps*. Chicago: Aermotor Co., 1947. TxLT-NWP.

———. *Aermotor Running Water for Farm Profit and Household Convenience*. Publication 7-56-110M. Form No. 262-1. Chicago: Aermotor Co., [ca. 1956]. Folder. NbHiK, TxTLB.

———. *Aermotor Selling Points[,] No. 2[,] October 25, 1906*. Chicago: Aermotor Co., 1906. Private collection.

———. *Aermotor Special Towers[,] TV Antenna Towers[,] Radio Towers[,] Observation Towers[,] Transmission Towers[,] Siren Towers[,] "Bilby" Triangulation Towers*. Chicago: Aermotor Co., [ca. 1957]. Folder. Private collection.

———. *Aermotor Stock Watering Troughs*. Chicago: Aermotor Co., [ca. 1905]. Folder. TxLT-NWP.

———. *Aermotor System Air Pressure Water Supply*. Chicago: Aermotor Co., [ca. 1910]. Folder. TxCaP.

———. *Aermotor Towers Are Towers of Strength[,] Trussed Tripod Towers[,] Head Room[,] Pump Room[,] Tank Room[,] Stock Room[,] Stock Proof*. Publication T.T.3. Chicago: Aermotor Co., [ca. 1903]. Folder. TxCaP.

———. *Aermotor Towers Mean Extra Sales and Extra Profits*. Publication W-D 4. Chicago: Aermotor Co., 1949. Handbill. TxTLB.

———. *Aermotor Trussed Tripod Towers[,] Head Room[,] Pump Room[,] Stock Room[,] Tank Room[,] Bull-Proof*. Publication T.T.5. Chicago: Aermotor Co., [ca. 1905]. Folder. TxCaP.

———. *Aermotor Windmills and Towers[,] the Aermotor[,] 4-Post Towers[,] 3-Post Towers[,] Wide Spread Towers[,] Suburban Outfits[,] Miscellaneous*. Chicago: Aermotor Co., [ca. 1957]. Private collection.

———. *Aermotor Windmills and Water Systems Price List[,] April 19, 1937*. Des Moines, IA: Aermotor Co., 1937. TxCaP.

———. *Aermotor Windmills and Water Systems[,] Price List[,] April 19, 1937[,] All Prices F.O.B. Kansas City[,] (Subject to change without notice)*. Kansas City, MO: Aermotor Co., 1937. TxLT-NWP.

———. *All Prices F.O.B. Chicago[,] Aermotor Electric Water Systems Price List 52-A[,] Aermotor Co.[,] 2500 Roosevelt Road[,] Chicago, Illinois*. Chicago: Aermotor Co., [ca. 1940]. TxTLB.

———. *Announcement[,] New Aermotor Branch House at Dallas, Texas*. Chicago: Aermotor Co., 1953. Mimeograph. TxCaP.

———. *Announcing the 6-Foot Auto-Oiled Aermotor*. Chicago: Aermotor Co., 1925. Handbill. TxTLB.

———. *Another First for Aermotor[,] New 'A' Series Submersible*. Form No. 330-REV1. Chicago: Aermotor Co., [ca. 1953]. Brochure. Private collection.

——. *Auto-Oiled Aermotor with Duplicate Gears Running in Oil and the Easy-To-Build-Up Tower.* Publication B.U.1. Chicago: Aermotor Co., [ca. 1915]. MdCCG, CU-AH.

——. *Auto-Oiled Aermotor[,] This Cut Shows the Gear Case of the Auto-Oiled Aermotor[,] the Ring Oiler.* Chicago: Aermotor Co., 1920. Handbill. TxTLB.

——. *Be Thrifty—Pump with Wind.* Publication W-4X. Chicago: Aermotor Co., [ca. 1940]. Folder. Private collection.

——. *Chicago, September 28, 1933[,] to Aermotor Dealers.* Chicago: Aermotor Co., 1933. Handbill. Private collection.

——. *Crated Motors for Auto-Oiled Aermotors.* Publication S.M.1. Chicago: Aermotor Co., [ca. 1940]. Handbill. TxTLB.

——. *Electric Aermotor.* Publication E.A.1. Chicago: Aermotor Co., [ca. 1920]. Folder. TxLT-NWP.

——. *Eleventh Annual Descriptive Catalogue, March 1899.* Chicago: Aermotor Co., 1899. Private collection.

——. *Eleventh Annual Price List, October 1, 1899.* Chicago: Aermotor Co., 1899. Private collection.

——. *Engines for Pumping and Power.* Publication G.P.7. Chicago: Aermotor Co., [ca. 1910]. Handbill. Private collection.

——. *Facts about Aermotor Water Systems.* Chicago: Aermotor Co., 1936. Brochure. MdCCG.

——. *The Following Is a Contract in Which It Is Agreed to Pay $2 as an Advertising Fee on Each Aërmotor Sold before Jan. 1, 1890.* Chicago: Aermotor Co., [ca. 1899]. Postcard. TxTLB.

——. *Galvanized after Completion Wheels and Towers.* Form No. 365. Chicago: Aermotor Co., [ca. 1894]. Handbill, typed photocopy. TxTLB.

——. *Galvanized Steel Aermotor with Removable Shaft Arms.* Publication R.T.1. Chicago: Aermotor Co., [ca. 1905]. Broadside. CU-AH.

——. *A Great Line of Towers[;] The 1912 Model[;] Aermotor 4-Post Towers.* Publication R.T.2. Chicago: Aermotor Co., 1912. Folder. Private collection.

——. *The Great Strength of the Aermotor Wheel.* Publication A.703. Chicago: Aermotor Co., [ca. 1933]. Folder. TxCaP, WyU-AHC, TxTLB.

——. *How to Choose Pumping Equipment.* Chicago: Aermotor Co., [ca. 1935]. OkShaMus, TxTLB, WyU-AHC.

——. *How to Choose Your Water Pumping System.* Publication 10M-8-54. Chicago: Aermotor Co., 1954. Brochure. NbHiK.

——. *Improved Aermotor[,] Details of Construction Shown in Sectional Views.* Publication A.704. Chicago: Aermotor Co., [ca. 1940]. Folder. TxCaP.

——. *The Improved Auto-Oiled Aermotor with Adjustable Stroke.* Publication A.O.11. Chicago: Aermotor Co., [ca. 1928]. Folder. CU-AH.

——. *Improved Self-Oiling Windmill with Double Gears Running in Oil[,] the Auto-Oiled Aermotor.* Publication A.O.9. Chicago: Aermotor Co., [ca. 1928]. Folder. TxTLB, TxCaP, WyU-AHC.

——. *Information Blank for Electric Pumps.* Chicago: Aermotor Co., [ca. 1935]. Folder. TxLT-SW.

——. *Instructions for Assembling Aermotor, Model 702.* Publication 20M-7-37. Chicago: Aermotor Co., 1937. Folder. TxTLB.

——. *Instructions for Oiling and Care of Aermotor, 702 Model[,] (Tack This Card Up Where You Will See It).* Publication 15M 6-33. [Chicago]: Aermotor Co., [ca. 1933]. Wall hanger. TxTLB.

——. *Instructions for Oiling and Care of Auto-Oiled Aermotor[,] (Tack This Card Up Where You Will See It).* Chicago: Aermotor Co., [ca. 1924]. Wall hanger. MdCCG.

———. *In the Present Removable Arms That Carry the Shafts with the Babbitt Cast Solid in Them, We Have Probably Six to Eight Times the Wearing Qualities Found in Any Windmill for the Following Reasons.* Publication R.A.1. Chicago: Aermotor Co., [ca. 1905]. Folder. TxCaP.

———. *Memorandum to Salesmen[,] Re: Complaints of a Clicking Noise in the #702 Model Aermotor.* Mimeograph. TxLT-NWP.

———. *This National Advertising of Our Pre-Sells Aermotor to Your Customers.* Chicago: Aermotor Co. [ca. 1945]. Folder. TxTLB.

———. *The New Improved, Model 702 Aermotor.* Publication A.702. Chicago: Aermotor Co., [ca. 1933]. Folder. CU-AH, WyU-AHC, TxTLB.

———. *Notice to All Branches[,] Re: Oil Leaks Down Posts of 6-foot Mills.* Chicago: Aermotor Co., 1953. Mimeograph. TxLT-NWP.

———. *Oil Hole Relocated in Pillar of Main Frame of Windmill.* Chicago: Aermotor Co., 1955. Mimeograph. TxLT-NWP.

———. *The Oil Lifter[;] el Alzador de Aceite.* Chicago: Aermotor Co., [ca. 1915]. Handbill. Private collection.

———. *Oiling System Auto-Oiled Aermotor.* Form O.S.2. Chicago: Aermotor Co., [ca. 1925]. Handbill. TxCaP.

———. *Pan American Airways Insure Pacific Water Supply with Improved Aermotor[,] the World[']s Most Dependable Windmills.* Chicago: Aermotor Co., [ca. 1937]. Brochure. TxTLB.

———. *Power Aermotors Bulletin P-22.* Chicago: Aermotor Co., [ca. 1922]. TxCaP, WyU-AHC, TxTLB.

———. *Power Aermotors Keep the Boys on the Farm.* Chicago: Aermotor Co., [ca. 1903]. Folder. TxTLB.

———. *Price List[,] Aermotor Gasoline Engines For Pumping and Power[,] November 1, 1911.* Chicago: Aermotor Co., 1911. TxTLB.

———. *Price List of Parts for Auto-Oiled Aermotors and Easy-to-Build-Up Towers.* Chicago: Aermotor Co., [ca. 1916]. TxTLB.

———. *Pump with Wind[,] It's Free! Aermotor.* Publication W-3X. Chicago: Aermotor Co., [ca. 1940]. Folder. TxTLB.

———. *Put the Wind to Work with Aermotor[,] Farm Proved for 50 Years.* Publication W-1X. Chicago: Aermotor Co., [ca. 1940]. Folder. CU-AH, TxTLB.

———. *A Real Self-Oiling Windmill with Duplicate Gears Running in Oil[,] the Auto-Oiled Aermotor[,] Every Working Part is Constantly and Completely Oiled.* Publication B.U.5. Chicago: Aermotor Co., [ca. 1920]. Folder. TxCaP, WyU-AHC.

———. *Reduction in Prices on Painted Goods[;] All Prices of Galvanized Goods Are Hereby Cancelled.* Chicago: Aermotor Co., 1915. Folder. Private collection.

———. *"Runs in Less Wind" Improved Aermotor.* Publication X-4. Chicago: Aermotor Co., [ca. 1940]. Folder. TxCaP.

———. *The Storm-Defying Auto-Oiled Aermotor.* Publication A.O.14. Chicago: Aermotor Co., [ca. 1928]. Folder. TxTLB.

———. *Thirteenth Annual Descriptive Catalogue. What a Buyer Should Know about Windmills. 1901.* Chicago: Aermotor Co., 1901. MdCCG.

———. *This Is the Windmill Which Has Captured the Windmill Business of the World.* Publication A.O.6. Chicago: Aermotor Co., [ca. 1925]. Folder. TxCaP.

———. *To Aermotor Agents. Bulletin No. 1. Oct. 16, 1893.* Chicago: Aermotor Co., 1893. Photocopy. TxCaP.

———. *To Aermotor Agents. Bulletin No. 2. Nov. 15, 1893.* Chicago: Aermotor Co., 1893. Photocopy. TxCaP.

———. *To Aermotor Agents. Bulletin No. 3. Dec. 15, 1893.* Chicago: Aermotor Co., 1893. Photocopy. TxCaP.

———. *To Aermotor Agents. Bulletin No. 4. Jan. 15, 1894.* Chicago: Aermotor Co., 1894. Photocopy. TxCaP.

———. *To Aermotor Agents. Bulletin No. 5. April 1, 1894.* Chicago: Aermotor Co., 1894. Photocopy. TxCaP.

———. *To Aermotor Agents. Bulletin No. 6. Nov. 1, 1894.* Chicago: Aermotor Co., 1894. Photocopy. TxCaP.

———. *To Aermotor Dealers.* Chicago: Aermotor Co., 1922. Handbill. TxTLB.

———. *To Aermotor Dealers.* Chicago: Aermotor Co., 1924. Handbill. TxTLB.

———. *To Put the Auto-Oiled Aermotor on an Old Aermotor 4-Post Tower.* Chicago: Aermotor Co., [ca. 1920]. Handbill. TxCaP.

———. *Twelfth Annual Descriptive Catalogue[,] Power Aermotor Edition[,] September 1900.* Chicago: Aermotor Co., 1900. MdCCG.

———. *Twentieth Annual Aermotor Repair List[,] January 1, 1908.* Chicago: Aermotor Co., 1908. MdCCG.

———. *Water Supply Bulletin.* Chicago: Aermotor Co., [ca. 1912]. TxCaP.

———. *Wide Spread Aermotor Towers.* Publication W.S.1. Chicago: Aermotor Co., [ca. 1947]. Folder. TxCaP.

———. *You Can Cut Your Pumping Cost with an Aermotor!* [Chicago]: Aermotor Co., [ca. 1940]. Folder. TxCaP, TxTLB.

Aermotor Division, Braden-Aermotor Corporation. *Aermotor Antenna Towers for Use by: Radio Hams—Industry—TV Fringe Areas—TV and Radio Stations—Citizen Band—2-Way Short Wave Radio—Lighting Towers.* Broken Arrow, OK: Aermotor Division of Braden-Aermotor Corp., [ca. 1967]. Looseleaf notebook page. TxLT-SW, TxTLB.

———. *Aermotor Galvanized Steel Towers.* Broken Arrow, OK: Aermotor Division of Braden-Aermotor Corp., [ca. 1967]. Looseleaf notebook page. TxLT-SW, TxTLB.

———. *Aermotor Observation Tower Price List Effective October 1, 1968[,] Net Prices—FOB Broken Arrow, Okla.* Broken Arrow, OK: Aermotor [Division of Braden-Aermotor Corp.], 1968. Broadside photocopy. TxTLB.

———. *Aermotor Repairs and Replacement Units.* Publication 4-1-66. Broken Arrow, OK: Aermotor Division of Braden-Aermotor Corp., [ca. 1966]. TxCaP.

———. *General Letter No. 75-67[,] To: All Branches and Salesman.* Broken Arrow, OK: Aermotor Division of Braden-Aermotor Corp., 1967. MdCCG.

———. *It's New! Aermotor Model Windmill.* Broken Arrow, OK: Aermotor Division of Braden-Aermotor Corp., 1967. Handbill. NbAMC.

———. *Windmills . . . Lowest-Priced Pumping Power on Earth.* Publication No. 6-66-25M. Form No. 314-1. Broken Arrow, OK: Aermotor Division of Braden-Aermotor Corp., [ca. 1966]. Brochure. CU-AH.

Aermotor Division, Motor Products Corporation. *How to Choose Your Aermotor Water System for Suburban or Rural Homes, Farms, Cottages, Resorts, Motels, Service Stations, Restaurants.* Form No. 256-E-1 4-60-50M. Chicago: Aermotor Division of Motor Products Corp., [ca. 1960]. Brochure. NbHiK, TxTLB.

Aermotor Division, Nautec Corporation. *Aermotor Everpure Chlorination-Dechlorination Equipment for Safe, Bacteria Free Water[,] Sulphur Elimination[,] Iron Removal[,] Acid Water Treatment[,] Dechlorination-Filtration of City Water.* Form 298. Chicago: Aermotor Division of Nautec Corp., [ca. 1961]. Folder. NbHiK.

———. *Announcing.* Broken Arrow, OK: Aermotor Division of Nautec Corp., 1965. Folder. OkHi, TxLT-SW, TxTLB.

Aermotor Division, Valley Industries, Inc. *Aermotor Wind Energy Price List Effective 2/15/82 Revision 1*. Conway, AR: Aermotor[,] Division of Valley Industries, 1982. TxCaP.

——. *Installation Instructions[,] Aermotor Windmill Towers[,] Standard 4-Post Tower*. Publication No. 28176. Conway, AR: Aermotor Division of Valley Industries, [ca. 1978]. Manual. Private collection.

——. *Installation, Operating and Maintenance Instructions[,] Aermotor Windmills*. Publication No. 28140. Conway, AR: Aermotor Division of Valley Industries, [ca. 1978]. Manual. TxTLB.

Aermotor Windmill Corporation. *Aermotor Windmills[,] Price List[,] Complete Windmills*. San Angelo, TX: Aermotor Windmill Corp., 1992. TxTLB.

Aerodyne Company. *Electricity from Wind![,] "Wherever the Wind Blows the Aerodyne Goes."* Minneapolis, MN: Aerodyne Co., [ca. 1927]. Folder. MdCCG.

American Wind Power Center. *American Wind Power Center[,] Lubbock, Texas[,] A Museum of Wind Power*. Lubbock, TX: American Wind Power Center, [ca. 2010]. Brochure. MdCCG.

Baker Manufacturing Co. *Gentlemen:—We Regret That We Are Compelled to Make Another Slight Advance on Our Windmill Goods January 1st*. Evansville, WI: Baker Manufacturing Co., 1915. Handbill. Private collection.

Braden-Aermotor Corporation. *Braden-Aermotor Corporation[,] Formerly Braden Winch and Aermotor Divisions of Nautec Corporation[,] Announces the Merging and Consolidation of Both Divisions into a Separate Corporate Entity[,] the Braden-Aermotor Corporation*. [Broken Arrow, OK]: Braden-Aermotor Corp., [September 1965]. Advertising card. TxCaP.

Burke-Bollmeyer Oiler Company. *Red King Windmills[,] the Burke-Bollmeyer Manufacturing Co.[,] Wauseon, Ohio[,] U.S.A.* Wauseon, OH: The Burke-Bollmeyer Manufacturing Co., [ca. 1903]. Private collection.

——. *You Can't Afford To Be Without It*. Wauseon, OH: Burke-Bollmeyer Oiler Co., [ca. 1902]. Handbill. InKendF&W.

Challenge Wind Mill and Feed Mill Co. *The Dandy Steel Wind Mill Is the Strongest and Simplest Made . . .* Batavia, IL: Challenge Wind Mill and Feed Mill Co., [ca. 1892]. Broadside. DSImce.

E. C. Leffel. *Croft's Improved Iron Wind Engine. Manufactured by E. C. Leffel, Springfield, O.* Springfield, OH: E. C. Leffel, [ca. 1880]. Folder. TxSaW.

Elgin Wind Power and Pump Company. *1882 Up-to-Date Windmills 1906* [cover title; no title page]. Elgin, IL: Elgin Wind Power & Pump Company, 1906. TxTLB.

——. *Catalog Number Twenty-One of the Elgin Wind Power and Pump Co.* Elgin, IL: Elgin Wind Power and Pump Co., [ca. 1912]. TxCaP, NbHiK.

Farm Light and Power Year Book[,] Dealers' Catalog and Service. New York: Farm Light and Power Publishing Co., 1922. MdCCG.

Felizardo Elizondo Guajardo, S.A. de C.V. *Aermotor[,] La Fabrica de Papalotes mas Grande y Moderna de Mexico desde 1953*. Apodaca, Mexico: Felizardo Elizondo Guajardo, S.A. de C.V., [ca. 2000]. Folder. TxTLB.

Flint and Walling Manufacturing Company. *A Triumph of Inventive Ingenuity[,] Star Wind Mills Fitted with the Burke-Bollmeyer Oiler*. Kendallville, IN: Flint and Walling Manufacturing Co., [ca. 1902]. InKendF&W.

Heller-Aller Company. *December 1983[,] Dear Dealer: The Heller-Aller Company Has Constantly Strived to Offer You, Our Customer, the Very Best Equipment at Competitive Prices*. Napoleon, OH: Heller-Aller Co., 1983. Handbill. TxTLB.

Herbert E. Bucklen Corporation. *Power and Light from the Free Wind[,] Cook—Light and Freeze Direct from the Breeze*. Elkhart, IN: Herbert E. Bucklen Corp., [ca. 1927]. Folder. MdCCG.

———. *Where Service Is a Problem the United States Government Uses a HEBCO*. Elkhart, IN: Herbert E. Bucklen Corp., [ca. 1927]. Folder. MdCCG.

Illinois State Gazetteer and Business Directory for the Years 1864–65. Chicago: J.C.W. Bailey, at the Directory Office, 1864. DLC.

Illinois State Gazetteer and Business Directory, 1878. Detroit: R. L. Polk & Co., 1878. DLC.

Iron Man Windmill Co. *The Perfect Solution for Providing a Lifetime of Free Water*. Wuhan, Hubei, China: Iron Man Windmill Co., 2010. Accessed January 26, 2012. http://www .ironmanwindmill.com/images/brochure/brochure.pdf. Printout. MdCCG.

James Martin & Company. "James Martin's Aermotor All-Steel Self-Regulating Galvanised Windmill." *Our New Harvesting Machinery Catalogue*. Sydney, Australia: James Martin & Company, [ca. 1895]. MdCCG.

John Danks & Son Pty. Ltd. *The "Coo-ee" Auto Oiled Encased Gear Windmill*. Melbourne, Australia: John Danks & Son Pty. Ltd., [ca. 1935]. Folder. TxTLB.

John Mullens & Sons. *The Divining-Rod: Its History, Truthfulness and Practical Utility*. Bath, UK: J. & H. W. Millins, 1927. MdCCG.

Kirkwood Manufacturing Company. *Descriptive Catalogue of the Kirkwood Iron Wind Engine, for Pumping Water from Wells, Springs, and Brooks, to Supply Railroads, Residences, Stock, and for Irrigating Purposes*. Arkansas City, KS: Traveler Print, [ca. 1888]. KArMus.

Mast, Foos and Company. *The Improved Iron Turbine Wind Engine, Manufactured by Mast, Foos & Co., Springfield, Ohio*. Form 6, 4-84-12,000. Springfield, OH: Mast, Foos & Co., [ca. 1884]. Folder. DSImce.

———. *Mast, Foos & Co., Springfield, Ohio, U.S.A., Manufacturers of Iron Turbine Wind Engines, Buckeye Force Pumps, Buckeye Iron Fence, Buckeye Senior & Junior Lawn Mowers, Etc.* Springfield, OH: Winters Print & Litho Co., [ca. 1890]. Folder. TxTLB.

Mid-America Windmill Museum. *Mid-America Windmill Museum*[,] *Kendallville, Indiana*[,] *Take a Look at the Past!* Kendallville, IN: Mid-America Windmill Museum, [ca. 2010]. Brochure. MdCCG.

Molinos de Viento, S.A. *6 Tamaros Diferentes*. Chihuahua, Mexico: Molinos de Viento, S.A., [ca. 1978]. Folder. TxCaP.

Noyes, La Verne W. *A Score of Ways of Looking at a Thing, including Fourteen Prize Essays Answering the Question: "Why Should I Use a Dictionary Holder?"* Chicago: La Verne W. Noyes, 1888. ICHi.

O'Brock Windmill Distributors. *E-Z Air Windmill Compressor*. North Benton, OH: O'Brock Windmill Distributors, [ca. 2000]. Handbill. MdCCG.

———. *"Harness the Wind" with a Water Pumping Windmill*. North Benton, OH: O'Brock Windmill Distributors, 2000. MdCCG.

Potomac Wind Energy. *American Classic Wind Power Operator Manual*. Dickerson, MD: Potomac Wind Energy, 2011. MdCCG.

Red Cross Manufacturing Company. *Red Cross Line of Wind Mills*[,] *Towers, Tanks, Pumps, Cylinders, Pipe and Water Supply Goods, Manufactured by the Red Cross Mf'g. Co.*[,] *Bluffton, Ind., U.S.A.* [General Catalogue No. 22]. Bluffton, IN: Red Cross Mf'g. Co., [ca. 1905]. TxCaP, MsSM.

Redhead, Norton, Lathrop & Co., Wholesale Stationers & Jobbers. Advertisement. *Bushnell's Des Moines City Directory, 1889–90*. Des Moines, IA: Des Moines Directory Co., Publishers, 1889. DLC.

Sears, Roebuck and Company. *Agriculture Implements*[:] *Windmills*[,] *Hay Presses*[,] *Incubators*[,] *Steel Tanks*. 1st ed. Publication 124E. Chicago: Sears, Roebuck & Co., [ca. 1906]. TxTLB.

——. *Sears, Roebuck and Co.*[,] *Incorporated*[,] *Cheapest Supply House on Earth*[,] *Consumer Guide Catalogue No*[.] *104*. Chicago: Sears, Roebuck and Co., 1897. Abridged rpt. ed. Edited by Fred L. Israel. New York: Chelsea House Publishers, 1976. TxTLB.

Second Wind Windmill Service. *The Wind Engine 702*. Publication 06/01/2000. Fort Worth, TX: Second Wind Windmill Service, 2000. Handbill. TxTLB.

Smith and Winchester. *Price List and Testimonial Circular of the Aërmotor as Arranged for Pumping and Power Purposes*. Boston: Smith & Winchester, [ca. 1892]. Private collection.

——. *Smith & Winchester, Illustrated Catalogue of Steel, Iron and Wood Wind Engines*[,] *Wood, Iron, Brass and Copper Pumps*[,] *Artesian Well Tools and Supplies*[,] *Steam Boilers and Engines, Steam Pumps, Wrought Iron Pipe Fittings, Brass Goods, Hose, Belting, Etc., Etc.* Buffalo, NY: Matthews, Northrup & Co., 1890. DeGE, TxCaP.

Southwestern Pioneer Windmill Association. *Southwestern Pioneer Windmill Association*[,] *J. B. Buchanan Vintage Windmill Collection*[,] *Spearman, Texas*. Spearman, TX: Southwestern Pioneer Windmill Association, [ca. 2012]. Brochure. MdCCG.

United Pump & Power Co. *Fresh Water on the Farm*. Form U.0.1. Milwaukee, WI: United Pump & Power Co., [ca. 1918]. Folder. MdCCG.

——. *Perry Water System*[,] *Installation Plans*. Chicago: United Pump & Power Co., [ca. 1910]. Brochure. MdCCG.

Valley Pump Group Aermotor-Weinman-Midland. *Valley Pump Group Aermotor-Weinman-Midland Product Capabilities*. Conway, AR: Valley Pump Group Aermotor-Weinman-Midland, [ca. 1982]. Folder. TxTLB.

Wincharger Corporation. *Every Farm Home Can Now Enjoy "Big City" Radio Reception*. Sioux City, IA: Wincharger Corp., [ca. 1925]. Brochure. MdCCG.

——. *Light Your Farm for 50¢ a Year Power Operating Cost*. Sioux City, IA: Wincharger Corp., [ca. 1930]. Brochure. MdCCG.

U.S. Wind Engine and Pump Company. *Descriptive Catalogue and Price List of Halladay's Patent Pumping and Power Wind Mills, Double and Single Acting Force Pumps, Farm Pumps, Halladay & Ruggles' Patent I.X.L. Feed Mills, Railroad Tank Spouts, Outlet Valves, Etc., Etc., Manufactured by United States Wind Engine and Pump Co., Batavia, Kane Co., Illinois*. Chicago: Culver, Page, Hoyne & Co., [ca. 1871]. MWA.

——. *Descriptive Catalogue of U.S. Wind Engine & Pump Co.*[,] *Batavia, Illinois. Manufacturers of Halladay's Standard Wind Mills, Double and Single Acting Pumps, the IXL Feed Grinders, Halladay's Celebrated Outlet Valves, Railroad Tanks, Drop Pipes, Goosenecks, Well Augers, Etc.* Chicago: Rand, McNally & Co., [ca. 1876]. DLC, ICHi.

——. *Halladay Standard Pumping & Geared Wind Mills. Pumps, Tanks. Etc. Manufactured by U.S. Wind Engine & Pump Co.*[,] *Batavia, Ill.*[,] *U.S.A.* Buffalo, NY: Cosack & Co., [ca. 1885]. Advertising card. Private collection.

——. *Halladay Standard Wind Mill as Shown When Not at Work, T. A. Morrison, Agent, Homer, Champaign Co., Ill.* Chicago: Rand, McNally & Co., [ca. 1876]. Advertising circular. ICHi.

——. "Halladay's Improved Wind Engine for Pumping Water and Grinding Grain" (advertisement). *H. C. Chandler & Co.'s Railway Business Directory and Shippers' Guide, for the State of Illinois*. Indianapolis: H. C. Chandler & Company, 1868. DLC.

——. *Is the Halladay Mill Durable?: The First Halladay Ever Sold* [Batavia, IL: U.S. Wind Engine and Pump Company, 1876]. Handbill. MiDbEI.

W. D. Moore & Company. *The Cheapest Power for Pumping Water*[,] *the "Aermotor" Windmill*. O'Connor, Australia: W. D. Moore & Company, [ca. 1953]. Folder. TxTLB.

Wind-Power Manufacturing Company. *WinPower*[,] *the Wind Plant with the Power Ring*. Newton, IA: Wind-Power Manufacturing Co., [ca. 1930]. Folder. MdCCG.

"W. I. Tustin, Patentee and Sole Proprietor. The Economy Wind-Mill." *Langley's San Francisco Directory for the Year Commencing April, 1880*. San Francisco: Francis, Valentine & Co., 1880. SFPL.

GOVERNMENT DOCUMENTS

Aermotor Division, Valley Industries, Inc. *Development of a 1kW High-Reliability Wind Turbine Generator*[,] *Contract Proposal*. 3 parts. Request for Proposal No. PF64410F to Rockwell International, June 20, 1977. Conway, AR: Aermotor Division of Valley Industries, 1977.

Anderson, John W., Robert V. Brulle, Edwin B. Birchfield, and William D. Duwe. *Development of a 40 kW Giromill*[,] *Phase I*. Vol. 2, *Design and Analysis*. Prepared for Rockwell International Corporation, Energy Systems Group, Rocky Flats Plant, Wind Systems Program, Golden, Colorado, as Part of the US Department of Energy, Division of Distributed Solar Technology, Federal Wind Energy Program. St. Louis: McDonnell Aircraft Co., 1979.

Australia. Department of Patents. Commonwealth of Australia [Patents on windmills]. Victoria, Australia.

Barbour, Erwin Hinckley. *The Homemade Windmills of Nebraska*. US Agricultural Experiment Station of Nebraska, Bulletin 59. Lincoln: University of Nebraska, 1899.

———. *Wells and Windmills in Nebraska*. US Department of the Interior, Geological Survey, Water-Supply and Irrigation Paper No. 29. Washington, DC: Government Printing Office, 1899.

Bowman, Isaiah. *Well-Drilling Methods*. US Department of the Interior, Geological Survey, Water-Supply Paper No. 257. Washington, DC: Government Printing Office, 1911.

Brulle, R. V. *Feasibility Investigation of the Giromill for Generation of Electrical Power*[,] *Final Report for the Period April 1975–April 1976*. 2 vols. Prepared for the US Energy Research and Development Administration, Division of Solar Energy. St. Louis: McDonnell Aircraft Co., 1977.

Day, P. C. "The Winds of the United States and Their Economic Uses." In *Yearbook of the United States Department of Agriculture*, 337–50. Washington, DC: US Department of Agriculture, 1911.

Dubois, Coert. *Systematic Fire Protection in the California Forests*. Washington, DC: US Department of Agriculture, Forest Service, 1914.

Fuller, Myron L. *Underground Waters for Farm Use*. US Department of the Interior, Geological Survey, Water-Supply and Irrigation Paper No. 255. Washington, DC: Government Printing Office, 1910.

Hamilton, C. L., and Hans G. Jepson. *Stock-Water Developments: Wells, Springs and Ponds*. Farmers' Bulletin No. 1859. Washington, DC: US Department of Agriculture, July 1940.

Illinois Workers' Compensation Commission. *Chronology of Workers' Compensation Legislation in Illinois*. Chicago, 2011. Accessed April 24, 2013. http://www.iwcc.il.gov/chronology.pdf.

Leigh, George E. "Bilby Towers." *NOAA Celebrates 200 Years of Science, Service, and Stewardship*. Accessed May 23, 2013. http://celebrating200years.noaa.gov/magazine/bilby/.

Moran, W. A. *Giromill Wind Tunnel Test and Analysis*. Vol. 2, *Technical Discussion*[,] *Final Report for the Period July 1976–October 1977*. Prepared for the US Energy Research and Development Administration, Division of Solar Energy. St. Louis: McDonnell Aircraft Co., 1977.

Murphy, Edward Charles. *Windmills for Irrigation*. US Department of the Interior, Geological Survey, Water-Supply and Irrigation Paper No. 8. Washington, DC: Government Printing Office, 1897.

Newell, F. H. *Letter of Transmittal, July 15, 1898*. US Department of the Interior. Water-Supply and Irrigation Papers of the United States Geological Survey, No. 20. Washington, DC: Government Printing Office, 1899.

Perry, Thomas O. *Experiments with Windmills*. US Department of the Interior, Geological Survey, Water-Supply and Irrigation Paper No. 20. Washington, DC: Government Printing Office, 1899.

Rohwer, Carl, and M. R. Lewis. *Small Irrigation Pumping Plants*. Farmers' Bulletin No. 1857. Washington, DC: US Department of Agriculture, December 1940.

Royaume de Belgique Brevet D'Invention. Patents on windmills. Le Ministre de L'Industrie et du Travail, Brussels, Belgium.

US Department of Agriculture. Forest Service Equipment Development Center. *Range Water Pumping Systems State-of-the-Art-Review*, by Dan W. McKenzie. Project Report 8522 1201. February 1985. San Dimas, CA: Forest Service Development Center, 1985.

US Department of Agriculture. Natural Resources Conservation Service. "Environmental Quality Incentives Program." Accessed January 8, 2014. http://www.nrcs.usda.gov/wps/portal/nrcs/main/national/programs/financial/eqip/.

——. "Farm Bill 2008 Fact Sheet: Environmental Quality Incentives Program." Accessed January 8, 2014. http://www.nrcs.usda.gov/Internet/FSE_DOCUMENTS/stelprdb1042024.pdf.

——. Rural Electrification Administration. *A Brief History of the Rural Electric and Telephone Programs*. Washington, DC: US Department of Agriculture, 1982. Accessed June 27, 2013. http://www.rurdev.usda.gov/rd/70th/rea-history.pdf.

US Department of Commerce, Bureau of the Census. *Abstract of the Census of Manufactures, 1919*. Washington, DC: Government Printing Office, 1923.

——, Bureau of Foreign and Domestic Commerce. *Statistical Abstract of the United States, 1921*. Washington, DC: Government Printing Office, 1921.

——. *Statistical Abstract of the United States, 1922*. Washington, DC: Government Printing Office, 1922.

——. *Statistical Abstract of the United States, 1931*. Washington, DC: Government Printing Office, 1931.

——, Coast and Geodetic Survey. *Bilby Steel Tower for Triangulation*. Rev. ed. by Jasper S. Bilby. Special Publication No. 158. Washington, DC: US Government Printing Office, 1940.

——, Patent Office. "Aermotor Company, of Chicago, Illinois. Trade-Mark for Windmills." Statement and Declaration No. 56,596. Registered October 9, 1906. US Patent and Trademark Office, Washington, DC.

——. Patents on windmills filed according to patent number. US Patent and Trademark Office, Washington, DC.

US Department of Commerce and Labor, Bureau of Statistics. *Windmills in Foreign Countries*. Special Consular Reports, Vol. 31. 58th Congress, 3rd Session, House Document No. 73 (Serial 4836). Washington, DC: Government Printing Office, 1904.

US Department of the Treasury, Bureau of Statistics. *Statistical Abstract of the United States, 1902*. Washington, DC: Government Printing Office, 1903.

US National Archives and Records Administration. "National Industrial Recovery Act (1933)." Accessed June 21, 2013. http://www.ourdocuments.gov/doc.php?doc=66.

Wallace, Henry C. "The Year in Agriculture: The Secretary's Report to the President[,] Washington, D.C., November 16, 1921." *United States Department of Agriculture Yearbook of Agriculture 1921*. Washington, DC: Government Printing Office, 1922.

INDEX

OTHER BOOKS

in the Tarleton State University Southwestern Studies in the Humanities Series

Pastoral Vision of Cormac McCarthy
Georg Guillemin

Exploding the Western: Myths of Empire on the Postmodern Frontier
Sara L. Spurgeon

Undaunted: A Norwegian Woman in Frontier Texas
Charles H. Russell

Lost Years of William S. Burroughs: Beats in South Texas
Robert E. Johnson

Birth of a Texas Ghost Town: Thurber, 1886–1933
Mary J. Gentry and Lindsay T. Baker

Alexandre Hogue: An American Visionary—Paintings and Works on Paper
Susie Kalil

Fritos® Pie: Stories, Recipes, and More
Kaleta Doolin

Faded Glory: A Century of Forgotten Texas Military Sites, Then and Now
Thomas E. Alexander and Dan K. Utley

Marfa Flights: Aerial Views of Big Bend Country
Paul V. Chaplo